Chaudhery Mustansar Hussain, Rüstem Keçili, Chaudhery Ghazanfar Hussain (Eds.)
Carbon-Based Molecularly Imprinted Polymers

Also of interest

Superabsorbent Polymers.
Chemical Design, Processing and Applications
Sandra Van Vlierberghe, Arn Mignon (Eds.), 2021
ISBN 978-1-5015-1910-9, e-ISBN (PDF) 978-1-5015-1911-6,
e-ISBN (EPUB) 978-1-5015-1171-4

Physical Chemistry of Polymers.
A Conceptual Introduction
2nd Edition
Sebastian Seiffert, 2023
ISBN 978-3-11-071327-5, e-ISBN (PDF) 978-3-11-071326-8,
e-ISBN (EPUB) 978-3-11-071339-8

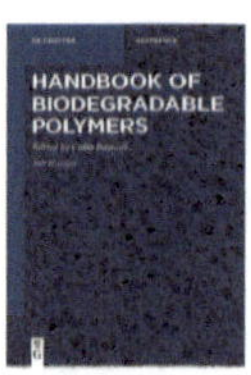

Handbook of Biodegradable Polymers
Catia Bastioli (Ed.), 2020
ISBN 978-1-5015-1921-5, e-ISBN (PDF) 978-1-5015-1196-7,
e-ISBN (EPUB) 978-1-5015-1198-1

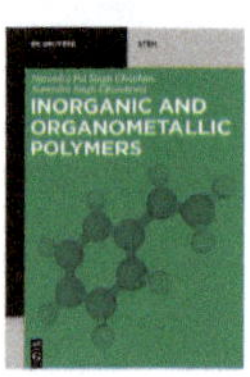

Inorganic and Organometallic Polymers
Narendra Pal Singh Chauhan, Narendra Singh Chundawat, 2019
ISBN 978-1-5015-1866-9, e-ISBN (PDF) 978-1-5015-1460-9,
e-ISBN (EPUB) 978-1-5015-1479-1

Carbon-Based Molecularly Imprinted Polymers

Smart Nanomaterials

Edited by
Chaudhery Mustansar Hussain, Rüstem Keçili
and Chaudhery Ghazanfar Hussain

DE GRUYTER

Editors

Prof. Chaudhery Mustansar Hussain
Department of Chemistry and Environmental Science
New Jersey Institute of Technology
Newark, NJ 07102
USA
chaudhery.m.hussain@njit.edu

Chaudhery Ghazanfar Hussain
Computer Science and Technology
Department of Education Lahore
Defence Road
Punjab, Lahore
Pakistan
ur1947@gmail.com

Rüstem Keçili
Yunus Emre Vocational School of Health Services
Anadolu University
26470 Eskişehir
Türkiye
rkecili@anadolu.edu.tr

ISBN 978-3-11-150302-8
e-ISBN (PDF) 978-3-11-150332-5
e-ISBN (EPUB) 978-3-11-150367-7

Library of Congress Control Number: 2025939114

Bibliographic information published by the Deutsche Nationalbibliothek
The Deutsche Nationalbibliothek lists this publication in the Deutsche Nationalbibliografie; detailed bibliographic data are available on the Internet at http://dnb.dnb.de.

Cover image: Love Employee/iStock/Getty Images Plus
Typesetting: Integra Software Services Pvt. Ltd.
Printing and binding: CPI books GmbH, Leck

www.degruyterbrill.com
Questions about General Product Safety Regulation:
productsafety@degruyterbrill.com

Preface

The field of nanomaterials has witnessed transformative advancements, with carbon-based materials leading the way due to their unique properties and versatile applications. Among these, **molecularly imprinted polymers (MIPs)** represent a breakthrough in creating highly selective, smart materials capable of mimicking biological recognition systems. This book offers an in-depth exploration of these innovative materials, detailing their fundamentals, performance evaluation, diverse applications, and future prospects.

Structured into five comprehensive parts, the book provides a systematic understanding of carbon-based MIPs. **Part 1: Fundamentals of molecular imprinting-based smartcarbon nanomaterials** introduces the foundational concepts and polymerization techniques underlying these materials. Chapters explore the basics of molecular imprinting technology, the integration of carbonaceous nanomaterials, and the methodologies for synthesizing these advanced polymers.

Part 2: Performance evaluation of molecular imprintingbasedsmart carbon nanomaterials delves into characterization methods, offering insights into the structural and functional evaluation of carbon-based MIPs. These methods are critical for understanding the efficiency and reliability of these smart materials.

Part 3: Applications of molecular imprinting-based smart carbon nanomaterials highlights the versatility of these materials across various domains. Chapters cover environmental remediation, biomedical innovations, analytical and industrial advancements, sensing in agriculture, and forensic science. Each application showcases the potential of carbon-based MIPs to address pressing challenges with precision and efficiency.

Part 4: Case studies of molecular imprinting-based smartcarbon nanomaterials provides a focused examination of binding kinetics and thermodynamics, demonstrating the practical implications and mechanistic insights of these materials in real-world scenarios.

Finally, **Part 5: Various aspects and future prospects of molecularimprinting-based smart carbon nanomaterials** addresses the sustainability of these materials, exploring green synthesis approaches and highlighting the challenges and future directions in this rapidly evolving field.

This book is designed to cater to researchers, practitioners, and students interested in nanotechnology, materials science, and applied chemistry. By offering a blend of theoretical knowledge, experimental techniques, and practical applications, it aims to inspire further advancements in the field of carbon-based molecularly imprinted polymers. We hope this work serves as a valuable resource, fostering innovation and progress in creating smarter, sustainable materials for a better future.

This compendium is an invaluable resource for researchers, practitioners, and students aiming to contribute to the advancement of molecular imprinting technology. It reflects the collective expertise and vision of a community dedicated to har-

https://doi.org/10.1515/9783111503325-202

nessing the power of nanotechnology for a sustainable and technologically advanced future. On behalf of Walter de Gruyter Publishers, we are very thankful to all contributors for their special hard work in making of this book. Gratitude extended to Helene Chavaroche (Senior Acquisitions Editor in Materials Science, Walter de Gruyter Publishing) for this project. Special thanks to Melanie Götz (Content Editor, Walter de Gruyter Publishing) for dedicated support and help during this project. In the end altogether appreciations to Walter de Gruyter Publishers for publishing the book.

CHAUDHERY MUSTANSAR HUSSAIN, RÜSTEM KEÇILI & CHAUDHERY GHAZANFAR HUSSAIN (Editors)

Acknowledgments

We would like to acknowledge this book to Our Lovely Sons & Daughters SARIM HUSSAIN CHAUDHERY, AZLAN HUSSAIN CHAUDHERY, MAIDA ALI, HAFIZA SAFA HUSSAIN CHAUDHERY, HAFIZA DUA HUSSAIN CHAUDHERY, AIRA HUSSAIN CHAUDHERY for their dedicated support during the compiling this book.

(Editors: Chaudhery Mustansar Hussain & Chaudhery Ghazanfar Hussain)

I would like to acknowledge this book to My Lovely Son Little MERIÇ for his dedicated support during the compiling this book.

(Editor: Rüstem Keçili)

https://doi.org/10.1515/9783111503325-203

Contents

Part 3: Applications of molecular imprinting-based smart carbon nanomaterials

Contributing authors

Chapter 1
Long Wu
Hainan University
School of Food Science and Engineering
Key Laboratory of Tropical Fruits and Vegetables
Quality and Safety for State Market Regulation
Haikou 570228
PR China
E-Mail: longquan.good@163.com

Jing Wei
Hainan University
School of Food Science and Engineering
Key Laboratory of Tropical Fruits and Vegetables
Quality and Safety for State Market Regulation
Haikou 570228
PR China

Chapter 2
Vaishali Goel
School of Pharmacy
Graphic Era Hill University
Dehradun
Uttarakhand
India

Deepika Raina
Vaishali Goel
School of Pharmacy
Graphic Era Hill University
Dehradun
Uttarakhand
India
E-Mail: deepika709@gmail.com

Prachi Mall
School of Pharmacy
Graphic Era Hill University
Dehradun
Uttarakhand
India

Chapter 3
Vishnu Biju
Center for Advanced Functional Materials (CAFM)
(Research Center affiliated to University of Kerala)
Department of Physics
Bishop Moore College
Mavelikara
Kerala 690110
India

A V Mahadev
Aryabhatta Research Institute of Observational Sciences
Division of Astronomy
Nainital
Uttarakhand 263001
India

Arun Aravind
Center for Advanced Functional Materials (CAFM)
(Research Center affiliated to University of Kerala)
Department of Physics
Bishop Moore College
Mavelikara
Kerala 690110
India
E-Mail: arun@bishopmoorecollege.org

Chapter 4
Dr. Neha Sen
M.L.S.M College Sundernagar
Distt. Mandi
Himachal Pradesh 175018
India
Email: neharikasen007@gmail.com

https://doi.org/10.1515/9783111503325-206

Chapter 5
Ziyin Mei
China University of Geosciences
State Key Laboratory of Geomicrobiology and Environmental Changes
Engineering Research Center of Nano-Geomaterials of the Ministry of Education
Faculty of Materials Science and Chemistry
Wuhan 430074
China
E-Mail: meiziyin@cug.edu.cn

Martins Lateef Yusuf
China University of Geosciences
State Key Laboratory of Geomicrobiology and Environmental Changes
Engineering Research Center of Nano-Geomaterials of the Ministry of Education
Faculty of Materials Science and Chemistry
Wuhan 430074
China
E-Mail: martins_yusuf@cug.edu.cn

Jiamei Liu
China University of Geosciences
State Key Laboratory of Geomicrobiology and Environmental Changes
Engineering Research Center of Nano-Geomaterials of the Ministry of Education
Faculty of Materials Science and Chemistry
Wuhan 430074
China
E-Mail: 1202320155@cug.edu.cn

Shaoguang Li
China University of Geosciences
State Key Laboratory of Geomicrobiology and Environmental Changes
Engineering Research Center of Nano-Geomaterials of the Ministry of Education
Faculty of Materials Science and Chemistry
Wuhan 430074
China
E-Mail: lisg@cug.edu.cn

Hui Li
China University of Geosciences
State Key Laboratory of Geomicrobiology and Environmental Changes
Engineering Research Center of Nano-Geomaterials of the Ministry of Education
Faculty of Materials Science and Chemistry
Wuhan 430074
China
E-Mail: lihui-chem@cug.edu.cn

Fan Xia
China University of Geosciences
State Key Laboratory of Geomicrobiology and Environmental Changes
Engineering Research Center of Nano-Geomaterials of the Ministry of Education
Faculty of Materials Science and Chemistry
Wuhan 430074
China
E-Mail: xiafan@cug.edu.cn

Chapter 6
Khasan Berdimuradov
Karshi State Technical University
Shahrisabz Faculty of Food Engineering
19 Khanabad Branch Street
Karshi City
Kashkadarya Region
180119
Uzbekistan
khasanberdimuradov@gmail.com;
elyor170690@gmail.com

Akbarali Rasulov
National University of Uzbekistan
Department of Physical chemistry
Tashkent
100034
Uzbekistan

Husan Yaxshinorov
National University of Uzbekistan
Department of Physical chemistry
Tashkent
100034
Uzbekistan

Javokhir Abdisattorov
National University of Uzbekistan
Department of Physical chemistry
Tashkent
100034
Uzbekistan

Elyor Berdimurodov
New Uzbekistan University
Chemical & Materials Engineering
54 Mustaqillik Ave.
Tashkent
100007
Uzbekistan

National University of Uzbekistan
Faculty of Chemistry
Tashkent
100034
Uzbekistan

Omar Dagdag
Gachon University
Department of Mechanical Engineering
Seongnam
13120
Republic of Korea

Mohamed Rbaa
The Higher Institute of Nursing Professions and
Health Techniques of Casablanca
P.O. Box 20250
Casablanca
Morocco

Mohammed V University in Rabat
Laboratory of Materials
Nanotechnology and Environment
Faculty of Sciences
P.O. Box 1014
Agdal-Rabat
Morocco

Chapter 7

Akbarali Rasulov
National University of Uzbekistan
Department of Physical chemistry
Tashkent
100034
Uzbekistan

Husan Yaxshinorov
National University of Uzbekistan
Department of Physical chemistry
Tashkent
100034
Uzbekistan

Javokhir Abdisattorov
National University of Uzbekistan
Department of Physical chemistry
Tashkent
100034
Uzbekistan

Elyor Berdimurodov
New Uzbekistan University
Chemical & Materials Engineering
54 Mustaqillik Ave.
Tashkent
100007
Uzbekistan

National University of Uzbekistan
Faculty of Chemistry
Tashkent
100034
Uzbekistan

Omar Dagdag
Gachon University
Department of Mechanical Engineering
Seongnam
13120
Republic of Korea

Mohamed Rbaa
The Higher Institute of Nursing Professions and
Health Techniques of Casablanca
P.O. Box 20250
Casablanca
Morocco

Mohammed V University in Rabat
Laboratory of Materials, Nanotechnology, and Environment, Faculty of Sciences
P.O. Box 1014
Agdal-Rabat
Morocco

Khasan Berdimuradov
Karshi State Technical University
Shahrisabz Faculty of Food Engineering
19, Khanabad Branch Street
Karshi City
Kashkadarya Region, 180119
Uzbekistan
khasanberdimuradov@gmail.com;
elyor170690@gmail.com

Chapter 8

Dr. Hemlata Vashisht
University of Delhi
Department of Chemistry
Kirori Mal College
Delhi
110007
India
hemlata.vashisht@gmail.com

Prof. Sudershan Kumar
University of Delhi
Department of Chemistry
Hindu College
Delhi
110007
India

Dr. Ajai Kumar
University of Delhi
Department of Chemistry
Hindu College
Delhi
110007
India

Chapter 9

Milad Tavassoli
Yasuj University of Medical Sciences
Student Research Committee
Yasuj
Iran

Yasuj University of Medical Science
Department of Nutrition
Faculty of Health and Nutrition Sciences
Yasuj
Iran
E-Mail: mtavassoli2006@gmail.com

Reza Abedi-Firoozjah
Kermanshah University of Medical Sciences
Department of Food Science and Technology
School of Nutrition Sciences and Food Technology
Kermanshah
Iran
E-Mail: r.abedi1878@gmail.com

Behnam Bahramian
Tabriz University of Medical Sciences
Department of Food Science and Technology
Faculty of Nutrition and Food Sciences
Tabriz
Iran
E-Mail: bahramianbehnam1372@gmail.com

Ali Ehsani
Tabriz University of Medical Sciences
Department of Food Science and Technology
Faculty of Nutrition and Food Sciences
Tabriz
Iran
E-Mail: ehsani@tbzmed.ac.ir

Ehsan Sadeghi
Kermanshah University of Medical Sciences
Department of Food Science and Technology
School of Nutrition Sciences and Food Technology
Kermanshah
Iran

Kermanshah University of Medical Sciences
Research Center for Environmental
Determinants of Health (RCEDH)
Health Institute
Kermanshah
Iran
E-Mail: ehsan.sadeghi59@yahoo.com

Chapter 10

Arunima Reghunadhan
Cochin University of Science and Technology
Materials Science and Nano Engineering Lab
(MSNE-Lab)
Department of Polymer Science and Rubber
Technology
Kochi

University of Limerick
Bernal Institute
Castletroy
Limerick
Republic of Ireland

Viswanathan Pillai Aromal
Cochin University of Science and Technology
Materials Science and Nano Enginering Lab
(MSNE-Lab)
Department of Polymer Science and Rubber
Technology
Kochi

Kerala University of Health Sciences
Government Medical College-Pariyaram
Kannur
Kerala
India

Linu Tresa K G
Cochin University of Science and Technology
Materials Science and Nano Enginering Lab
(MSNE-Lab)
Department of Polymer Science and Rubber
Technology
Kochi
India

Yadhu Krishna M. A
Cochin University of Science and Technology
Materials Science and Nano Enginering Lab
(MSNE-Lab)
Department of Polymer Science and Rubber
Technology
Kochi
India

Cochin University of Science and Technology
Department of Ship Technology
Kochi
India

Satheesh Babu
Cochin University of Science and Technology
Department of Ship Technology
Kochi
India

Sari P. Sasidharan
Cochin University of Science and Technology
Department of Polymer Science and Rubber
Technology
Kochi
India

Prasanth Raghavan
Cochin University of Science and Technology
Materials Science and Nano Enginering Lab
(MSNE-Lab)
Department of Polymer Science and Rubber
Technology
Kochi
India

Gyeongsang National University (GNU)
Department of Materials Engineering and
Convergence Technology
Jinju-daero
Republic of Korea

Mahatma Gandhi University
International and Inter University Centre for
Nanoscience and Nanotechnology (IIUCNN)
Kottayam
Kerala
India

Sindhu Karthika Ammini
Kinder Multispecialty Hospital
Department of Paediatrics
Kalamassery
Kochi
Kerala
India
dr.sindhukrishna@gmail.com

Chapter 11

Jemmyson Romário de Jesus
Federal University of Viçosa
Research Laboratory in bionanomaterials
LPbio
Department of Chemistry
36570-900 Viçosa
Minas Gerais
Brazil
jemmyson.jesus@ufv.br

Tatianny de Araujo Andrade
Federal University of Viçosa
Research Laboratory in bionanomaterials
LPbio
Department of Chemistry
36570-900 Viçosa
Minas Gerais
Brazil

Chapter 12

Jonathan Tersur Orasugh
University of Johannesburg
Department of Chemical Sciences
Doorfontein
Johannesburg 2028
South Africa

Centre for Nanostructures and Advanced Materials
DSI-CSIR Nanotechnology Innovation Centre
Council for Scientific and Industrial Research
Pretoria 0001
South Africa
E-Mail: calunivjto@gmail.com

Lesego Tabea Temane
University of Johannesburg
Department of Chemical Sciences
Doorfontein
Johannesburg 2028
South Africa

Centre for Nanostructures and Advanced Materials
DSI-CSIR Nanotechnology Innovation Centre
Council for Scientific and Industrial Research
Pretoria 0001
South Africa

Suprakash Sinha Ray
University of Johannesburg
Department of Chemical Sciences
Doorfontein
Johannesburg 2028
South Africa

Centre for Nanostructures and Advanced Materials
DSI-CSIR Nanotechnology Innovation Centre
Council for Scientific and Industrial Research
Pretoria 0001
South Africa
E-Mail: rsuprakas@csir.co.za

Chapter 13

Smrita Singh
IFTM University
School of Biotechnology
Moradabad
Uttar Pradesh
India

Tanzeel Ahmed
IFTM University
School of Biotechnology
Moradabad
Uttar Pradesh
India

Ashutosh Singh Chauhan
Air Pollution and stone conservation Laboratry
Archaeological Survey of India
Agra
Uttar Pradesh
India
E-Mail: ashu.chauhan1979@gmail.com

Part 1: Fundamentals of molecular imprinting-based smart carbon nanomaterials

Long Wu* and Jing Wei

Chapter 1
Introduction to molecular imprinting technology

Abstract: Molecular imprinting technology (MIT) is a cutting-edge method in polymer science that creates specific sites in polymers to recognize target molecules. The cornerstone of MIT is the creation of cavities in the polymer that match the target molecule in shape, size, and function, achieved by assembling a complex of the template molecule, monomers, and a cross-linker, which, after polymerization and template removal, leaves a customized site for molecule recognition. The applications are broad in biosensing, drug delivery, and environmental monitoring, thanks to its high selectivity and affinity for specific molecules. The evolution of the technology has enhanced the robustness and selectivity of the polymers while also integrating new materials into the approach. However, challenges still remain, such as imprinting large biomolecules and scaling up from labs to industry. Future research aims to overcome these hurdles, using interdisciplinary approaches to expand its applications and enhance its efficiency. By developing new materials with high specificity and sensitivity, MIT has significant potential to impact various sectors, particularly in healthcare and environmental science, offering great significance and potential in the advancing materials science and its interdisciplinary applications.

Keywords: Imprinted polymers, molecular recognition, nanotechnology, polymer materials

1.1 Introduction

Molecular imprinting technology (MIT) is a cutting-edge technique used to create polymer matrices with specific predetermined binding sites for a target molecule, akin to a "molecular lock-and-key" mechanism [1–3]. The fundamental concept of molecular imprinting involves the creation of template-shaped cavities in polymer matrices that have a memory of the template molecule, which is typically achieved by polymerizing functional and cross-linking monomers in the presence of a target molecule, also

*Corresponding author: Long Wu, School of Food Science and Engineering, Key Laboratory of Tropical Fruits and Vegetables Quality and Safety, State Administration for Market Regulation, Hainan University, Haikou 570228, PR China, e-mail: longquan.good@163.com
Jing Wei, Institute of Food Testing, Hainan Academy of Inspection and Testing, Key Laboratory of Tropical Fruits and Vegetables Quality and Safety, State Administration for Market Regulation, Hainan University, Haikou 570228, PR China

https://doi.org/10.1515/9783111503325-001

named as the template [4, 5]. Once the polymerization is complete, the template molecule is removed, leaving behind a cavity that is complementary in shape and functional group orientation to the template [6]. These cavities are thus highly specific to the template molecule and can selectively rebind to it in the presence of other substances.

In the realm of MIT, molecularly imprinted polymers (MIPs) stand out for their array of advantageous properties. For example, MIPs are not only cost-effective and straightforward to synthesize but also exhibit remarkable stability under stringent chemical and physical conditions [7]. Their reusability further enhances their appeal, making them a sustainable choice for various applications [8]. The utility of MIPs extends beyond their inherent qualities, as they have garnered significant interest across a spectrum of fields [9]. Particularly noteworthy is their application as selective adsorbents in solid-phase extraction (SPE) processes, where they demonstrate an exceptional ability to target and concentrate specific analytes from complex mixtures [10]. This selective adsorption is a critical step in sample preparation for subsequent analysis, ensuring the purity and integrity of the samples.

In addition to SPE, MIPs are also making a mark in the domain of chromatographic separation. Here, they serve as stationary phases that can be tailored to interact with target molecules through specific binding sites, leading to highly efficient and selective separations. This capability is particularly valuable in the purification and isolation of biomolecules, pharmaceuticals, and other compounds of interest [11–13]. Furthermore, the realm of chemical sensors has also benefited from the advent of MIPs [14]. These polymers can be integrated into sensor devices to provide selective detection of target analytes with high sensitivity and specificity, which is particularly important in environmental monitoring, food safety, and medical diagnostics, where the accurate identification and quantification of specific substances are paramount [15]. The versatility of MIPs is further highlighted by their potential for customization. By selecting appropriate template molecules and functional monomers, MIPs can be designed to recognize a wide range of target molecules, from small organic compounds to large biomolecules [16–18]. This tailor-made approach allows for the creation of highly specific recognition elements, which are invaluable in the development of advanced analytical techniques and devices.

The combination of MIPs with nanomaterials has shown potential in improving adsorption capacity, selectivity, and mass transfer rate for analytes [19]. For example, nanomaterials such as quantum dots [20], upconversion materials [21], plasmonic nanoparticles [22, 23], and magnetic nanoparticles [24] are being integrated with MIPs to leverage their unique physical responses for high-sensitivity detection signals. However, these nanomaterials often lack inherent molecular recognition capabilities, necessitating surface functionalization with biological molecules like antibodies or lectins, which can be challenging due to structural and stability differences. To address the limitations of traditional MIPs, surface molecular imprinting has emerged as an alternative method [25]. This involves anchoring the shell to a substrate like nanoparticles, which can mitigate some of the traditional MIPs' drawbacks and improve performance or provide new functionalities [26]. This method, particularly when applied to carbon-based MIPs, offers a robust platform for the selective recognition and binding of various molecules. In the context of carbon-based MIPs, the use of carbon materials, such as carbon nanotubes [27], graphene [28], or carbon dots [29], adds enhanced physical and chemical stability, increased surface area, and improved conductivity to the polymers [30]. These merits make carbon-based MIPs particularly valuable in applications where robustness and specificity are critical, such as in sensing, separation, and catalysis.

The introduction to MIT typically covers the historical development of the field, key milestones, and the evolution of various imprinting strategies. It delves into the different types of monomers and cross-linkers used, the polymerization techniques employed (such as free radical polymerization or controlled polymerization), and the methods for template removal, which are crucial steps in creating effective MIPs. Additionally, this chapter explores the theoretical aspects underpinning the molecular recognition mechanisms in MIPs, such as the thermodynamics and kinetics of the imprinting and rebinding processes, providing the necessary background to appreciate the intricacies and potential applications of this fascinating technique.

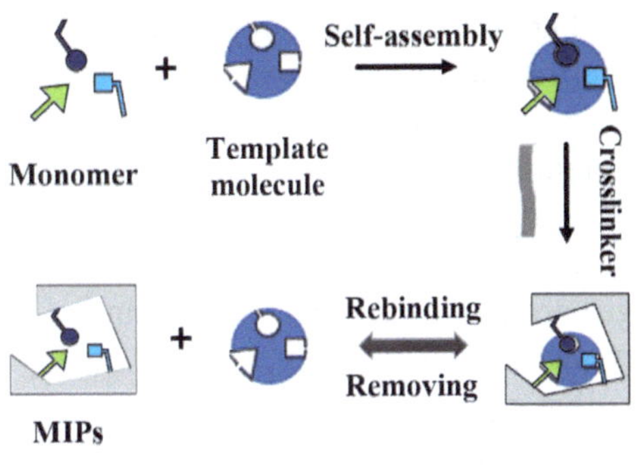

Figure 1.1: Schematic presentation of MIT.

1.2 Synthesis of MIPs

MIPs are polymer materials prepared through imprinting technology that have spatial structures and binding sites that perfectly match the template molecules, thus possessing specific recognition functions for the template molecules. In 1972, Wulff [31] first proposed the concept of MIPs and used polymers with enzyme-analogous structures for the resolution of racemates. In 1993, Vlatakis's group [32] reported on the molecular imprinting for creating selective recognition sites in synthetic polymers to prepare polymers that mimic antibody combining sites. Since then, MIPs have developed rapidly, with significant progress made in the principles of imprinting technology, the types of imprinted polymers, preparation methods, and applications [33, 34]. Due to the perfect match between the spatial structure and binding sites of the imprinted polymers and the template molecules, they exhibit specific or selective recognition functions for the template molecules (see Figure 1.1).

1.2.1 Principle of MIPs

MIPs are a type of recognition functional material that not only possesses the adsorption properties of adsorbent materials but also exhibits excellent selectivity [35]. The specific recognition function of MIPs is closely related to their unique spatial structure and binding sites (intermolecular interactions) (Figure 1.2). The process involves several steps that lead to the formation of a polymer matrix with cavities that are complementary in shape, size, and chemical functionality to the template [36]. During the preparation of MIPs, an expanded explanation of the molecular imprinting principle can be described as follows:

1) Template-monomer interaction: The process begins with the interaction between the template molecule and functional monomers. These monomers have functional groups that can form various types of intermolecular interactions with the template based on covalent interaction and non-covalent assembly such as hydrogen bonding, electrostatic interactions, and van der Waals forces. This interaction is crucial as it sets the stage for the selective binding sites within the polymer.
2) Formation of the pre-polymerization complex: Template molecules are mixed with the functional monomers to form a pre-polymerization complex. This complex is stabilized by the interactions between the template and the monomers, ensuring that the monomers are arranged around the template in a manner that will later translate into a complementary cavity in the polymer.
3) Cross-linking and polymerization: Cross-linking agent is introduced to the mixture. This agent reacts with the functional monomers to form a network of polymer chains, effectively creating a three-dimensional matrix. The cross-linking is essential for maintaining the structural integrity of the polymer and the shape of the binding sites after the template is removed.
4) Curing and template removal: The mixture is then subjected to conditions that initiate the polymerization reaction, such as heat or UV light, causing the monomers to polymerize and form a solid, cross-linked network. Once the polymerization is complete, the template molecules are removed from the polymer matrix, typically through a washing or elution process. This step leaves behind cavities within the polymer that are complementary to the template molecules.
5) Resulting MIP structure: The resulting MIP has a network of cavities that are specific to the template molecule. These cavities have the same shape, size, and chemical functionality as the template, allowing them to selectively rebind the target molecule with high specificity and affinity.

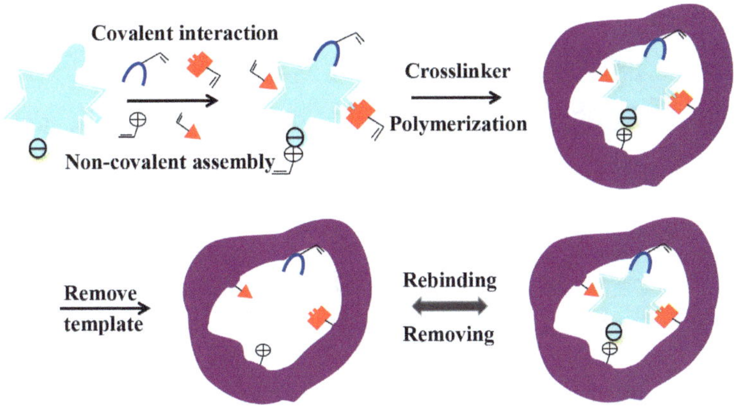

Figure 1.2: Scheme of the recognition principle of MIPs.

MIPs are used in various applications due to their high selectivity and affinity for the target molecules. They are particularly useful in analytical chemistry for sample preparation, separation techniques, and as recognition elements in various sensors. Additionally, they are employed in the pharmaceutical industry for controlled drug release and in environmental science for selective extraction and detection of analytes. This overview provides a comprehensive look at the molecular imprinting process, from initial interactions between the template and monomers to the final application of the MIPs. The technique continues to evolve, offering new possibilities for the development of highly selective materials in various fields.

1.2.2 Classification of MIPs

Normally, MIPs can be categorized based on the type of interaction between the template molecule and the functional monomers used in their synthesis [37]. They are covalent MIPs, where a chemical bond is formed between the template and the monomers (preassembled approach), and non-covalent MIPs, where interactions such as hydrogen bonding, electrostatic forces, or van der Waals forces are used (assembled approach). From the perspective of substrate recognition, MIPs can also be classified based on the type of substrate they are designed to recognize. There are MIPs for ion recognition, which are particularly useful in environmental monitoring and analytical chemistry. Organic small molecule imprinted polymers are widely used for selective separation and detection in various industries. Additionally, there are MIPs for biomacromolecule recognition, which are crucial in the fields of biomedicine and diagnostics.

The location of the template molecules within the MIP structure can also determine the classification. There are three main types based on this criterion (Figure 1.3).

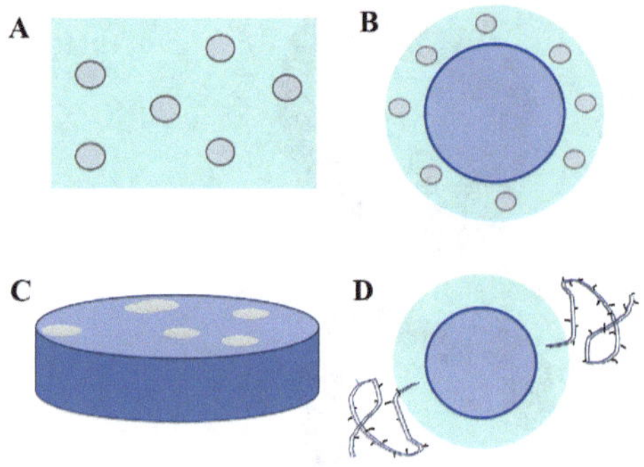

Figure 1.3: Schematic presentation of the classification of MIPs (BIPs, SIPs, and EIPs).

The first one is bulk imprinted polymers (BIPs). In this type, template molecules are dispersed throughout the volume of the polymer matrix (Figure 1.3A). The distribution is random, and template molecules located deeper within the matrix are more challenging to remove and have slower recognition rates. BIPs are typically used for the recognition of small molecules and metal ions. The second one is surface imprinted polymers (SIPs). Here, the template molecules are situated within the top layer or surface of the polymer (Figure 1.3B and C). This arrangement allows for faster template removal and higher recognition rates, making SIPs suitable for a broader range of substrates, including larger molecules. The third one is epitope imprinted polymers (EIPs). In this specialized type, only a part of the template molecule, typically a specific epitope, is bound to the surface of the polymer (Figure 3D). This method results in very rapid removal and recognition rates, which is particularly advantageous for the recognition of biomacromolecules, where high specificity and speed are required.

The distribution of the template molecules within the MIP has a significant impact on the practical application of the material. The ease of template removal and the rate of target molecule recognition are critical factors that determine the suitability of an MIP for specific applications. Ongoing research is focused on improving the performance of MIPs, such as enhancing their stability, selectivity, and the ability to imprint larger or more complex molecules. There is also a significant effort to scale up the production of MIPs for industrial applications while maintaining their high recognition properties.

1.2.2.1 Bulk imprinted polymers

BIPs are a class of advanced materials developed through a process that involves the creation of a network with specific recognition sites for target molecules. The synthesis of these polymers is achieved by thoroughly mixing a template molecule, which is the analyte of interest, with a functional monomer capable of forming a complex with the template through non-covalent interactions. A cross-linking agent is then added to ensure the stability and structural integrity of the polymer network. The mixture is subjected to a polymerization reaction, which can be initiated by heat or radiation, resulting in a cross-linked polymer matrix. After the reaction, the template molecules are removed, leaving behind cavities that are complementary in shape and function to the original template. This process, known as imprinting, endows the polymer with selective binding properties. The resulting bulk MIP is then purified through drying, crushing, grinding, and sieving to obtain a material that can be used for various applications, such as selective adsorption, catalysis, and sensor development.

For example, Hussain et al. [38] studied an innovative synthesis of siloxane-decorated polyurethane triazole (Si-PUT) coatings, which were developed via a solvent-free and catalyst-free bulk polymerization process involving azide-alkyne cycloaddition. The materials were designed to exhibit excellent surface hydrophobicity, high transparency, and significant anticorrosion properties. The methods described the functional modification of aminopropyl trimethoxysilane and the synthesis of azide-terminated urethane monomers, followed by their cycloaddition to form highly cross-linked Si-PUT matrices. The application of the coatings is demonstrated through their performance as corrosion-resistant coatings, with the potential to formulate environmentally friendly and multifunctional high-solid protective coatings. The study included a comprehensive assessment of the physicochemical properties of the polymers and their evaluation as corrosion-resistant coatings using electrochemical methods, showcasing the potential of Si-PUTs as promising materials for sustainable and protective applications in the coatings industry.

In addition, Nageib et al. [39] presented a comprehensive study on the computational modeling, analysis, and synthesis of MIPs, focusing on their application in detecting sulfamethoxazole (SMX), a widely used antibiotic in the dairy industry. The research aimed to address the need for a rapid and efficient method to monitor SMX levels in milk due to the potential health risks and antimicrobial resistance posed by its overuse. The study employs two functional monomers, methacrylic acid (MAA) and acrylamide (AA), in the bulk polymerization process to create MIPs with specific binding to SMX. Computational predictions using PyRx-Virtual Screening Tool were utilized to analyze the binding affinity between SMX and the monomers, with MAA showing a higher affinity. This was experimentally verified via the synthesis of MIPs and non-imprinted polymers, followed by batch rebinding experiments in ultrapure water and phosphate buffer solution. The results demonstrated that MIPs synthesized with MAA exhibited a higher imprinting factor, indicating a superior binding capacity for SMX compared to those with AA. The study concludes that MIP-MAA can be effectively used for the selective detection of SMX in complex matrices, offering a simple, stable, and cost-effective alternative to conventional detection methods such as immunoassays, gas chromatography/mass spectrometry (MS), and high-performance liquid chromatography (HPLC).

1.2.2.2 Surface imprinted polymers

SIPs are a subtype of MIPs that feature the recognition sites or cavities on the surface of the material rather than throughout the bulk [40]. This allows for easier access to the target molecules and often results in improved interaction kinetics. The imprinted sites are located on the surface of the material, which can facilitate rapid binding and release of the target molecules. Since the templates are removed from the surface, it can be easier to regenerate the SIPs without significant loss of binding sites. SIPs are typically synthesized on a solid support, which can provide mechanical stability and allow for easy handling and reuse. In this regard, SIPs are ideal for applications requiring rapid interaction with the target, such as sensors, diagnostic devices, and certain separation techniques.

As an example, Cheng et al. presented a study on the development of a surface MIP (MIP-IL@UIO-66) for the rapid and highly selective adsorption of SMZ in environmental samples [41]. The polymer was synthesized by polymerization on the surface of a metal-organic framework (UiO-66), which served as a carrier due to its large specific surface area. The resulting MIP-IL@UIO-66 demonstrated excellent adsorption capacity (284.66 mg g^{-1}), rapid equilibrium rate (10 min), and high selectivity (selectivity coefficient of 11.36) for SMZ. The adsorption process was primarily driven by hydrogen bonds, electrostatic and π-π interactions, and the molecular memory effect. The MIP-IL@UIO-66 was also found to be reusable, maintaining its performance over five cycles, which was successfully applied as an SPE adsorbent and coupled with HPLC to detect SMZ in various samples, including water, soil, egg, and pork, with recoveries between 91% and 106%. The study highlights the potential of MIP-IL@UIO-66 as an efficient, selective, and reusable adsorbent for environmental applications.

1.2.2.3 Epitope imprinted polymers

EIPs are a special type of MIPs designed to mimic the binding properties of antibodies, specifically recognizing epitopes (specific sites) on protein surfaces, and they are synthesized to mimic the high specificity of antibodies for their antigens (epitopes) [42]. EIPs are used as the target molecule is too large to be effectively imprinted or when only a specific part of the molecule is of interest. The process involves the use of a template molecule that corresponds to the epitope or a part of the protein. This template is then imprinted into the polymer matrix, creating specific binding sites that can recognize the epitope with high specificity. The binding sites in the polymer are complementary to the shape, size, and chemical functionality of the target epitope. This allows the polymer to selectively bind to the epitope as if it were an antibody.

EIPs are particularly useful in the field of biochemistry and medical diagnostics, which can be employed in the detection and analysis of proteins, in immunoassays, and for the development of synthetic substitutes for antibodies in various analytical and bioanalytical applications [43]. For instance, Xing et al. [44] outlined a method known as the dual MIP-based plasmonic immunosandwich assay (duMIP-PISA) for the specific and sensitive detection of protein biomarkers in complex biological samples. The method utilized C-terminal epitope-imprinted gold nanoparticle monolayers on a glass slide as a plasmonic substrate for the extraction of target proteins, and N-terminal epitope-imprinted Raman-responsive silver-silica nanoparticles as nanotags for the specific labeling of captured proteins. The duMIP-PISA is demonstrated with neuron-specific enolase as a model protein biomarker in human serums, allowing the differentiation of small cell lung cancer patients from healthy individuals. Compared to traditional enzyme-linked immunosorbent assay (ELISA), the duMIP-PISA offers a simpler procedure, faster speed, lower sample volume requirement, and a wider linear range, showcasing its potential in disease diagnosis and other important applications.

1.2.3 Construction of MIPs

The preparation of MIPs involves several key steps and can be conducted using various polymerization techniques. The conventional preparation methods mainly include free radical polymerization, bulk polymerization, solution polymerization, emulsion polymerization, and suspension polymerization. This section details the methods of MIP preparation, from conventional to modern approaches, highlighting the evolution and current trends in the field.

1.2.3.1 Free radical polymerization

The cornerstone of MIP synthesis is free radical polymerization, which involves the copolymerization of functional monomers and cross-linkers around a template molecule. The functional monomers chosen are capable of interacting with the template through various non-covalent interactions, such as hydrogen bonding, van der Waals force, or electrostatic interactions. To obtain complex polymer architectures using radical polymerization of multivinyl monomers, Gao et al. [45] suggested that conventional free radical polymerization is a widely used method for synthesizing industrially important vinyl polymers, and it hints at future directions in the field.

It has long been an important subject to access functional materials in free radical polymerization by designing novel initiators and monomers as well as developing new polymerization techniques.

Liu [46] discussed an extended model of radical polymerization of acrylates and methacrylates at high temperatures, showing a departure from classic free-radical kinetics. Biobased polymer molecules are a goal for the future, and functionality now plays a much greater role in various applications. Pirman et al. [47] outlined the future of free radical polymerizations, indicating the potential for new developments in the synthesis of vinyl polymers. It is important to highlight the synthesis of complex polymer architectures, the use of visible light to initiate polymerization, and the development of novel methods for controlling the polymerization process to produce polymers with specific properties and applications.

1.2.3.2 Bulk polymerization

In bulk polymerization, the template, functional monomers, cross-linkers, and initiators are mixed and polymerized in the absence of a solvent. This method is straightforward but may result in a heterogeneous distribution of the template within the polymer matrix. Jašo et al. [48] described the materials and methods used in a study of the bulk polymerization of methyl methacrylate (MMA). The process involved washing the MMA monomer and initiator, preparing a solution, and conducting the polymerization under isothermal conditions. The results showed three characteristic moments in the differential scanning calorimetry thermograms: the onset of acceleration, the maximum of the polymerization rate, and the end of polymerization. Suzuki et al. investigated phase separation during the bulk free-radical polymerization of MMA using the redox reaction of benzoyl peroxide with *N,N*-dimethyl-*p*-toluidine (DMT) for initiation at room temperature. It was found that phase separation into monomer-rich and polymer-rich phases occurs simultaneously with the onset of the Trommsdorff effect [49].

1.2.3.3 Solution polymerization

To overcome the limitations of bulk polymerization, solution polymerization is employed, where all components are dissolved in a suitable solvent. This technique allows for better control over the polymerization process and can lead to a more homogeneous distribution of the template.

By employing dissipative particle dynamics simulations combined with stochastic polymerization models, Shi et al. [50] utilized dissipative particle dynamics simulations combined with stochastic polymerization models to conduct a detailed study on a coarse-grained model that closely resembles the real monomer structure. With an incorporation of bending angle potentials into coarse-grained models representing supramolecular reactive monomers, the study achieved simulations for various types of multiple hydrogen bonds. The simulation results indicated that for the interfacial polymerization system, the volume of the monomer caused a significant steric hindrance effect, which in turn led to a lower average degree of polymerization of the product examined in the relationship between the solvent-to-monomer ratio and the resin's appearance and water solubility.

Anderson et al. [51] presented a breakthrough in the synthesis of ultrahigh-molecular-weight (UHMW) polymers via topochemical polymerization. They demonstrated the polymerization of *para*-azaquinodimethane compounds, which can undergo visible light and thermally initiated polymerization in the solid state. This process allowed the production of polymers with high molecular weights, a challenge previously faced in the field. The study showed that the polymers are soluble and can be functionalized post-polymerization, which broadens their potential applications. Notably, the polymers exhibit excellent dielectric properties, making them promising candidates for energy storage applications. The use of cryoelectron microscopy for the first time to determine the crystal structure of a topochemical polymer product is another highlight of this work. This method offered a practical alternative to traditional X-ray diffraction, which required larger single crystals. The findings paved the way for the design and synthesis of a wide range of functional UHMW polymers with tailored properties for various applications.

1.2.3.4 Emulsion and suspension polymerization

Emulsion and suspension polymerization are variations that involve the dispersion of the monomer droplets in an aqueous phase. Emulsion polymerization uses surfactants to stabilize the droplets, while suspension polymerization forms larger, more stable particles. Both methods can produce MIPs with well-defined structures and improved accessibility to the binding sites.

Jahanzad et al. [52] described the production of polymer beads through suspension polymerization, which often results in a broad size range of particles. The control over particle size is maintained by the stirring system and the stabilizer system. The article further explained the formation of different particle types such as bead particles, satellite particles, and emulsion particles, and the conditions leading to their formation. It further discusses the role of initiators in the formation of emulsion particles and the influence of secondary particle formation via micellar nucleation. The article also outlined the stages of a typical suspension polymerization reaction and the impact of an inhibitor on the kinetics of polymerization.

Chaudhary and Sharma [53] described different heterogeneous polymerization techniques, focusing on suspension polymerization. They discussed the synthesis of polystyrene cross-linked with divinylbenzene copolymer through suspension polymerization. The study examined the effects of different synthesis parameters, including agitation speed, temperature, initiator, cross-linker, and diluent, on particle surface morphology, size and distribution, surface area, and cross-linking density. Also, the review provided a detailed understanding of how suspension polymerization works, including the formation of organic and aqueous phases, the role of stirrers and suspension agents, and the conversion of monomer droplets into polymer particles.

1.2.3.5 Thermal and radiation-induced polymerization

Conventional thermal initiation using an initiator and heat is a common method to start the polymerization. However, this approach can be time-consuming and energy-intensive. To address the issues, radiation-induced polymerization techniques, such as UV light, gamma rays, electron beams, and microwave irradiation, have been explored. These methods offer faster reaction times, lower energy consumption, and in some cases, the elimination of the need for initiators.

Pola [54] provided a comprehensive review of the research on the thermal reactive modification of polymer surfaces using infrared (IR) laser radiation, which focused on the comparison between ablative and non-ablative laser irradiations and their effects on polymer surfaces. The study revealed that IR laser radiation can lead to unique thermal processes that result in the formation of polymer films with specific structural and compositional changes. The authors highlight that the efficiency of the laser degradation process depends on factors such as the absorbance of the polymer at the laser wavelength, thermal diffusivity, and the interaction between the polymer and the substrate. It offered a different approach compared to conventional heating, with the potential to achieve new goals in polymer research.

Due to its high energy and ability to penetrate solid materials, ionizing radiation is an effective tool for the preparation of nanomaterials. It can generate free radicals homogeneously in the media, allowing for the irradiation of monomers or polymers at or below room temperature. In the case of polymer-based nanomaterials, radiation is used to create nanoparticles, nanogels, and polymer-clay nanocomposites. Güven [55] discussed the synthesis of protein-based nanoparticles for drug delivery and the creation of thermoresponsive cell culture surfaces using radiation-induced grafting. The use of radiation-induced grafting was discussed for the functionalization of track-etched membranes, which can be used in energy conversion and storage. It is supposed that the radiation engineering is a promising green technology for the preparation and modification of nanomaterials, with potential to replace conventional techniques as new nanomaterials are developed.

1.2.3.6 Green solvent techniques

Considering environmental and safety concerns, there is a growing interest in using environmentally friendly solvents in MIP synthesis. Supercritical carbon dioxide (CO_2), ionic liquids, and deep eutectic solvents are emerging as green alternatives that can facilitate the polymerization process while reducing the environmental impact.

To underscore the importance of green synthesis methods in the development of MIPs for SPE, focusing on the use of environmentally benign materials and processes to create highly selective and efficient sorbents that meet the challenges of modern analytical chemistry, Ma et al. [56] highlighted the development and application of green synthesized MIPs as sorbents in SPE techniques. MIPs are tailor-made polymers with molecular recognition capabilities, offering high selectivity for target analytes. The green synthesis of MIPs involves the use of nontoxic metal ions, natural ligands, environmentally friendly solvents, and economic synthesis routes, which align with the principles of green chemistry and green sample preparation. Martins et al. [57] discussed the development of greener approaches to synthesize MIPs, which are used for the molecular recognition capabilities in various analytical techniques. Traditional MIP synthesis often involves toxic reagents and solvents, leading to environmental concerns. The study highlighted the use of green monomers, porogenic solvents, and template molecules that are less harmful to the environment. The study emphasized the application of dummy templates to replace toxic and carcinogenic templates, reducing health risks and environmental impact. Additionally, it pointed out the use of computational model to optimize MIP synthesis, thus minimizing waste and resource consumption. The integration of green MIPs in chromatographic and MS applications was discussed, showing their potential in creating more sustainable and eco-friendly analytical techniques.

Besides, electropolymerization (EP) is another method used to prepare polymers by the application of an electric current, which facilitates the polymerization of monomers at the electrode surface (Figure 1.4). This method is particularly useful for the synthesis of conducting polymers and MIPs with electrical conductivity. In the process of EP, a potential is applied across an electrode (usually a working electrode in an electrochemical cell) and a counter electrode [58]. The monomers are oxidized or reduced at the electrode surface, leading to the formation of a polymer film [59]. This film can be used for various applications, including sensors, capacitors, and as a stationary phase in chromatography. EP allows for the in situ formation of MIPs on electrode surfaces for use in sensor development. The technique can provide a stable and conductive MIP layer that can selectively bind target molecules, making it useful for detecting specific analytes in complex samples.

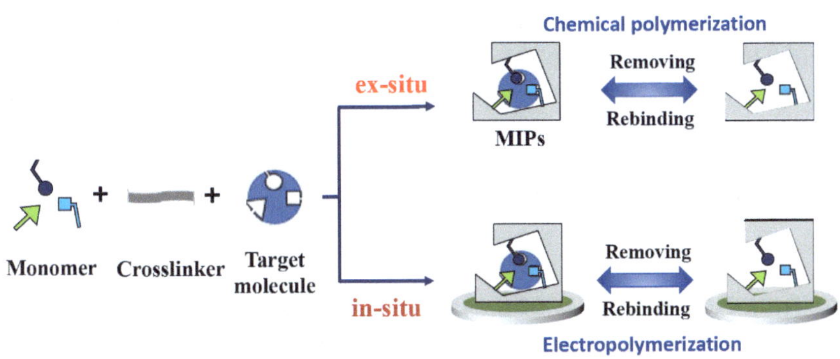

Figure 1.4: Schematic view of MIP synthesis via a chemical/electropolymerization method. Reproduced with permission from Zhou et al. [15]. Copyright 2022, MDPI publisher.

1.3 Applications

Currently, MIPs have been widely applied in various fields such as analytical detection, separation and extraction, environmental protection, and controlled drug release [60–62]. With the advancement of technology and the expansion of application fields, the requirements for the structural design and preparation methods of MIPs are becoming increasingly high, and an increasing number of MIPs with different structures are being prepared using various methods [63]. In a word, molecular imprinting is a versatile and powerful tool with a wide range of applications across various fields. Its ability to create highly specific artificial receptors makes it a valuable technique for many analytical and synthetic challenges.

1.3.1 Analytical and environmental science

Molecular imprinting is used to create materials with high specificity for target molecules, which is critical in the analysis of complex samples. Besides, MIPs are employed in SPE to selectively concentrate and purify target analytes from environmental and biological samples. To effectively enhance sample extraction, MIPs along with associated techniques are designed to have specialized binding capacities for specific target analytes, offering high selectivity in analytical and environmental science.

Xu et al. [64] presented a comprehensive review on the integration of MIPs with paper-based analytical devices (PADs) for various applications. The review highlights the application of MIP-PADs in on-site environmental analysis, food safety monitoring, point-of-care detection, biomarker detection, and exposure assessment. MIP-PADs have been used to detect a range of pollutants and contaminants, including nitrophenols, pentachlorophenol, neonicotinoids, and heavy metal ions. They offer advantages such as low cost, rapid detection, and ease of operation, making them suitable for resource-limited settings. Gkika et al. [65] summarized the application of MIPs as tools for environmental separation. The review emphasized the importance of MIPs in sample preparation, highlighting their high selectivity, reusability, and sensitivity. Also, it outlined the use of MIPs in SPE for the pre-concentration of target pollutants in water samples, and discussed the development of MIP-based sensors for detecting environmental contaminants and biomarkers, showcasing the versatility of MIPs in analytical chemistry and environmental science.

1.3.2 Pharmaceutical and medical applications

Due to the advantages of selective recognition, enhanced drug loading, sustained release, and robustness in harsh conditions, MIPs as versatile drug delivery materials have been widely studied [66]. MIPs are utilized in high-throughput screening for drug candidates, offering a stable and cost-effective alternative to biological receptors [67]. In drug delivery, molecular imprinting can also be used to design systems that release therapeutic agents selectively at the site of action.

For example, Hasanah and Susanti [68] reported a comprehensive review on the design and synthesis of MIPs for the separation and analysis of impurities in pharmaceutical products. In the study, the critical role of MIPs was highlighted in ensuring drug safety by selectively separating impurities that can compromise the pharmacological and toxicological profile of active pharmaceutical ingredients. Besides, it underlined the potential of MIPs in advancing pharmaceutical impurity analysis and separation, and indicated further research to expand MIPs to other types of impurities and to optimize the synthesis methods for improved efficiency and cost-effectiveness, thus contributing to the safety and quality of pharmaceutical products.

1.3.3 Food safety and control

MIP-coupled techniques play an important role in safeguarding the integrity of the food safety [69]. MIPs are designed to selectively bind and identify trace amounts of harmful substances, thus enhancing the performance of existing analytical methods. A critical application of MIPs in the food industry is the detection and quantification of pesticide residues on agricultural products. These residues can be hazardous to human health, and their levels are strictly regulated by safety standards. MIPs provide a robust platform for the extraction and concentration of these residues from complex matrices, allowing for sensitive and accurate measurement. By facilitating compliance with these standards, MIPs contribute to the reduction of health risks associated with pesticide exposure.

Moreover, MIPs can be synthesized to target a broad spectrum of contaminants, including heavy metals, toxins, and microbial pathogens [70], which are prevalent concerns in the food industry. The technology is also valuable in the authentication of food products, ensuring their genuineness and preventing economic fraud. With the potential to be integrated into portable detection devices and smart packaging, MIPs are set to revolutionize on-site monitoring and real-time food safety assessment, offering consumers and manufacturers greater assurance and control over the quality and safety of the food products they consume and handle.

1.3.4 Biosensing and diagnostics

MIPs are synthesized to have a high affinity for target molecules, such as proteins, small molecules, or even whole cells, making them ideal for the selective detection of biomarkers [71]. MIPs can be integrated into various detection platforms, including optical, electrochemical, and piezoelectric systems, to transduce the binding event into a measurable signal. The use of MIPs in biosensors allows for the detection of trace amounts of target molecules in complex biological samples, such as blood, urine, or tissue homogenates, with minimal interference from other components.

Early diagnosis is another area where MIPs show great promise. The ability to identify and quantify biomarkers at an early stage of disease is crucial for effective treatment and improved patient outcomes. MIPs can be designed to recognize specific biomarkers associated with various diseases, including cancer, cardiovascular diseases, and neurodegenerative disorders [72]. By incorporating MIPs into diagnostic assays, it is possible to achieve high sensitivity and selectivity, which are essential for the accurate detection of disease biomarkers at low concentrations. Furthermore, MIPs offer several advantages over traditional biological receptors, such as antibodies. They are more stable, resistant to harsh conditions, and have a longer shelf life, making them suitable for repeated use and storage.

Additionally, MIPs do not suffer from the risk of lot-to-lot variability and can be produced in large quantities at a lower cost compared to antibodies. The development of point-of-care (POC) devices using MIPs is another exciting development [73]. The devices can be used for rapid, on-site testing, providing immediate results to patients and healthcare professionals. The simplicity, portability, and cost-effectiveness of MIP-based POC devices make them particularly attractive for use in resource-limited settings or for large-scale screening programs.

MIT is a versatile and powerful tool in the development of specific biosensors and diagnostic tools. Its application in early disease detection and POC diagnostics has the potential to transform healthcare, enabling more effective disease management and improved patient outcomes. As research in this field continues to advance, it is expected that MIPs will play an increasingly important role in the prevention, diagnosis, and treatment of various diseases.

1.3.5 Catalysis and chemical synthesis

MIT is a versatile approach that has been extended to catalysis field, particularly in the creation of enantioselective catalysts. These catalysts are designed to differentiate between enantiomers, the mirror-image isomers of chiral molecules, which is crucial in the pharmaceutical industry, where the desired biological activity often resides in just one enantiomer [74]. MIPs have shown potential in providing a stable and selective environment for enantioselective catalysis, mimicking the action of naturally occurring enzymes but with the added benefit of being more robust and resistant to various reaction conditions.

The integration of MIPs into materials that exhibit specific binding or release behavior is another significant application in chemical synthesis. These materials can be tailored to recognize and selectively bind target molecules, facilitating a controlled release of active ingredients or a selective catalysis process [75]. At the nanoscale, molecular imprinting is employed to fabricate nanoparticles with tailored recognition properties. These nanoparticles can have a significant surface-area-to-volume ratio, enhancing their interaction with target molecules and making them highly effective in catalysis and separation processes [76]. The unique size and shape of these imprinted nanoparticles allow for advanced applications, such as in sensors for detecting trace amounts of specific compounds or in targeted drug delivery systems that can release their payload upon encountering specific biological markers. Moreover, the use of MIPs in catalysis and chemical synthesis is not limited to enantioselective reactions [77]. They are also explored in various other types of selective catalysis, including the synthesis of complex organic molecules, where the imprinted sites can provide the necessary steric and electronic environment to facilitate the desired reaction.

In conclusion, MIT offers a promising avenue for the development of advanced catalysts and materials for chemical synthesis. The ability to create MIPs with tailored recognition properties allows for greater control over reaction efficiency, leading to more precise and environmentally friendly synthetic processes. As research progresses, it is anticipated that MIPs will continue to expand its impact in catalysis and chemical synthesis, contributing to the development of innovative materials and processes that meet the challenges of modern chemistry.

1.3.6 Forensic and security applications

MIPs have been increasingly recognized for its potential in forensic and security applications, particularly in the detection of specific controlled substances, explosives, and other compounds of interest, which offer a synthetic alternative to biological receptors, providing a high degree of specificity and selectivity for target molecules, which is crucial in forensic analysis and security screening [78]. In forensic science, the identification and analysis of controlled substances like drugs or their metabolites are paramount for legal proceedings. Specifically, MIPs can be designed to recognize these substances with high precision, serving new functions in the development of sensors and assay systems that can detect and quantify even trace amounts of these compounds in complex samples, such as blood, urine, or hair [79]. This capability is particularly valuable in cases where the presence of a controlled substance must be established beyond reasonable doubt.

In addition, MIPs can be easily prepared to selectively bind to the unique chemical signatures of various explosives, enabling the creation of portable detection devices that can be deployed at points of entry, such as airports and border crossings [80]. The devices can provide a rapid and reliable means of screening for explosives, enhancing security measures and potentially thwarting dangerous situations. Furthermore, MIPs can also be utilized in the detection of chemical warfare agents and toxic industrial chemicals, which are of significant concern in both military and civilian settings. The high selectivity of MIPs allows for the differentiation between similar compounds, ensuring that only the compound of interest is detected, reducing the likelihood of false positives.

The development of MIP-based sensors for forensic and security applications is an active area of research, with ongoing efforts to improve the sensitivity, stability, and response time of these devices. As the technology matures, it is expected that MIPs will play an increasingly prominent role in the forensic and security sectors, providing a reliable and effective means of identifying and analyzing compounds of interest.

1.4 Conclusions and perspectives

MIT is an avant-garde approach in polymer science that fabricates polymers with specific binding sites tailored to recognize target molecules. The essence of MIT lies in the formation of cavities within the polymer matrix that are a mirror image of the target molecule in terms of shape, size, and functionality. This is achieved by polymerizing functional and cross-linking monomers in the presence of the target molecule, known as the template. Post-polymerization, the template is extracted, leaving behind a cavity that is highly specific for the template molecule, enabling selective rebinding amid other substances.

MIPs are noteworthy for their advantageous properties, such as cost-effectiveness, ease of synthesis, and remarkable stability under harsh chemical and physical conditions. Their reusability adds to their sustainability, making them appealing for a variety of applications. MIPs have been instrumental as selective adsorbents in SPE processes, showing an exceptional ability to target and concentrate specific analytes from complex mixtures. They play a significant role in chromatographic separation as stationary phases, offering efficient and selective separations, and in the realm of chemical sensors, providing selective detection of target analytes with high sensitivity and specificity. The integration of MIPs with nanomaterials has shown promise in enhancing adsorption capacity, selectivity, and mass transfer rate for analytes. Materials such as quantum dots, upconversion materials, plasmonic nanoparticles, and magnetic nanoparticles are being combined with MIPs to leverage their unique physical responses for high-sensitivity detection signals. Surface molecular imprinting has emerged as an alternative method to address the limitations of traditional MIPs, involving the anchoring of the shell to a substrate like nanoparticles, which can improve performance or provide new functionalities.

Looking ahead, the field of MIT is poised for further advancements. Challenges such as imprinting large biomolecules and scaling up from lab to industry remain. Future research aims to overcome these hurdles through interdisciplinary approaches, expanding the applications of MIPs, and enhancing their efficiency. The development of new materials with high specificity and sensitivity will significantly impact various sectors, particularly healthcare and environmental science. As the technology evolves, it holds great significance and potential in advancing materials science and its interdisciplinary applications, offering new possibilities for the development of highly selective materials in various fields.

References

[1] Song J., Lin X., Ee L. Y., & Li S. F. (2024). Research progress on molecularly imprinted materials for the screening and identification of organic pollutants, *Chemosensors*. 12(5), 72.

[2] Ostovan A., Arabi M., Wang Y., Li J., Li B., Wang X., & Chen L. (2022). Greenificated molecularly imprinted materials for advanced applications, *Advanced Materials*. 34(42), 2203154.

[3] Wang X., Lyu H., Hu Z., & Shen B. (2024). Application of molecular imprinting for targeted removal of organic contaminants and resistance genes from water: A review, *Journal of Environmental Chemical Engineering*. 12(2), 112068.

[4] Arabi M., Ostovan A., Li J., Wang X., Zhang Z., Choo J., & Chen L. (2021). Molecular imprinting: Green perspectives and strategies, *Advanced Materials*. 33(30), 2100543.

[5] Chen R. N., Kang S. H., Li J., Lu L. N., Luo X. P., & Wu L. (2021). Comparison and recent progress of molecular imprinting technology and dummy template molecular imprinting technology, *Analytical Methods*. 13(39), 4538–4556.

[6] Reville E. K., Sylvester E. H., Benware S. J., Negi S. S., & Berda E. B. (2022). Customizable molecular recognition: Advancements in design, synthesis, and application of molecularly imprinted polymers, *Polymer Chemistry*. 13(23), 3387–3411.

[7] Wei J., Liu C., Wu T., Zeng W., Hu B., Zhou S., & Wu L. (2022). A review of current status of ratiometric molecularly imprinted electrochemical sensors: From design to applications, *Analytica Chimica Acta*. 1230, 340273.

[8] Kong J., Xu X., Ma Y., Miao J., & Bian X. (2023). Rapid and sensitive detection of sulfamethizole using a reusable molecularly imprinted electrochemical sensor, *Foods*. 12(8), 1693.

[9] Wu L., Li G., Xu X., Zhu L., Huang R., & Chen X. (2019). Application of nano-ELISA in food analysis: Recent advances and challenges, *TrAC Trends in Analytical Chemistry*. 113, 140–156.

[10] Wan Q., Liu H., Deng Z., Bu J., Li T., Yang Y., & Zhong S. (2021). A critical review of molecularly imprinted solid phase extraction technology, *Journal of Polymer Research*. 28, 1–16.

[11] Song Z., Li J., Lu W., Li B., Yang G., Bi Y., . . . & Chen L. (2022). Molecularly imprinted polymers based materials and their applications in chromatographic and electrophoretic separations, *TrAC Trends in Analytical Chemistry*. 146, 116504.

[12] Bouvarel T., Delaunay N., & Pichon V. (2021). Molecularly imprinted polymers in miniaturized extraction and separation devices, *Journal of Separation Science*. 44(8), 1727–1751.

[13] Song Z., Song Y., Wang Y., Liu J., Wang Y., Lin W., . . . & Chen L. (2022). Chromatographic performance of zidovudine imprinted polymers coated silica stationary phases, *Talanta*. 239, 123115.

[14] Wu L., Zhang C., Long Y., Chen Q., Zhang W., & Liu G. (2022). Food additives: From functions to analytical methods, *Critical Reviews in Food Science and Nutrition*. 62(30), 8497–8517.

[15] Zhou S., Liu C., Lin J., Zhu Z., Hu B., & Wu L. (2022). Towards development of molecularly imprinted electrochemical sensors for food and drug safety: Progress and trends, *Biosensors*. 12(6), 369.

[16] Xu J., Miao H., Wang J., & Pan G. (2020). Molecularly imprinted synthetic antibodies: From chemical design to biomedical applications, *Small*. 16(27), 1906644.

[17] Sarvutiene J., Prentice U., Ramanavicius S., & Ramanavicius A. (2024). Molecular imprinting technology for biomedical applications, *Biotechnology Advances*. 71, 108318.

[18] Liu Y., Wang L., Li H., Zhao L., Ma Y., Zhang Y., . . . & Wei Y. (2024). Rigorous recognition mode analysis of molecularly imprinted polymers – Rational design, challenges, and opportunities, *Progress in Polymer Science*. 150, 101790.

[19] Cg A. M., & Varghese A. (2023). Recent advances in nanomaterials based molecularly imprinted electrochemical sensors, *Critical Reviews in Analytical Chemistry*. 53(1), 88–97.

[20] Mantashloo R., & Bahar S. (2023). Synthesis of magnetic graphene quantum dots based molecularly imprinted polymers for fluorescent determination of quercetin, *Microchemical Journal*. 185, 108233.

[21] Mao T., Shi X., Liu M., Cheng Y., Fang C., & Luo X. (2024). Rapid and efficient detection of perfluorooctanoic acid by up-conversion molecular imprinting fluorescent probe, *Journal of Physics and Chemistry of Solids*. 187, 111879.

[22] Wu L., Yan H., Li G., Xu X., Zhu L., Chen X., & Wang J. (2019). Surface-imprinted gold nanoparticle-based surface-enhanced Raman scattering for sensitive and specific detection of patulin in food samples, *Food Analytical Methods*. 12, 1648–1657.

[23] Yin W., Wu L., Ding F., Li Q., Wang P., Li J., . . . Han H. (2018). Surface-imprinted SiO_2@Ag nanoparticles for the selective detection of BPA using surface enhanced Raman scattering, *Sensors and Actuators B: Chemical*. 258, 566–573.

[24] Wu L., Li Y., Han Y., Liu X., Han B., Mao H., & Chen Q. (2024). A dual-mode optical sensor for sensitive detection of saxitoxin in shellfish based on three-in-one functional nanozymes, *Journal of Food Composition and Analysis*. 130, 106190.

[25] Zhu Y., Wu L., Yan H., Lu Z., Yin W., & Han H. (2020). Enzyme induced molecularly imprinted polymer on SERS substrate for ultrasensitive detection of patulin, *Analytica Chimica Acta*. 1101, 111–119.

[26] Wang L., Yu J., Wang X., Li J., & Chen L. (2023). Molecular imprinting-based nanocomposite adsorbents for typical pollutants removal, *Journal of Hazardous Materials Letters*. 4, 100073.

[27] Su J., & Su X. (2024). A multi-walled carbon nanotubes-based molecular imprinting electrochemical sensor for the detection of Aspartame in sports beverage, *Journal of Food Measurement and Characterization*. 18(1), 618–624.

[28] Soman G., Vandana M., & Hegde G. (2023). Molecularly imprinted graphene based biosensor as effective tool for electrochemical sensing of uric acid, *Sensors International*. 4, 100243.

[29] Keçili R., Hussain C. G., & Hussain C. M. (2023). Emerging trends in green carbon dots coated with molecularly imprinted polymers for sensor platforms, *TrAC Trends in Analytical Chemistry*. 166, 117205.

[30] Siripongpreda T., Hoven V. P., Narupai B., & Rodthongkum N. (2023). Emerging 3D printing based on polymers and nanomaterial additives: Enhancement of properties and potential applications, *European Polymer Journal*. 184, 111806.

[31] Wulff G. (1972). The use of polymers with enzyme-analogous structures for the resolution of racemates, *Angewandte Chemie International Edition*. 11, 341.

[32] Vlatakis G., Andersson L. I., Müller R., & Mosbach K. (1993). Drug assay using antibody mimics made by molecular imprinting, *Nature*. 361(6413), 645–647.

[33] McCluskey A., Holdsworth C. I., & Bowyer M. C. (2007). Molecularly imprinted polymers (MIPs): Sensing, an explosive new opportunity?, *Organic & Biomolecular Chemistry*. 5(20), 3233–3244.

[34] Huang Y., & Wang R. (2018). Review on fundamentals, preparations and applications of imprinted polymers, *Current Organic Chemistry*. 22(16), 1600–1618.

[35] Cui Y., He Z., Xu Y., Su Y., Ding L., & Li Y. (2021). Fabrication of molecularly imprinted polymers with tunable adsorption capability based on solvent-responsive cross-linker, *Chemical Engineering Journal*. 405, 126608.

[36] Ertürk G., & Mattiasson B. (2017). Molecular imprinting techniques used for the preparation of biosensors, *Sensors*. 17(2), 288.

[37] Fresco-Cala B., Batista A. D., & Cárdenas S. (2020). Molecularly imprinted polymer micro-and nano-particles: A review, *Molecules*. 25(20), 4740.
[38] Hussain S. S., Sykam K., Narayan R., & Basak P. (2023). A route to high-solid coatings: Hydrophobic polyurethane triazoles via bulk polymerization of azide-alkyne cycloaddition, *Progress in Organic Coatings*. 179, 107529.
[39] Nageib A. M., Halim A. A., Nordin A. N., & Ali F. (2023). Computational modelling analysis and synthesis of molecularly imprinted polymers (MIPs) with two functional monomers using bulk polymerization, *Materials Today: Proceedings*. Available at https://doi.org/10.1016/j.matpr.2023.04.680
[40] Schirhagl R., Ren K., & Zare R. N. (2012). Surface-imprinted polymers in microfluidic devices, *Science China Chemistry*. 55, 469–483.
[41] Cheng G., Li X., Li X., Chen J., Liu Y., Zhao G., & Zhu G. (2022). Surface imprinted polymer on a metal-organic framework for rapid and highly selective adsorption of sulfamethoxazole in environmental samples, *Journal of Hazardous Materials*. 423, 127087.
[42] Teixeira S. P., Reis R. L., Peppas N. A., Gomes M. E., & Domingues R. M. (2021). Epitope-imprinted polymers: Design principles of synthetic binding partners for natural biomacromolecules, *Science Advances*. 7(44), eabi9884.
[43] Han Y., Tao J., Ali N., Khan A., Malik S., Khan H., & Mohamed A. A. (2022). Molecularly imprinted polymers as the epitome of excellence in multiple fields, *European Polymer Journal*. 179, 111582.
[44] Xing R., Wen Y., Dong Y., Wang Y., Zhang Q., & Liu Z. (2019). Dual molecularly imprinted polymer-based plasmonic immunosandwich assay for the specific and sensitive detection of protein biomarkers, *Analytical Chemistry*. 91, 9993.
[45] Gao Y., Zhou D., Lyu J., Xu Q., Newland B., Matyjaszewski K., . . . & Wang W. (2020). Complex polymer architectures through free-radical polymerization of multivinyl monomers, *Nature Reviews Chemistry*. 4(4), 194–212.
[46] Liu S. (2024). The future of free radical polymerizations, *Chemistry of Materials*. 36(4), 1779–1780.
[47] Pirman T., Ocepek M., & Likozar B. (2021). Radical polymerization of acrylates, methacrylates, and styrene: Biobased approaches, mechanism, kinetics, secondary reactions, and modeling, *Industrial & Engineering Chemistry Research*. 60(26), 9347–9367.
[48] Jašo V., Stoiljković D., Radičević R., & Bera O. (2013). Kinetic modeling of bulk free-radical polymerization of methyl methacrylate, *Polymer Journal*. 45(6), 631–636.
[49] Suzuki Y., Cousins D. S., Shinagawa Y., Bell R. T., Matsumoto A., & Stebner A. P. (2019). Phase separation during bulk polymerization of methyl methacrylate, *Polymer Journal*. 51(4), 423–431.
[50] Shi M. Y., Li C. X., Song W. Y., Liu H., Xue Y. H., & Wang Y. (2022). Understanding of supramolecular solution polymerization and interfacial polymerization via forming multiple hydrogen bonds: A computer simulation study, *Soft Matter*. 18(29), 5446–5458.
[51] Anderson C. L., Li H., Jones C. G., Teat S. J., Settineri N. S., Dailing E. A., . . . & Liu Y. (2021). Solution-processable and functionalizable ultra-high molecular weight polymers via topochemical synthesis, *Nature Communications*. 12(1), 6818.
[52] Jahanzad F., Sajjadi S., & Brooks B. (2013). Suspension polymerisation in the presence of an inhibitor; prolonged transition stage and suppressed emulsion particle formation, *Polymer*. 54(1), 16–23.
[53] Chaudhary V., & Sharma S. (2019). Suspension polymerization technique: Parameters affecting polymer properties and application in oxidation reactions, *Journal of Polymer Research*. 26(5), 102.
[54] Pola J. (2023). Thermal reactive modifications of polymer surfaces by infrared laser radiation, *Journal of Analytical and Applied Pyrolysis*. 169, 105819.
[55] Güven O. (2021). Radiation-assisted synthesis of polymer-based nanomaterials, *Applied Sciences*. 11(17), 7913.

[56] Ma X., Wang L., He Q., Sun Q., Yin D., & Zhang Y. (2023). A review on recent developments and applications of green sorbents-based solid phase extraction techniques, *Advances in Sample Preparation*. 6, 100065.
[57] Martins R. O., Bernardo R. A., Machado L. S., Junior A. C. B., Maciel L. Í. L., De Aguiar D. V. A., . . . Chaves A. R. (2023). Greener molecularly imprinted polymers: Strategies and applications in separation and mass spectrometry methods, *TrAC Trends in Analytical Chemistry*. 168, 117285.
[58] Zidarič T., Majer D., Maver T., Finšgar M., & Maver U. (2023). The development of an electropolymerized, molecularly imprinted polymer (MIP) sensor for insulin determination using single-drop analysis, *Analyst*. 148(5), 1102–1115.
[59] Buensuceso C. E., Tiu B. D. B., Lee L. P., Sabido P. M. G., Nuesca G. M., Caldona E. B., . . . & Advincula R. C. (2022). Electropolymerized-molecularly imprinted polymers (E-MIPS) as sensing elements for the detection of dengue infection, *Analytical and Bioanalytical Chemistry*. 414, 1347–1357.
[60] Pandey A., Singh S., & Singh S. (2023). Molecularly imprinted polymers: Applications and challenges in biological and environmental sample analysis, *Molecularly Imprinted Polymers (MIPS)*. 321–344.
[61] Bouvarel T., Delaunay N., & Pichon V. (2021). Molecularly imprinted polymers in miniaturized extraction and separation devices, *Journal of Separation Science*. 44(8), 1727–1751.
[62] He S., Zhang L., Bai S., Yang H., Cui Z., Zhang X., & Li Y. (2021). Advances of molecularly imprinted polymers (MIP) and the application in drug delivery, *European Polymer Journal*. 143, 110179.
[63] Tse Sum Bui B., Mier A., & Haupt K. (2023). Molecularly imprinted polymers as synthetic antibodies for protein recognition: The next generation, *Small*. 19(13), 2206453.
[64] Xu S., Xu Z., & Liu Z. (2022). Paper-based molecular-imprinting technology and its application, *Biosensors*. 12(8), 595.
[65] Gkika D. A., Tolkou A. K., Lambropoulou D. A., Bikiaris D. N., Kokkinos P., Kalavrouziotis I. K., & Kyzas G. Z. (2024). Application of molecularly imprinted polymers (MIPs) as environmental separation tools, *RSC Applied Polymers*. 2, 127–148.
[66] Liu R., & Poma A. (2021). Advances in molecularly imprinted polymers as drug delivery systems, *Molecules*. 26(12), 3589.
[67] Yuksel N., & Tektas S. (2022). Molecularly imprinted polymers: Preparation, characterisation, and application in drug delivery systems, *Journal of Microencapsulation*. 39(2), 176–196.
[68] Hasanah A. N., & Susanti I. (2023). Molecularly imprinted polymers for pharmaceutical impurities: Design and synthesis methods, *Polymers*. 15(16), 3401.
[69] Huang C., Wang H., Ma S., Bo C., Ou J., & Gong B. (2021). Recent application of molecular imprinting technique in food safety, *Journal of Chromatography A*. 1657, 462579.
[70] Ayerdurai V., Lach P., Lis-Cieplak A., Cieplak M., Kutner W., & Sharma P. S. (2024). An advantageous application of molecularly imprinted polymers in food processing and quality control, *Critical Reviews in Food Science and Nutrition*. 64(11), 3407–3440.
[71] Denmark D. J., Mohapatra S., & Mohapatra S. (2020). Point-of-care diagnostics: Molecularly imprinted polymers and nanomaterials for enhanced biosensor selectivity and transduction, *The EuroBiotech Journal*. 4(4), 184–206.
[72] Rajpal S., & Mishra P. (2022). Next generation biosensors employing molecularly imprinted polymers as sensing elements for in vitro diagnostics, *Biosensors and Bioelectronics: X* 11, 100201.
[73] Park R., Jeon S., Jeong J., Park S. Y., Han D. W., & Hong S. W. (2022). Recent advances of point-of-care devices integrated with molecularly imprinted polymers-based biosensors: From biomolecule sensing design to intraoral fluid testing, *Biosensors*. 12(3), 136.
[74] Liu Y., Li Z., & Jia L. (2020). Synthesis of molecularly imprinted polymer modified magnetic particles for chiral separation of tryptophan enantiomers in aqueous medium, *Journal of Chromatography A*. 1622, 461147.

[75] Chen Q., Wen G., Liang A., & Jiang Z. (2024). A dimode scattering method for ultratrace dinitrofuran detection with nanopalladium molecularly imprinted polymer nanocatalytic probe, *Langmuir*. 40, 5753–5763.
[76] Wei Z., Cao H., Mao J., Chen Z., Wu X., Yuan M., . . . & Xu F. (2023). Enhancement of the sensing performance and stability of a MOF based-molecularly imprinted polymer by utilizing dual-ligands and triethanolamine catalysis, *Talanta*. 258, 124459.
[77] Sharifzadeh Z., Razavi S. A. A., & Morsali A. (2023). Construction of hierarchically chiral metal–organic frameworks for fast and mild asymmetric catalysis, *Green Chemistry*. 25(21), 8661–8678.
[78] Ferreira J. B., De Jesus Macrino C., Dinali L. A. F., Filho J. F. A., Silva C. F., Borges K. B., & Romão W. (2021). Molecularly imprinted polymers as a selective sorbent for forensic applications in biological samples – A review, *Analytical and Bioanalytical Chemistry*. 413(24), 6013–6036.
[79] Díez-Pascual A. M. (2023). Perspectives of polymers in forensic analysis, *Macromol*. 3(2), 108–119.
[80] Martins B. O. (2023). Security knowledges: Circulation, control, and responsible research and innovation in EU border management, *Science as Culture*. 32(3), 435–459.

Vaishali Goel, Deepika Raina*, and Prachi Mall

Chapter 2
Fundamentals and primer of molecular imprinting technology

Abstract: Molecular imprinting technology is a technique used to develop false recognition sites within polymer matrices that are compatible in size, shape, and spatial arrangement with the functional groups of a template. It provides a flexible method for constructing customized molecular recognition sites within polymeric matrices, based on the lock-and-key model. This primer explains the fundamentals of molecular imprinting technology, detailing its principles, methodologies, and applications. The process involves creating molecularly imprinted polymers in which template molecules are inserted into a polymer matrix and then extracting them, thereby exposing complementary recognition sites on the target polymer. The procedure includes selecting a functional monomer and template, polymerization methods, and a strategy to remove the template. The technology finds its applications in various fields such as drug delivery, chromatography, catalysis, and sensors. By thoroughly understanding the concepts and methodologies, the potential of this technology can be harnessed to synthesize more efficient materials with precise molecular identification capabilities, leading to new opportunities in areas like environmental monitoring and biotechnology. Recent technological advancements have led to the development of novel approaches such as stimuli-responsive MIPs, surface imprinting, and nanoimprinting, which have increased the technology's adaptability and usefulness. These developments demonstrate the ongoing progress of molecular imprinting technology and open the door for the creation of smart materials with specialized molecular recognition properties for a wide range of biosensing, drug delivery, and nanotechnology applications.

Keywords: Molecular recognition, artificial antibodies, electropolymerization, surface imprinting, biosensors

*__Corresponding author: Deepika Raina__, School of Pharmacy, Graphic Era Hill University, Dehradun, Uttarakhand, India, e-mail: deepika709@gmail.com
Vaishali Goel, Prachi Mall, School of Pharmacy, Graphic Era Hill University, Dehradun, Uttarakhand, India

https://doi.org/10.1515/9783111503325-002

2.1 Introduction

Molecular imprinting technology (MIT) refers to a technique that involves the development of polymers that completely matches with a specific target in the functional group site and three-dimensional arrangement. Molecularly imprinted polymers (MIPs) are also a novel way of customized material that incorporates molecular recognition sites, enabling specific binding to natural antigens and antibodies, thereby exhibiting encoded selectivity and identification [1].

The substrates may cause hindrance in the identification and recognition of target molecules in real samples. Several organic biological molecule pairings, such as those between an enzyme and a substrate or an antigen and an antibody, are extensively utilized in the recognition of target molecules due to their strong affinity and specificity for their intended targets. However, limited stability, difficulty in modification, susceptibility to denaturation, and high cost are common challenges associated with natural molecular recognition. Additionally, moderate conditions restrict their utilization. Thus, it is vital to investigate molecular recognition systems with enhanced stability and specificity [2].

The prototype fragments were first utilized in the 1980s for the production of polymers exhibiting pores, which were further involved in qualitative analysis, leading to the first reports of MIP as a "biomimetic material." Since then, the development of biosensing technologies has generated significant interest in MIPs from both the scientific and industrial communities [3]. Molecules with multifunctional binding sites of various sizes can be achieved efficiently using the process of molecular imprinting. Furthermore, molecular imprinting has become a practical technique for simulating natural enzymes due to the inherent benefits of MIPs, which include cost-effectiveness, precision, thermal stability, and the absence of a need for protocols based on biological principles [4].

For an extended period, scientists have dedicated themselves to developing innovative methods for medication delivery, such as controlled release and targeted distribution. However, some successes have been achieved. For instance, the typical products for targeted delivery systems and controlled-release systems are doxorubicin hydrochloride liposome injection (Doxil®) and glipizide extended-release tablets (GLUCOTROL XL®). New technologies are being applied in conjunction with functional carriers or materials to create novel delivery systems that have distinct advantages. The use of carriers has resulted in targeted drug delivery along with the reduced nonspecific absorption and prolonged half-life of drugs in serum. Many studies on drug delivery systems have been conducted with the goal of improving the therapeutic efficacy of medications with minimal side effects. Only a small number of them, meanwhile, have clinical use approval. The search for viable medicine delivery methods has never stopped, which resulted in unearthing of MIP. The unique benefits of MIPs, commonly referred to as "artificial antibodies," have drawn the interest of pharmaceutical researchers. MIPs are artificial materials created using an MIT that have a specific strength and spatial organization [5].

2.2 Historical background

The field of MIPs began to take an interest in research due to Wulff and Sarhan's 1972 description, regardless of the concept of molecular imprinting which was first proposed in 1930. This was due to the promising attributes of MIPs, which include high binding affinity and specific selectivity for target molecules, as well as their simplicity, robustness, and stability [6]. It has advanced significantly in the last several years as a successful method for a various uses, such as solid-phase extraction (SPE), cell recognition, catalysis of chemical reactions, chromatographic separation, and biosensing of biomacromolecules. By producing MIPs with specific molecular recognition capabilities, a technique known as "molecular imprinting" is employed to incorporate recognition features into artificial networks [7].

MIT has been inspired by biomimetic approach, which allows synthetic and semisynthetic materials to replicate the molecular identification. Molecular recognition or identification exhibits molecular complementary designs, which involves various interactions such as covalent interactions and the non-covalent interactions including van der Waals forces, ionic interactions, hydrogen bonding, and hydrophobic interactions [8].

The graph showing an upsurge has been observed over the past several years, giving rise to the importance and activity of research into the design of artificial blocks possessing same natural recognition progressions. One approach used to create materials having specific selectivity, similar to biological systems (such as antibodies and biological receptors), is molecular imprinting. This method is considered flexible and promising, as it can be applied to both living and biochemical substances, including peptides and building blocks of proteins, derivatives of the nucleotides, contaminants, medications, and food. Additionally, food has applications in various fields such as separation science (including purification), biological antibodies and receptor systems, chemical sensors, catalytic processes, and drug delivery. In contrast to organic systems like proteins and macromolecules, imprinted polymers offer greater corporeal resilience, tolerance to extreme pressures and temperatures, and resistance toward biological solvents, bases, acids, and metal ions. Furthermore, the cost of synthesizing these polymers is lower, and they have a very long shelf life, maintaining their recognition capabilities for several years at ambient temperature [9].

2.3 Preparation techniques of molecularly imprinted polymers

The preparation and synthesis of MIPs refer to a simple and efficient process, which involves number of components such as: (a) template molecule or target molecule, (b) one or more functional monomers, (c) initiators, (d) cross-linkers, (e) porogenic solvent, and (f) a solvent required for template extraction post-polymerization [10].

The selection of the most suitable monomer, cross-linker, initiator, and the solvent for a template molecule has regarded as the biggest challenge in the development of MIPs with high specificity as well as the selectivity. Some considerations for different components are used as follows:

First, the **template** plays a pivotal role in all molecular imprinting procedures by guiding the prearrangement of functional groups that adhere to functional monomers. Second, the **functional monomer** is selected during the pre-polymerization stage, which aims to establish the proper interface between the monomers and the functional groups of the template molecule. Afterward, **cross-linkers** play a crucial role in the synthesis of MIPs, as they are responsible for the formation of polymer backbone surrounding the template molecule and act as multifunctional monomers. Careful consideration must be given to choose the cross-linking agent so as to maintain the profile of the template molecule having specific voids. In order to facilitate template movement both inside and outward while preserving the MIP molecule's structure or in other words to enhance its mechanical stability, the degree of cross-linking needs to be optimized. Then, the type of **solvents** called porogenic solvent is required for creating pores in macroporous polymers, which in turn leads to the development of polymers having porous structures and high specific surface areas. Thermodynamically good solvents are generally used as porogenic solvents. Finally, **initiators** also play a critical role as they are responsible for the formation of complexes or initiation of complexation in the presence of a template. For example, when the formation of complex is mediated by hydrogen bonding, decreased levels of temperature are desirable, which is effectively achieved by using photochemically active initiators as they perform well at low temperatures [11].

Some examples of functional monomers, initiators, cross-linking agents, and porogenic solvents used in the synthesis of MIPs are listed in Table 2.1

2.3.1 Synthesis of MIP

In its traditional form, initially the functional monomer consisting of functional group is allowed to interact with the template molecule or target molecule, which is accompanied by certain forces such as covalent bond or non-covalent bond including hydrogen bonds, metal-ligand interactions, or hydrophobic bonds. The cross-linkers and ini-

Table 2.1: Examples of functional monomers, initiators, cross-linking agents, and porogenic solvents [11, 12].

Types of components	Examples
Functional monomer	1-Vinyl imidazole Methacrylic acid (MAA) Itaconic acid Vinyl benzoic acid
Initiators	Ammonium persulfate (APS) Azobisisobutyronitrile (AIBN) Azobisdimethylvaleronitrile (ABV) Benzoylperoxide (BPO)
Cross-linking agent	*N,N′*-Ethylenebisacrylamide (EBA) *N,N′*-Methylenebisacrylamide (MBA) *p*-Divinylbenzene (DVB) Ethyleneglycol dimethacrylate (EGDMA)
Porogenic solvents	2-Methoxyethanol Acetonitrile Chloroform Cyclohexane Dichloroethane

tiators are then involved for copolymerization with the succeeding monomer-template complexes in an appropriate porogenic solvent. After the completion of polymerization process, there occurs the formation of cross-linked polymer network, followed by the detachment of template molecule from the network and thereby results in the development of the template having specificity in regard to dimensions and the functional group distribution. This process results in development of MIPs having the ability to selectively bind the template [4, 13]. The diagrammatic representation of MIP synthesis is shown in Figure 2.1.

2.3.2 Strategies used in MIP synthesis

Basically, during the synthesis of MIP, the template molecule and the functional monomer undergo an association, which forms the basis for the establishment of two approaches to molecular imprinting techniques, that is, covalent interactions and non-covalent interactions.

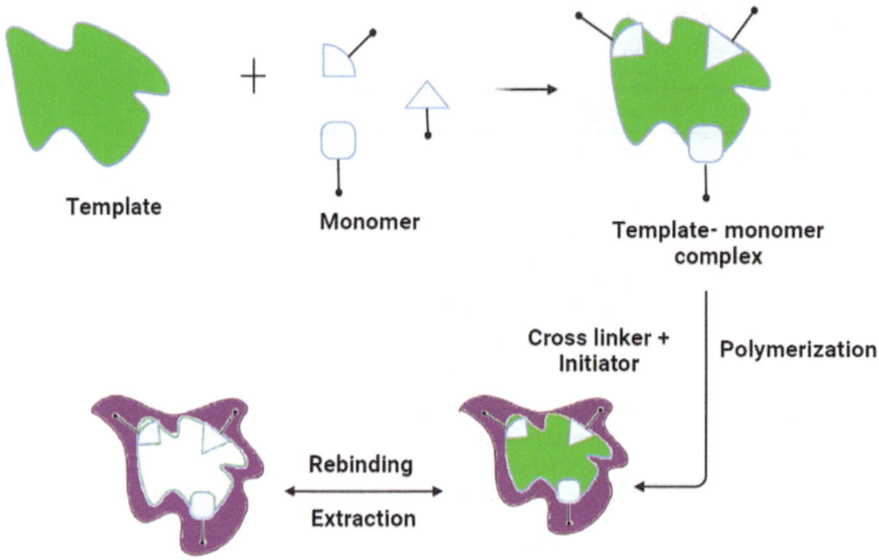

Figure 2.1: Diagrammatic representation of the synthesis of molecularly imprinted polymers [10, 13].

2.3.2.1 Covalent interactions approach

In this interaction, the template molecule and functional monomers are bonded to each other by covalent bonds. Afterward, polymerization occurs in the presence of cross-linker and the template is extracted out from the extremely cross-linked polymer by a chemical method. For example, once MIP was synthesized using amino acid derivatives and polymerizable vinylphenylboronate via covalent bonds, and subsequently, the template was removed by the hydrolysis of amino acids followed by utilization of the resultant polymer possessing specific binding. In MIPs, which involved the use of the covalent approach, the functional monomer and the template molecule are firmly bonded together and therefore covalent imprinting is more appropriate to achieve or establish stringent imprinting. These covalent approaches between functional and template monomer are reversible, and on the other hand, the number of possible templates and the variety of reversible covalent interactions significantly restricted and also frequently necessitate the hydrolysis process in the presence of acid for breakage of the bonds connecting the functional and the template monomer.

2.3.2.2 Non-covalent approach

In this interaction, the template molecule and functional monomer are bonded together by non-covalent bonds like hydrogen bonding, ionic attractions, hydrophobic forces, and van der Waals forces. This approach is much preferred due to the below-mentioned reasons:

a) The laborious pre-polymerization complex production is avoided, as the method is much easier to be followed.
b) The non-covalent interactions can be disrupted simply by washing the polymer with an appropriate solvent to remove the template.
c) It has the advantage of not requiring the synthesis of an adduct between the functional monomer and the template, which enables imprinting to be achieved across a wide range of systems with minimal restrictions on available options. Non-covalent methods also allow for a greater variety of functionality to be introduced into the MIP binding site, along with the continuous removal of the template [13, 14].

2.3.3 Different preparation methods involved in MIP synthesis

The preparation of MIP generally involves free radical polymerization, which allows abundant monomers to be polymerized, for example, methyl methacrylate, ethylene, and styrene. Free radical polymerization can be conducted under low temperature and atmospheric conditions. It is basically carried out in three steps: The first step is initiation, where the initiator is decomposed and produces a free radical that breaks up the double bond of the monomer. The opened double bond of monomer allows it to join their structure and form other compound. The second step involves the propagation by a free radical and allows the chain to grow to form a polymer linear chain. The last and third step involves the termination of high-molecular-weight polymer chain formed during the process [11, 15]. The different methods involved in MIP synthesis are described further.

2.3.3.1 Bulk polymerization

This method involves the template molecule and functional monomer forming a complex during pre-polymerization, which is followed by the polymerization of a complex in the presence of small amount of solvent, that is, porogen and cross-linkers. It is a quick and easy process. After polymerization, porous polymer monoliths are grinded and sieved along with the removal of monomer by washing with a solvent. This method makes the weak selectivity and repeatability because particles obtained are uneven in size and shape due to the crushing and grinding of polymer monoliths and thereby reducing the binding capacity [16].

2.3.3.2 Precipitation polymerization

The key element of this method is the use of a larger amount of solvent compared to bulk polymerization, and thereby it is also known as homogeneous solution polymerization. All the components, including template, cross-linkers, initiators, and functional monomers in low quantity, are added in a sizable amount of solvent, resulting in the development of a polymer. These compounds are added in the precipitation medium and the insoluble nature causes the polymer to precipitate out at the bottom. MIPs are obtained by the continuous extraction of a template molecule that is entrapped in the interconnected polymer chain network, resulting in the generation of a template molecule having well-defined dimensioned cavities. This method provides MIPs that are spherical in shape, and the size obtained ranges from nano- to sub-micrometers [17].

2.3.3.3 Suspension polymerization

As the name suggests, the suspension polymerization method involves polymerization in the presence of a stabilizer that is added to the solvent along with other components, that is, functional monomer, template molecule, initiator, cross-linker, and certain amount of porogenic solvent if the required MIP should be porous in nature. Due to the addition of the stabilizer, the resulting MIP got suspended in the solution rather than being precipitated. This technique is typically employed to generate MIP spheres having a more uniform sphere and better size distribution and it can be regulated by adjusting some variables like the quantity of cross-linkers or stirring speed [5, 18].

2.3.3.4 Emulsion polymerization

This method involves the addition of a surfactant to the aqueous medium along with the template molecule, initiator, and functional monomer at a concentration above the critical micelle concentration. During this process, the functional monomer disperses in the medium/solvent, the initiator dissolves in the solvent, moves from the external phase into the micelle, and triggers the polymerization reaction. Nevertheless, the interactions between the functional monomer and the template molecule may be impacted by the incorporation of surfactants, which in turn affects the productivity and effectiveness of procedure by causing hindrance to remove the surfactant at the washing stage. However, the MIPs obtained from this method are nanometers in size [12, 19].

2.3.3.5 Electropolymerization

This method involves the discharge of an electric potential in a solution containing a template molecule, a porogenic solvent, a functional monomer, and a cross-linker. The generation of electric potential around the electrode, which is immersed in the solvent containing the template and monomer, creates a thin layer around the exterior of the electrode and initiates the polymerization reaction. The strict regulation of polymer thickness can be done by utilizing different conductive materials of various sizes and shapes. Electropolymerization results in the generation of both conducting and nonconducting MIPs that have both advantages and disadvantages. The nonconducting MIPs are used for capacity chemosensors that can detect the analyte at the nanoscale level, while conducting MIPs are suitable for the fabrication of a sensing membrane of ion-selective electrode due to its high resistance [20, 21]. The electropolymerization technique that is extensively used for the synthesis of MIPs is cyclic voltammetry. This method provides the authority to monitor the width and density of deposited polymer and alterations in the number of scan rates and scan cycles [22].

2.3.3.6 Surface imprinting

The preparation procedures discussed above, such as bulk polymerization or precipitation polymerization, involve certain processes of crushing and grinding of the obtained polymer monoliths, leading to reduced particle size and also binding efficiency, thereby making their usefulness limited. Such limitations can be overcome by surface imprinting technology as polymer monoliths produced by this method possess greater selectivity and more binding sites along with better kinetics. The preparation method involves the use of solid substrate or matrix, for example, silica (SiO_2), alumina, and titanium dioxide (TiO_2). Among the mentioned matrices, the most preferred is titanium dioxide, owing to its low cost, easy availability, and nontoxicity. The preparatory method is divided into three stages. In the first stage, the template molecules are immobilized on the solid exterior of substrate or matrix via chemical interaction, including covalent and non-covalent interactions. In the second stage, the obtained immobilized solid substrate is further processed for polymerization reaction by mixing it in the solution containing functional monomers and cross-linking agents. In the final stage, after polymerization, the solid substrate breaks down into the solution with the help of reagent (e.g., hydrogen fluoride) and the template molecules are eluted to attain MIPs having specific binding site either close to or on the surface [23, 24].

2.3.3.7 Bulk imprinting

The bulk imprinting technique is considered the most beneficial method for imprinting low-molecular-weight MIPs. One of the most significant advantages of bulk imprinting is that it facilitates the formation of 3D binding sites for the entire protein. However, alongside its advantages, it also has some disadvantages such as (a) slow diffusion of macromolecules via highly cross-linked polymers, (b) the more time taken for binding equilibrium, and (c) the occurrence of configurational changes in the macromolecule due to the presence of nonphysiological circumstances. The bulk imprinting process involves the interaction of functional monomers (methacrylic acid or acrylamide) and template molecules along with the cross-linking agent (methylene-bisacrylamide) for imprinting. The template protein is subsequently washed around five times by elution with sodium dodecyl sulfate, which works by cleaving the non-covalent bonds and causes the diffusion out of the polymer network. However, there might reside some amount of sodium dodecyl sulfate during the washing phase, so one must ensure the removal of all washing compounds before recognition studies to maintain the integrity of binding sites [25, 26].

2.4 Applications of molecularly imprinted polymers

MIPs possess a wide variety of applications, featuring specific recognition binding sites that account for high specificity, strong binding affinity, and targeted drug delivery. Some of their applications are summarized below.

2.4.1 Cell concentration and separation

MIPs are widely used for capturing and separating bacterial spores. They have also been utilized in a comprehensive biological system designed for *Bacillus anthracis* species to trap and concentrate the spores. Robust spore interactions and a strong imprinting effect, which showed 25% more binding compared to non-imprinted controls, were demonstrated in using binding experiments. Methanotrophs from paddy soil were also selectively captured using surface imprinted polymers [27].

2.4.2 Drug delivery

MIPs were successfully engineered for transdermal drug delivery, involving the use of subcutaneous implant systems that provide controlled drug delivery. Molecularly imprinted hydrogels for ocular drug delivery have proven useful due to their sufficiently high binding affinity and improved sustained-release profiles of the drug. Thus, MIPs have been found to be effective in controlling the drug-release profile or rate, thereby providing better therapeutic effects [28].

2.4.3 MIPs as electrochemical biosensors

MIPs are widely used as sensing elements and are referred to as electrochemical biosensors. They have significant applications in sensing biomolecules such as proteins, hormones, pesticides, antibiotics, and neurotransmitters. The use of MIP-based electrochemical biosensors shows more accurate sensing performance compared to other analytical techniques by serving as specific recognition elements for templates and thereby generating the output as electrochemical signals [29–31].

2.4.4 MIPs with quantum dots

The combination of quantum dots (analytical tools) with MIPs has been used as a significant tool for imaging tumor biomarkers because it showed good biocompatibility and solubility, high specificity and selective binding affinity, and improved performance in crossing cell membranes. Due to their exceptional targeting ability, they are widely used to target tumor cells, thereby reducing the common side effects of traditional cancer therapies [32, 33].

2.4.5 MIPs in analytical chemistry

As MIPs show significant selectivity and specificity, they have been found to have substantial applications in the extraction and removal of pollutants and toxic compounds from environmental water, as well as in the extraction of naturally sourced bioactive compounds or essential nutrients across various fields such as pharmaceuticals, medicine, food, and chemical industry. They also exhibit durability, cost-effectiveness, simplicity of synthesis, and longevity across a variety of pH and temperature environments, making them suitable for use in combination with other analytical procedures [34].

2.4.6 Separation and purification

Various chromatographic techniques, including mixed-mode chromatography, capillary electrochromatography, and SPE, are mostly involved in separation and purification including MIPs. The separation field is becoming significant due to enlarged likelihood of market potential in chemical industry, pharmaceutical industry, and waste material treatment industries, where MIPs play a critical role because they can withstand the separation operating ecological conditions like low or high pH, elevated temperatures and pressures, organic solvents, and toxic environment. Another area for exploitation of MIPs is the distinction of chiral compounds forming the basis for fundamental investigation, medication design, and polymer chemistry [35, 36].

2.4.7 Diagnostic application

MIPs are now being used for diagnostic purposes as molecularly imprinted sorbent assays. The combination of MIP is done with fluorescent and enzyme-linked assays to detect the presence of antibodies and to estimate the levels of hormone. Dual-template magnetic MIPs were used in combination with graphene-Au-grafted recombinant apoferritin-encoded metal labels for the detection of carcinoembryonic antigen [37, 38].

2.4.8 Quality control in food industry

Several MIPs have been found to be effective in the food industry for the detection of various toxins, such as triazine, chloramphenicol, and clenbuterol, as well as several mycotoxins, such as moniliformin and patulin. MIPs also utilize various sophisticated analytical techniques, such as liquid chromatography (LC), for the targeted removal of analytes from complex food matrices. Paper-based MIP technology is also employed in the food industry for the purification of food from contaminants, such as viruses, bacteria, and heavy metal ions, due to their high reactivity, selectivity, and ease of operation [39, 40].

2.5 Conclusion

MIT has evolved significantly, turning into a highly specialized and adaptive tool for enhanced drug delivery approach by providing complementary design or molecular recognition sites that are analogous to precise attachment of natural antigens and antibodies. The efficacy of MIPs has been widened by invading a new perspective during the development of techniques such as electropolymerization, surface imprinting, and bulk imprinting. This approach thereby distended the application of MIPs, by intensifying it for targeted drug delivery, separation, and purification and for diagnostic purposes. Overall, MIPs have significant potential for building smart strategies, accelerating progress in a variety of academic and professional fields.

References

[1] Liu Y., Lian Z., Li F., Majid A., & Wang J. (2021). Review on molecular imprinting technology and its application in pre-treatment and detection of marine organic pollutants, *Marine Pollution Bulletin*. 169, 112541.

[2] Zhou Q., Xu Z., & Liu Z. (2022). Molecularly imprinting–aptamer techniques and their applications in molecular recognition, *Biosensors*. 12, 576.

[3] Li Y., et al. (2024). Recent advances in molecularly imprinted polymer-based electrochemical sensors, *Biosensors and Bioelectronics*. 249. 116018.

[4] Tian R., et al. (2022). Recent development in the design of artificial enzymes through molecular imprinting technology, *Journal of Materials Chemistry B*. 10. 6590–6606.

[5] He S., et al. (2021). Advances of molecularly imprinted polymers (MIP) and the application in drug delivery, *European Polymer Journal*. 143. 110179.

[6] Kadhem A. J., Gentile G. J., & De Cortalezzi M. M. F. (2021). Molecularly imprinted polymers (MIPs) in sensors for environmental and biomedical applications: A review, *Molecules*. 26, 6233.

[7] Nawaz N., Abu Bakar N. K., Muhammad Ekramul Mahmud H. N., & Jamaludin N. S. (2021). Molecularly imprinted polymers-based DNA biosensors, *Analytical Biochemistry*. 630, 114328.

[8] Bereli N., Yavuz H., & Denizli A. (2020). Molecular imprinting technology for biomimetic assemblies, *Hacettepe Journal of Biology and Chemistry*. Hacettepe University, 48, 575–601.

[9] Vasapollo G., et al. (2011). Molecularly imprinted polymers: Present and future prospective, *International Journal of Molecular Sciences*. 12. 5908–5945.

[10] Kosheleva R. I., Mitropoulos A. C., & Kyzas G. Z. (2019). New trends in molecular imprinting techniques, *Interface Science and Technology*. Elsevier, 30, 151–172.

[11] Cormack P. A. G., & Elorza A. Z. (2004). Molecularly imprinted polymers: Synthesis and characterisation, *Journal of Chromatography B: Analytical Technologies in the Biomedical and Life Sciences*. 804, 173–182.

[12] Yuksel N., & Tektas S. (2022). Molecularly imprinted polymers: Preparation, characterisation, and application in drug delivery systems, *Journal of Microencapsulation*. 39, 176–196.

[13] Komiyama M., Mori T., & Ariga K. (2018). Molecular imprinting: Materials nanoarchitectonics with molecular information, *Journal Bulletin of the Chemical Society of Japan*. 91, 1075–1111.

[14] Yan H., & Kyung H. R. (2006). Characteristic and synthetic approach of molecularly imprinted polymer, *Bulletin of the Chemical Society of Japan*. 7, 155–178.

[15] Chen L., Wang X., Lu W., Wu X., & Li J. (2016). Molecular imprinting: Perspectives and applications, *Chemical Society Reviews*. 45, 2137–2211.

[16] Włoch M., & Datta J. (2019). Synthesis and polymerisation techniques of molecularly imprinted polymers, *Current Analytical Chemistry*. 86, 17–40.

[17] Gkika D. A., et al. (2024). Application of molecularly imprinted polymers (MIPs) as environmental separation tools, *RSC Applied Polymers*. 2. 127–148.

[18] Chen L., Xu S., & Li J. (2011). Recent advances in molecular imprinting technology: Current status, challenges and highlighted applications, *Chemical Society Reviews*. 40, 2922–2942.

[19] De Middeleer G., Dubruel P., & De Saeger S. (2016). Characterization of MIP and MIP functionalized surfaces: Current state-of-the-art, *TrAC – Trends in Analytical Chemistry*. 76, 71–85.

[20] Frasco M. F., Truta L. A. A. N. A., Sales M. G. F., & Moreira F. T. C. (2017). Imprinting technology in electrochemical biomimetic sensors, *Sensors*. MDPI (Multidisciplinary Digital Publishing Institute Switzerland, 17, 523.

[21] Wang J., Liang R., & Qin W. (2020). Molecularly imprinted polymer-based potentiometric sensors, *TrAC – Trends in Analytical Chemistry*. 130, 115980.

[22] Yarman A., et al. (2017). Enzymes as tools in MIP-sensors, *Chemosensors*. 5, 11.

[23] Xu P., Xu W., Zhang X., Pan J., & Yan Y. (2009). Molecularly-imprinted material for dibenzothiophene recognition prepared by surface imprinting methods, *Journal Adsorption Science & Technology*. 27, 975–987.
[24] Dong C., et al. (2021). Molecularly Imprinted Polymers by the surface imprinting technique, *European Polymer Journal*. 145. 110231.
[25] Kryscio D. R., & Peppas N. A. (2012). Critical review and perspective of macromolecularly imprinted polymers, *Acta Biomaterialia*. 8, 461–473.
[26] Ansari S., & Masoum S. (2019). Molecularly Imprinted Polymers for capturing and sensing proteins: Current progress and future implications, *TrAC – Trends in Analytical Chemistry*. 114, 29–47.
[27] Piletsky S., Canfarotta F., Poma A., Bossi A. M., & Piletsky S. (2020). Molecularly Imprinted Polymers for cell recognition, *Trends in Biotechnology*. 38, 368–387.
[28] Luliński P. (2017). Molecularly Imprinted Polymers based drug delivery devices: A way to application in modern pharmacotherapy. A review, *Materials Science and Engineering C*. 76, 1344–1353.
[29] Gui R., Jin H., Guo H., & Wang Z. (2018). Recent advances and future prospects in molecularly imprinted polymers-based electrochemical biosensors, *Biosensors and Bioelectronics*. 100, 56–70
[30] Ahmad O. S., Bedwell T. S., Esen C., Garcia-Cruz A., & Piletsky S. A. (2019). Molecularly imprinted polymers in electrochemical and optical sensors, *Trends in Biotechnology*. 37, 294–309.
[31] Li R., Feng Y., Pan G., & Liu L. (2019). Advances in molecularly imprinting technology for bioanalytical applications, *Sensors*. MDPI (Multidisciplinary Digital Publishing Institute) Switzerland, 19, 177.
[32] Díaz-álvarez M., & Martín-Esteban A. (2021). Molecularly imprinted polymer-quantum dot materials in optical sensors: An overview of their synthesis and applications, *Biosensors*. 11, 39.
[33] Zaidi S. A. (2017). Molecular imprinting polymers and their composites: A promising material for diverse applications, *Biomaterials Science*. 5, 388–402.
[34] Malik M. I., Shaikh H., Mustafa G., & Bhanger M. I. (2019). Recent applications of molecularly imprinted polymers in analytical chemistry, *Separation and Purification Technology*. 48, 179–219.
[35] Yemiş F., Alkan P., Yenigül B., & Yenigül M. (2013). Molecularly Imprinted Polymers and their synthesis by different methods, *Polymer-Polymer Composite*. 21, 145–150.
[36] Shi W., Zhang S. Q., Li K. B., Jia W. P., & Han D. M. (2018). Integration of mixed-mode chromatography and molecular imprinting technology for double recognition and selective separation of proteins, *Separation and Purification Technology*. 202, 165–173.
[37] Bedwell T. S., & Whitcombe M. J. (2016). Analytical applications of MIPs in diagnostic assays: Future perspectives, *Analytical and Bioanalytical Chemistry*. 408, 1735–1751.
[38] Belbruno J. J. (2019). Molecularly imprinted polymers, *Chemical Reviews*. 119, 94–119.
[39] Hussain M. (2015). “Molecular imprinting” as multidisciplinary material science: Today and tomorrow, *International Journal of Advanced Materials Research*. 1, 132–154.
[40] Xu S., Xu Z., & Liu Z. (2022). Paper-based molecular-imprinting technology and its application, *Biosensors*. 12, 595.

Vishnu Biju, A.V. Mahadev, and Arun Aravind*

Chapter 3
Polymerization techniques of carbon-based molecularly imprinted polymers

Abstract: Polymerization techniques are essential in creating selective recognition sites within molecularly imprinted polymers (MIPs), particularly those based on carbon matrices. The ability to control the polymerization method directly impacts the efficiency and specificity of molecular recognition, making it a critical aspect of MIP synthesis.

Choosing an appropriate polymerization method involves carefully considering factors such as the target analyte, template molecule, functional monomers, and the desired application of the resulting MIP. The need for optimal imprinting efficiency and binding affinity guides this decision. The selected polymerization method significantly influences the properties of the resulting MIP, including morphology, surface area, porosity, and mechanical strength. Variations in polymerization conditions such as temperature, solvent, initiator, and reaction time further contribute to the tunability of MIP characteristics, thereby impacting their performance in various applications.

This chapter provides an in-depth analysis of different polymerization techniques commonly employed for the fabrication of carbon-based MIPs. These techniques contain conventional methods such as bulk precipitation, suspension polymerization, and advanced approaches like surface imprinting.

Through comprehensive study of the importance of polymerization techniques in creating selective recognition sites, the criteria for method selection, the effects of method choice on MIP properties, and a detailed description of various polymerization techniques are discussed. This chapter aims to offer valuable insights into the synthesis and optimization of carbon-based MIPs which are used for a variety of applications, from molecular sensing to separation science, wastewater treatment, chromatographic separation, and drug delivery.

Keywords: Molecularly imprinted polymers (MIPs, carbon-based nanomaterials, selective binding sites, biosensors, electrochemical sensors, polymerization techniques, free-radical polymerization, porosity, emulsion polymerization, precipitation poly-

***Corresponding author: Arun Aravind,** Center for Advanced Functional Materials (CAFM), Department of Physics, Bishop Moore College (Research Center Affiliated to University of Kerala), Mavelikara 690110, Kerala, India, e-mail: arun@bishopmoorecollege.org

Vishnu Biju, Laboratory for Electro Optics Systems (LEOS), Indian Space Research Organisation (ISRO), India

A. V. Mahadev, Division of Astronomy, Aryabhatta Research Institute of Observational Sciences, Nainital 263001, Uttarakhand, India

https://doi.org/10.1515/9783111503325-003

merization, binding affinity, covalent approach, non-covalent approach, functional monomers, cross-linker concentration, template molecules, hydrophobic interaction, electrostatic interaction, hydrogen bonding, Le Chatelier's principle

3.1 Introduction

Molecularly imprinted polymers (MIPs) that are derived using carbon-based nanomaterials are increasingly recognized for their potential in various applications. These solutions have specific binding sites that enable them to interact with target compounds selectively. MIPs function similarly to receptors, using these particular recognition sites to identify and bind to molecules within the polymer network [1]. Research has shown that incorporating highly selective nanoparticles into MIPs can further enhance their sensitivity [2]. The ability to bond with a wide range of other elements and the chemically stable nature makes carbon an excellent choice for such applications [3]. Carbon-based nanomaterials are used in many applications and are gaining attention for their effectiveness as electron mediators in the development of highly sensitive biosensors and electrochemical sensors [4]. There are researches on new analytical methods using carbon-based nanomaterials for detection of *Salmonella* species, a major food-poisoning organism, in water and soil resources [5]. Moreover, they bind to a variety of biological molecules due to their large surface area, the presence of many functional groups, and varied intramolecular mixes. By modifying or enhancing the surface characteristics of these carbon-based nanomaterials, their chemical characteristics and reactivity can be improved. These unique qualities have led to the identification of various types of carbon-based materials. These carbon-based nanomaterials might be modified with other materials or applied directly to increase their viability for application [6].

The methods used for the polymerization of MIPs play a crucial role in determining their physical and chemical properties. Different polymerization techniques can significantly affect the morphology, surface area, porosity, and mechanical strength of the resulting MIPs, influencing their performance in specific applications. The main aim of this chapter is to construct a detailed exploration of the various polymerization techniques used to synthesize carbon-based MIPs. It will discuss the criteria for selecting polymerization methods, the effects of these methods on MIP properties, and provide detailed descriptions of conventional and advanced polymerization techniques, supported by case studies and practical applications.

3.2 Polymerization methods

3.2.1 Significance of polymerization techniques in MIPs

Polymerization techniques are vital for the formation of selective binding sites within MIPs. For example, free-radical polymerization can create MIPs with precise recognition sites for small organic molecules, enhancing their application in sensors. The efficiency and specificity of MIPs are greatly induced by the selected polymerization method. Elements such as polymerization rate, temperature, and solvent can affect the formation and stability of the recognition sites, ultimately impacting the selectivity and binding affinity of MIPs. The polymerization technique affects the morphology of MIPs, including particle size and shape. The morphology affects the surface area and accessibility of binding sites, influencing the overall performance of the MIPs. The surface area of MIPs determines their binding capacity and interaction with target molecules. Higher surface areas allow for more binding sites and better performance in applications like sensing and chromatography. Polymerization methods like emulsion polymerization are known to produce MIPs with high surface areas due to the formation of microporous structures. Porosity is a critical factor for the diffusion of target molecules into the recognition sites. High porosity allows for efficient binding and faster response times. Precipitation polymerization, for instance, can create highly porous MIPs, enhancing their effectiveness in applications requiring rapid interaction with analytes. Mechanical strength is essential for the durability and stability of MIPs in practical applications. Polymerization conditions such as cross-linker concentration and reaction time can be adjusted to change the different properties of MIPs such as mechanical properties. This can be seen in the case if we use higher cross-linker concentration, it will generally lead to stronger polymers but may reduce porosity.

A regular MIP synthesis process requires a porogenic solvent, a polymerization initiator, a cross-linker, a functional monomer and a template. To create MIPs with specificity and affinity comparable to their biological counterparts is the ultimate foundation of molecular imprinting. This section will delve deeply into the important approaches to fabricating MIPs, how different factors affect the synthesis of MIPs, and how these factors influence to select the appropriate polymerization techniques to synthesize desirable MIPs.

3.2.2 Approaches to polymerization

3.2.2.1 Covalent approach

It is also referred to as the pre-organization method. This requires the creation of MIPs by combining functional monomers and templates through reversible covalent bonds. The tough nature of these covalent bonds, grants increased stability to the polymers, allowing for the organization of precise identification sites across the spatial structure [7]. Toward the final stages of the process, we incorporate chemical procedures, like pH adjustments that are important for the elimination of the templates embedded in imprinting layers. The selection of monomers is limited and the reaction conditions are hard. At present, the commonly reported functional monomers are borate silanes, chelate compounds, and Schiff bases. The property of derivatization is limited for different target molecules. But the relevance of covalent methods is reduced due to the fact that one or more polymerizable groups needed for participation in the imprinting process [8].

3.2.2.2 Non-covalent approach

In the non-covalent approach, the functional monomers and templates are arranged in advance by self-assembly via non-covalent interactions (such as coordination, metal chelation, hydrophobic interaction, electrostatic interaction, and hydrogen bonding. Understanding the choice of polymerization method has huge significance in the production of MIPs. A few of the key factors are discussed below.

3.2.2.2.1 Functional monomers

The way substrates and template interact depends on the selection of monomers. In a covalent approach various polymers, with different compositions containing increasing template amounts are tested to determine the optimal template/monomer ratio through empirical means. The formation of the solution complex between monomers and the template is governed by Le Chatelier's principle. Functional monomers play a role in the bonding interactions within the binding sites, in MIP binding sites general mechanism. In covalent molecular imprinting techniques, an excess of functional monomers compared to the moles of template is typically used to promote the formation of template functional monomer assemblies.

3.2.2.2.2 Template molecule

When identifying a template molecule for the polymerization of MIPs, several factors must be taken into account. In order to prevent reaction or interference with the polymerization process, the template must first of all be chemically inert under the circumstances of the process. It is imperative that the template be free of any polymerizable groups that might interfere with the polymerization process. Furthermore, the template should not have any features that could impede or delay the polymerization of free radicals. Another crucial factor is the template's stability, especially under moderately high temperatures or in the presence of UV light. Although imprinting small organic molecules – such as medications, insecticides, steroids, sugars, amino acids, peptides, nucleotide bases, and steroids – is a well-established and reasonably common process, employing larger templates is more difficult because of their less favorable nature.

3.2.2.2.3 Cross-linkers

The cross-linker serves a key role in regulating the polymer matrix's shape. It keeps the polymer matrix mechanically stable and helps to maintain the imprinted binding sites' capacity to recognize molecules.

In order to produce persistently porous (macroporous) materials with enough mechanical stability, high cross-link ratios are usually used. For this, trimethylolpropane trimethacrylate (TRIM) and ethylene glycol dimethacrylate (EGDMA) are frequently utilized. Certain studies have noted that the selection of the cross-linker has a major impact on the physical characteristics of the polymers, but less of an influence on particular interactions between the functional monomers and the template [9–11].

Compared to EGDMA, polymers using TRIM as a cross-linker have higher stiffness, structural order, and efficient binding sites. When polymerization is accomplished by precipitation, it has been observed that the binding characteristics of the polymer are enhanced and the degree of non-specific interactions is reduced by adjusting the amount of cross-linker and lowering the concentration of the template [12]. In another study, it has been observed that the type of cross-linker strongly influences the final size and yield of MIP nanoparticles [13].

3.2.2.2.4 Solvents

In the molecular imprinting process, the type and volume of the solvent are also important variables. For the fabrication of MIPs, toluene, chloroform, dichloromethane, or acetonitrile are the most often utilized solvents. The solvent is responsible for the growth of the pores in macroporous polymers and for bringing all the components (monomer, template, initiator, and cross-linker) into one phase during the polymerization process. To ensure strong flow through qualities of the resulting MIP, the solvent ought to create big pores. As the solvent volume increases, the polymer's pore volume increases as well. The solvent is hence frequently referred to as the "porogen." In a non-covalent imprinting polymerization, the porogen should also be selected with consideration for its contribution to the formation of the template-functional monomer complex: less polar solvents, like chloroform or toluene, promote the formation of the complex and facilitate polar non-covalent interactions, like hydrogen bonds; more polar solvents, especially protic solvents that allow for a high degree of disruption of the hydrogen bonds, tend to dissociate the non-covalent interactions in the prepolymer complex. However, since strong template-monomer interactions have been reported, several studies have produced effective MIPs in fairly polar solvents (such as acetonitrile/water or methanol/water). For instance, an acetonitrile/water 4/1 imprinted polymer that is both water-compatible and has a strong affinity for a polar 1-methyladenosine template has been produced and used successfully as a solid phase extraction sorbent for the extraction of analytes from human urine samples that have been spiked [14].

3.2.2.2.5 Polymerization conditions

Various studies have demonstrated that, in comparison to polymers made at higher temperatures, MIP polymerization at lower temperatures yields polymers with improved selectivity. The majority of researchers typically choose a polymerization temperature of 60 °C. Nevertheless, the polymerization reaction starts very quickly at this temperature, which makes it difficult to control and results in low MIP repeatability. Furthermore, the stability of complexes is negatively impacted by the relatively high temperatures, which also result in notable pressure reductions in the columns and less repeatability of the monolithic stationary phases [15]. Achieving high imprinting efficiency and binding affinity involves optimizing the polymerization conditions. Adjusting parameters such as temperature and solvent can enhance the interaction between monomers and the template, leading to better imprinting outcomes.

3.2.3 Polymerization methods

3.2.3.1 Bulk polymerization

Bulk polymerization has been one of the most prominent and widely used forms of free radical polymerization. It gets its significance from its simplicity, requiring only a monomer, a monomer-soluble initiator, and occasionally a chain transfer agent for monitoring molecular weight. This technique involves mechanical grinding and sieving steps to obtain small particles. This method requires a substantial quantity of template molecules, and the fabricated particles usually have a reduced MIP binding capacity in contrast to theoretical values [16]. The necessity to crush, grind, and sieve the developed polymer particles to an appropriate size is often time-consuming and would lower the polymer yield. The particles produced after the grinding operation mostly have uneven shapes and sizes. In the course of this operation, multiple high-affinity binding sites get damaged and transform into low-affinity sites.

3.2.3.2 Precipitation polymerization

It is possible to develop MIPs through the precipitation polymerization method. It entails the organization of imprinted beads with a similar reaction mixture employed in the bulk method except for the presence of a greater amount of porogen. Polymer chains would keep growing, precipitating only when they become large enough to be insoluble in the reaction mixture. Then the polymer beads are readily recovered by washing and centrifugation operations. This technique is straightforward, much quicker than the bulk method and delivers MIP beads in good yields. The particle's diameters decrease, rising porogen volume possibly due to the formation of less oligomers and nuclei under diluted conditions and with an increase of flow ability of the polymer, thus less monomers and cross-linkers diffuse to the surface and polymerize into the small particles.

Precipitation polymerization was utilized to develop graphene oxide (GO)-MIPs) by co-polymerization of GO along with EGDMA through the covalent reaction of protic functional groups [17]. The pre-polymerization system of norfloxacin and methacrylic acid (MAA) was linked to the GO complex by non-covalent bonding, and the imprinted microspheres were formed by heat initiation. After washing the template norfloxacin from the polymers, imprinted holes with complementary binding sites were left in the GO/MIPs [13, 18–20].

3.2.3.3 Emulsion polymerization

Emulsion polymerization refers to the surface modification of porous commercial membranes by light-initiated water-in-oil emulsion polymerization. The process firstly requires preparing a water-in-oil emulsion containing template molecules and functional monomers, and then immersing the membrane in the emulsion to initiate polymerization.

Ma et al. [21] discussed in detail on the formulation and application of MIPs and molecularly imprinted microspheres. The work was built on the selective recognition of compounds such as gatifloxacin, tryptophan, bilirubin, and lysozyme in aqueous media. Reverse microemulsion polymerization with b-cyclodextrin and acrylamide was employed to grow a new form of molecularly imprinted microsphere for gatifloxacin in aqueous media. It revealed excellent selectivity and effective separation from counterparts in high-performance liquid chromatography. Water/oil reverse microemulsion polymerization was utilized for the synthesis of molecularly imprinted microspheres. Combinatorial functional monomers such as acrylamide and acryloyl-β-CD (b-CD-A) were used along with *N,N′*-methylenebisacrylamide as the cross-linker.

3.2.3.4 In situ polymerization

Monolithic MIPs are a variety of MIPs that are developed as a single continuous and porous piece, instead of individual micro- or nanoparticles. They have good capability to detach isomers specifically as chiral stationary phase in high-performance liquid chromatography. Considering their sizes can be controlled between micro and macro, they are capable of providing good pressure reduction and higher flow rates.

With the heterogeneity of target molecules to be analyzed or purified using liquid chromatography, there is an increasing demand for the growth of molecular-specific separation media and their preparation procedures. The molecular imprinting technique is anticipated to be effective to develop new phase systems in liquid chromatography. But given that it is time-consuming to prepare the columns of MIPs, since polymer particles are obtained by grinding and sieving bulk polymers. Hence, a more innovative approach would be desirable for practical use. Novel separation media in the form of continuous rods of MIPs were fabricated by the in-situ method [22]. The components were placed in a stainless-steel column and heated for polymerization. The templates were effectively imprinted in the polymer matrix and the imprinted polymers formed were more stable complexes with the template compounds than their positional isomers in the matrix. This method eliminated the requirement for mechanical grinding and sieving of bulk polymers for the formulation of MIP separation media which also triggered the expulsion of some amount of polymer particles to be thrown away without being packed into a column tube due to being too big or too small for packing. In situ polymerization method allowed the entire number of reagents to be utilized for the preparation of separation media. It was also found that nearly all of the template molecules could be recovered and utilized again without much chemical changes, thereby rendering this method economically feasible.

3.2.3.5 Surface imprinting

Surface imprinting of MIPs upon performed beads could also be used to produce chromatography-grade imprinted polymers. Depending on the requirements, the template molecule is bound to the substrates surface via a non-covalent or covalent method. A functional monomer is thereafter added, or it could be introduced concurrently with the template. Polymerization then occurs, enabling thin polymer films to be grafted onto planar support substrates or spheres. Once the reaction is terminated, the template molecule is removed from the polymer matrix.

In a work by Ye et al. [23], methacrylate-silica was developed by drying LiChrospher Si 1000 in a three-necked flask under vacuum, subsequently, sodium dried toluene along with a mixture of 3-methacryloxy-propyltrimethoxysilane and triethylamine were added. The mixture was refluxed under stirring for 15 h while being continuously flushed with dry nitrogen. The product was then washed with toluene, acetone and diethyl ether before being put overnight to dry under vacuum resulting in the yield of 20.1 g of methacrylate-silica. After producing methacrylate-silica, polymer-coated silica particles were prepared by the addition of methacrylate-silica (2 g) to a mixture that contained *N,N'*-1, 4-phenylenediacrylamide, *N,N'*-methylenediacrylamide, water, methyl methacrylate, and ammonium persulphate. The mixture was mechanically mixed ahead of being placed in an ultrasonic bath, seales, and being kept at 40 ° C for 48 h. The resulting polymer-coated silica was later washed and suspended in several solvents for additional purification

3.2.3.6 Sol-gel imprinting

Sol-gel polymerization involves introducing template molecules into inorganic and organic-inorganic network structures. The process is usually done by mixing the template, cross-linkers, initiators, and other compounds to form a gel through a hydration condensation reaction. Later, molecularly imprinted membranes are prepared by the addition of functional monomers. Moderate reaction conditions and simple operation are among the main advantages of this technique. Furthermore, this method also provides control over properties such as pore size, and surface area of the membrane material through the alterations in components, concentrations, and pH of the reaction system during preparation [24]. Kalogiouri et al. [25] worked on the preparation and optimization of bisphenol A-imprinted sol-gel polymer for selective extraction, with the effectiveness of acetonitrile as a solvent being highlighted and emphasis being given to application of methods to real water samples. This method achieved high efficiency and selectivity for BPA. High efficiency and selectivity were achieved for BPA from this method. PheTES and 3-APTES were utilized to improve the functional selectivity of the synthesized polymer. Central composite designs along with response surface methodology were used to perform optimized extraction protocol [25].

References

[1] Pan J., et al. (2018). Molecularly imprinted polymers as receptor mimics for selective cell recognition, *Chemical Society Reviews*. 47(15), 5574–5587, Available at https://doi.org/10.1039/C7CS00854F.

[2] Fang L., et al. (2021). Efficient removal of norfloxacin in water using magnetic molecularly imprinted polymer, *Chemosphere*. 262, 128032, Available at https://doi.org/10.1016/j.chemosphere.2020.128032.

[3] Cruz-Cruz A., et al. (2022). Recent advances in carbon dots based biocatalysts for degrading organic pollutants, *Current Pollution Reports*. 8(4), 384–394, Available at https://doi.org/10.1007/s40726-022-00228-5.

[4] Sadik O. A., Aluoch A. O., & Zhou A. (2009). Status of biomolecular recognition using electrochemical techniques, *Biosensors and Bioelectronics*. 24(9), 2749–2765, Available at https://doi.org/10.1016/j.bios.2008.10.003.

[5] Bakhshandeh B., et al. (2022). New analytical methods using carbon-based nanomaterials for detection of *Salmonella* species as a major food poisoning organism in water and soil resources, *Chemosphere*. 287(Pt 3), 132243, Available at https://doi.org/10.1016/j.chemosphere.2021.132243.

[6] Ansari S. (2017). Combination of molecularly imprinted polymers and carbon nanomaterials as a versatile biosensing tool in sample analysis: Recent applications and challenges, *TrAC Trends in Analytical Chemistry*. 93, 134–151, Available at https://doi.org/10.1016/j.trac.2017.05.015.

[7] Dong C., et al. (2021). Molecularly Imprinted Polymers by the surface imprinting technique, *European Polymer Journal*. 145, 110231, Available at https://doi.org/10.1016/j.eurpolymj.2020.110231.

[8] Wang S., et al. (2014). Affinity-tunable specific recognition of glycoproteins via boronate affinity-based controllable oriented surface imprinting, *Chemical Science*. 5(3), 1135–1140, Available at https://doi.org/10.1039/C3SC52986J.

[9] Molinelli A., et al. (2005). Analyzing the mechanisms of selectivity in biomimetic self-assemblies via IR and NMR spectroscopy of prepolymerization solutions and molecular dynamics simulations, *Analytical Chemistry*. 77(16), 5196–5204, Available at https://doi.org/10.1021/ac050525f.

[10] Navarro-Villoslada F., Vicente B. S., & Moreno-Bondi M. C. (2004). Application of multivariate analysis to the screening of molecularly imprinted polymers for bisphenol A, *Papers Presented at the 2nd International Workshop on Moleculary Imprinted Polymers Montpellier, France, 15–19 September 2002*. 504(1), 149–162, Available at https://doi.org/10.1016/S0003-2670(03)00766-9.

[11] Shi X., et al. (2007). Development and characterisation of molecularly imprinted polymers based on methacrylic acid for selective recognition of drugs, *Biomaterials*. 28(25), 3741–3749, Available at https://doi.org/10.1016/j.biomaterials.2007.04.036.

[12] Ye L., Yu Y., & Mosbach K. (2001). Towards the development of molecularly imprinted artificial receptors for the screening of estrogenic chemicals, *The Analyst*. 126, 760–765, Available at https://doi.org/10.1039/B009048O.

[13] Yoshimatsu K., et al. (2007). Uniform molecularly imprinted microspheres and nanoparticles prepared by precipitation polymerization: The control of particle size suitable for different analytical applications, *Analytica Chimica Acta*. 584(1), 112–121, Available at https://doi.org/10.1016/j.aca.2006.11.004.

[14] Qiao F., et al. (2006). Molecularly imprinted polymers for solid phase extraction, *Chromatographia*. 64(11), 625–634, Available at https://doi.org/10.1365/s10337-006-0097-2.

[15] Kempe M., & Mosbach K. (1991). Binding studies on substrate- and enantio-selective molecularly imprinted polymers, *Analytical Letters*. 24(7), 1137–1145, Available at https://doi.org/10.1080/00032719108052959.

[16] Chen L., et al. (2016). Molecular imprinting: Perspectives and applications, *Chemical Society Reviews*. 45(8), 2137–2211, Available at https://doi.org/10.1039/C6CS00061D.

[17] Chen J., et al. (2022). Graphene oxide molecularly imprinted polymers as novel adsorbents for solid-phase microextraction for selective determination of norfloxacin in the marine environment, *Polymers*. 14(9), Available at https://doi.org/10.3390/polym14091839.

[18] Mayes A. G., & Mosbach K. (1996). 'Molecularly Imprinted Polymer beads: Suspension polymerization using a liquid perfluorocarbon as the dispersing phase', *Analytical Chemistry*. 68(21), 3769–3774, Available at https://doi.org/10.1021/ac960363a.

[19] Scorrano S., Longo L., & Vasapollo G. (2010). Molecularly imprinted polymers for solid-phase extraction of 1-methyladenosine from human urine, *Analytica Chimica Acta*. 659(1), 167–171, Available at https://doi.org/10.1016/j.aca.2009.11.046.

[20] Zourob M., et al. (2006). A micro-reactor for preparing uniform molecularly imprinted polymer beads, *Lab on a Chip*. 6(2), 296–301, Available at https://doi.org/10.1039/b513195b.

[21] Ma Y., et al. (2012). Efficient synthesis of narrowly dispersed molecularly imprinted polymer microspheres with multiple stimuli-responsive template binding properties in aqueous media, *Chemical Communications*. 48(50), 6217–6219, Available at https://doi.org/10.1039/C2CC31932B.

[22] Jun M., et al. (1993). Molecular recognition in continuous polymer rods prepared by a molecular imprinting technique, *Analytical Chemistry*. 65(17), 2223–2224, Available at https://doi.org/10.1021/ac00065a009.

[23] Ye L., Weiss R., & Mosbach K. (2000). Synthesis and characterization of molecularly imprinted microspheres, *Macromolecules*. 33(22), 8239–8245, Available at https://doi.org/10.1021/ma000825t.

[24] Yang Z., et al. (2021). Synthesis and characterization of $Ca_3Co_4O_9$ micro-nanopowders by sol–gel-molten salt method, *Journal of Materials Science: Materials in Electronics*. 32(24), 28730–28738, Available at https://doi.org/10.1007/s10854-021-07254-x.

[25] Kalogiouri N. P., et al. (2020). Synthesis and application of molecularly imprinted polymers using sol–gel matrix imprinting technology for the efficient solid-phase extraction of BPA from water, *Microchemical Journal*. 157, 104965, Available at https://doi.org/10.1016/j.microc.2020.104965.

Neha Sen

Chapter 4
Polymerization techniques of carbon-based molecularly imprinted polymers

Abstract: Carbon-based nanomaterials such as carbon nanotubes, graphene, and carbon dots are added to the molecularly imprinted polymers (MIPs) to give enhanced sensitivity to the sensor. It significantly modifies the electrochemical response toward selective targeted moieties such as environmental pollutants that mainly include pesticides, insecticides, heavy metals, and antibiotics. The large surface area provided by these nanomaterials has remarkably altered the number of binding sites that excellently modify the detection limits. In this chapter we highlight the developments in the field of carbon-based MIPs. They behave as potent real-time sensors in the detection of almost any predetermined analyte in biological, pharmaceutical, and environmental samples.

Keywords: Carbon-based molecularly imprinted polymers, MIP, polymerization, carbon-based nanomaterial, carbon nanotubes, graphene, carbon dots and sensors

4.1 Introduction

About 50 years back molecular imprinting technology (MIT) opened doors to emerging advents in the field of separation technology when Wulf published his work on molecularly imprinted polymers that were commonly abbreviated as MIPs [1]. MIPs are best known for their simplicity, facile synthesis, stability, sensitivity, and selectivity toward the targeted species by possessing morphologically identical sites, capable of target specific recognition. This have made them efficient scavengers or separating sensors for hazardous predetermined pollutants in a wide range of analytes preferably the pollutants, ions, antibodies, and other variety of molecules [2, 3]. Besides this, they are used as catalysis in immunoassay and as environmental monitoring devices [4, 5].

Most of the commonly used analytical techniques such as high-performance liquid chromatography (HPLC), gas chromatography (GC) with electron capture, GC-coupled mass spectrometry, thin-layer chromatography, enzyme-linked immunosorbent assay, and capillary electrophoresis require skilled technicians, lot of time, and cost for their operation. Traditional electrochemical sensors have disadvantage in re-

Neha Sen, M.L.S.M College, Sundernagar, Mandi 175018, Himachal Pradesh, India, e-mail: neharikasen007@gmail.com

https://doi.org/10.1515/9783111503325-004

lation to lack of sensitivity. The best alternative for the real-time detection of any analyte is the use of MIP-based electrochemical sensor that include quick detection, high selectivity, and excellent efficiency in terms of sensitivity which make them ideal for any analyte determination [6]. Synthesis of MIP through traditional bulk polymerization process involves three main steps:

1. Formation of pre-polymerization complex: It is the association of a template molecule and a functional monomer. A template refers to any atom, ion, molecule, macromolecule, metal complex, or even a microorganism like virus or bacteria that is used to generate a recognition site during the polymerization process and is eventually removed to produce the vacancy that will trap the similar analyte later on as suggested by Pauling in his work in 1940 [7]. The template will control the arrangement of functional groups on the monomer unit that causes the precise and selective binding of the target species. It should facilitate the polymerization process and maintain its stability during the entire working conditions [8].
   The choice of a monomer is crucial. The functional monomers are designed in such a way that the substituents present on them efficiently interact with the template molecules. It should contain the functional groups that control the overall functionality of the MIP. Stability of the template on the polymer depends on the type of monomer used. Acidic templates require basic monomers and vice versa. The signal quality of the MIP-based sensor is directly related to the proper configuration and number of the pores created by monomers. An appropriate template-to-monomer ratio required for stability and sensitivity of polymer is approximately 1:4 under optimal conditions. The most commonly used monomers for the MIP synthesis are pyrrole, methacrylic acid, and 4-vinylpyridine.
2. Polymerization: This is initiated thermally or photochemically in a suitable solvent by using pre-polymerization complex as synthesized in step 1, initiator, and a cross-linker. The cross-linker plays a role to determine the stability and morphology of MIP. It maintains the shape of binding site, adsorption capacity, and the pore size of the MIP. While aiming to prepare a strong and porous polymer, the concentration of the cross-linker is recommended to be kept high. The resulting polymer has target/template molecules still bound to it.
3. Elution or washing away of a target molecule from the polymer matrix to generate three–dimensional (3D) cavities: These 3D cavities so formed have specificity in their shape and size to selectively bind and extract the target molecules, which was otherwise a tough and a tedious process.

The polymer as discussed above is produced by the bulk polymerization process which has low efficiency. To solve the problem, other polymerization techniques were introduced which include suspension polymerization [9], dispersion polymerization [10], and precipitation polymerization [11]. Few other polymerization techniques are emulsion, electrochemical, swelling, sol-gel, ultrasound, microwave, and surface imprinting polymerization. These are summarized in Figure 4.1.

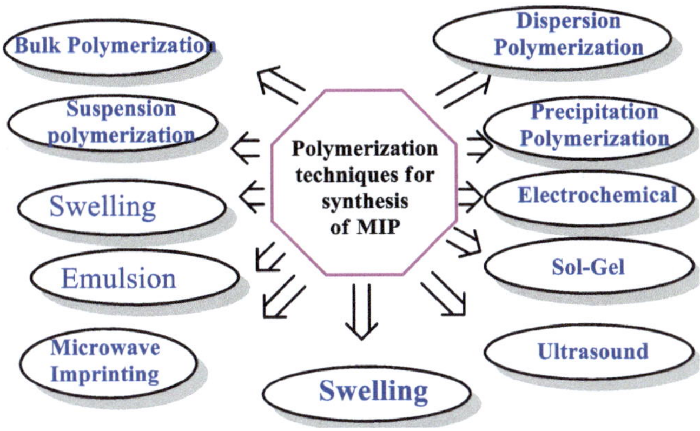

Figure 4.1: Various polymerization techniques for the synthesis of MIP.

MIPs so obtained have selective and sensitive recognition micro cavities for specific species, but they still suffer limitations including unevenly distributed binding sites, poor site accessibility for template molecules, slow rate of adsorption, and low diffusion across these sites. Most serious drawbacks were pertaining to their low conductivities and electrocatalysis. Among various polymerization techniques used so far for the synthesis of MIP, electropolymerization remains one of the most commonly used technique giving desirable detection limits. Since the 1970s, tremendous continuous efforts have been put forward to modify MIPs to update their binding capacities. Efforts were made to update and renew the MIPs by introducing controllable magnetic molecularly imprinted polymers (MMIPs) [12, 13].

In addition to their short pretreatment time, binding to analyte is fast and effective, making them efficient electrochemical sensors for the detection of various molecules. Going ahead, surface molecular imprinting technique (SMIT) [14], where polymerization process is carried on surface layer of a suitable substrate or matrix, has gained recent attentions. Here in SMIT, the recognition or binding sites are distributed on the surface layers avoiding the remote access to deep seated or embedded sites. It was after the work of Dickey in 1949 [15], SMIT attracted the attention of various researchers around the globe to be adopted as one of the leading separation technique.

He used methyl orange dye as the template that is imprinted on silica gel. SMIT permits outstanding results to be utilized as sensor, separator, purifier, and catalyst due to its preferred selectivity, high surface to volume ratio, more binding affinity, elimination of limitation pertaining to embedded binding sites, and binding kinetics.

During the formation of surface molecular imprint polymer (SMIP), (i) the template and the monomer are bonded mainly by covalent, semi-covalent, or non-covalent interactions [16, 17]; (ii) polymerization is carried on the surface of a suitable solid substrate in the presence of a cross-linker and an initiator; (iii) finally, the template is washed or eluted away with a suitable solvent to generate the predetermined cavities or pores for selective extraction process. Table 4.1 presents various possible interactions between a monomer and a template.

Table 4.1: Types of interactions between a monomer and a template in the MIP.

Covalent interactions	– Covalent interactions are best suited for monomer and template moieties with high formation kinetics leading to uniform distribution of binding sites. – High strength of covalent bond leads to the formation of a stable polymer. – It however poses a difficulty in removing the template during elution process. – Special reaction conditions are required to be maintained during removal of the template from binding site. – Highly specific and precise binding sites are formed in resulting MIP.
Non-covalent interactions	– During non-covalent interactions, week forces such as van der Waals forces, hydrogen bonds, dipole, or electrostatic interactions are responsible for holding the monomer and a template together. – This is the most preferred technique that involves self-assembly or template procedure of in situ non-covalent interaction between the monomer and the template. – Non-covalent process of polymerization is pretty fast that proceed under easy ambient conditions. However, special methods of polymerization are required for selection to minimize the formation of non-specific-binding sites. – The resulting SMIP however has less stability than those having covalent interactions.
Semi-covalent interactions	– Both covalent and non-covalent interactions prevail; monomer and template have covalent interactions during imprinting time, whereas non-covalent forces prevail during rebinding time. – Specificity of binding sites is maintained. – Special reaction conditions are required pertaining to covalent interactions.

4.2 Carbon-based molecular imprint polymer (CMIP)

Replacement of the planar surfaces by the nanostructured surfaces is one of the most attractive approaches for the development of MIPs. Most recently, development of carbon-based molecule imprint polymers (CMIPs) opened doors to recent events in separation research as promising solutions to maximize the properties and minimize the limitations associated with MIPs [18, 19]. Carbon and carbon-based nano materials such as carbon nanotubes (CNT), carbon nanoparticles (CNP), carbon electrodes (CE), graphene (G), graphene oxide (GO), and carbon dots (CDs) are attractive materials used to modify the properties of MIPs due to their large surface area, good thermal conductivity, high chemical and mechanical stability, fluorescence, and capability of bonding with a number of elements [20]. Among them, CNT and G are forefront nanomaterial used to make sensors. Nowadays work is heading toward the use of green carbon dots. Carbon-based nanomaterials are efficient electron mediators used to make efficient electrochemical biosensors [21, 22]. Their reactivities can be changed by modifying their surface properties. Imprinting of the polymer on the surface of substrate or a solid support (e.g., CNP) is done in two types: (i) surface imprinting: applying a layer of MIP on the outer surface of solid support; (ii) core-shell imprinting: MIP distributed throughout (surface and core) the solid support.

MIP finds one of the fascinating applications in the field of electrochemical sensors. MIP-based electrochemical sensors consist of (i) a sensor (which is MIP itself) used for selective sensing of an analyte under consideration and (ii) a transductor (which is an electrode) that converts the information gained from sensor into an electric signal [2]. The properties of carbon-based MIP sensors depend upon the surface properties of the nanomaterial under use, which are either used as such or fabricated with glass electrodes or any other membranes.

The chapter highlights the developments of highly specific and efficient CBMIP that mimic the antibiotics, biologically active molecules, or pollutants which are otherwise required to be extracted or selectively removed.

4.2.1 Carbon nanotube-based MIP (CNT-MIP)

An allotrope of carbon, CNTs are one of the most studies nanomaterial in last decades. They are hollow, cylindrical tubes of graphite with a diameter up to around 100 nm. Depending upon the number of graphite layers in a cylinder, CNT exits in two forms:

1. A single-layer hollow carbon sheet in the form of tubes called as single-walled carbon nanotubes (SWCNTs).
2. More than one layer hollow carbon sheets in the form of tubes that are held by weak van der Waals forces and called as multiwalled carbon nanotubes (MWCNT).

CNTs are excellent substrate and supporting material for MIPs proving it with large surface area, corrosion resistance, mechanical strength, electrical conductance, and biocompatibility. CNTs can be easily functionalized and modified by different organic functional groups. In most of the cases they are first functionalized with carboxylate group (–COOH) by treating with carboxylic group-based strong acids [23]. These are then grafted with the vinyl ends and coupled through silicon by using silicon coupling agent with an aim to anchor the monomer units. CNTs have the capacity of solid-phase extraction of various organic and inorganic moieties. In addition, they are now widely used to (a) modify the electrodes, (b) maximize the sensitivity by increasing the surface-to-volume ratio, (c) promote the electron transfer processes, and (d) high electrocatalytic activity. All these properties make them efficient materials for the analytical determination procedures in the development of sensitive and efficient biosensors. SWCNT and MWCNTs are now more widely used to modify the MIP-based electrodes.

A novel microextraction technique for electromembrane extraction (EME) procedure was developed by Tahmasebi et al. [24] by using MIP-coated MWCNTs. Naproxen, an acidic drug, was determined by this technique. EME extraction process generally faces a limitation of lack of selectivity that was overcome by introducing MIP-MWCNT -coupled EME. The MIP has low efficiency in aqueous phase; thus MIP-MWCNTs were protected by placing them between the pores of hollow fibrous polypropylene. The duel effect of EME and MIP led to high sensitivity, fast enrichment of EME, and easy cleaning of MIPs. This coupled process called MIP-MWCNTs-EME was used for the first time which not only enhanced the selectivity of EME but also isolated MIP from sample solution.

Alizadeh et al. [25] proposed a sensitive and selective electrochemical sensor to find ultra-trace levels of RDX (trinitroperhydro-1,3,5-triazine) on the basis of MIP technique. The carbon paste electrode is normally inactive against the RDX. However, on coupling with MIP, sites for RDX determination were generated on the glassy carbon electrode (GCE) together with MWCNTs. Use of MWCNTs enhanced the voltammetric signal of MIP-based glass electrode and acted as an ultrasensitive sensor for RDX.

A novel 3D MIP coated on carboxylated CNT acting as an amperometric sensor for profenofos (PFF), a common agricultural insecticide was proposed by Amatatongchai et al. [26]. Carboxylated CNT is coating with SiO_2 (to form CNT-SiO_2 using sol-gel technique) and vinyl end groups (to form CNT-SiO_2-vinyl). It was finally grafted with MIP resulting in CNT-SiO_2-vinyl-MIP or CNTs@MIP. The monomer, cross-linker, and initiator are used for generating MIP are methacrylic acid (MAA), ethylene glycol dimethacrylate (EGDMA), and 2,20-azobisisobutyronitrile (AIBN), respectively. The 3D CNTs@MIP is coated on the surface of a GCE. PFF templates are finally removed by thorough washing with acetic acid and methanol to form highly stable, efficient, and reproducible 3DCNTs@MIP/GCE sensor device for PFF. A diagrammatic representation of various steps for the synthesis of 3DCNTs@MIP/GCE is given in Figure 4.2.

The SEM and TEM studies show (i) the presence of uniformly distributed pores and (ii) the thickness of uncoated CNTs and coated 3D-CNTs@MIP to be approximately 14.3 and 78 nm, respectively. Step-wise confirmation of the development of sensor was done from FTIR studies. Cyclic voltammetry was carried out to see electrochemical behavior of PFF. The sensor CNTs-SiO_2 had no electrochemical peak current showing absence of PFF surface binding sites. The grafted CNTs@MIP sensor on other hand facilitated the electron transfer between PFF and the electrode surface and exhibited a well-defined oxidation peak owing to its good electrical conductivity.

A carboxyl-functionalized MWCNT was deposited on pencil graphite electrode (PGE) surface and a MIP film of pyrrole (monomer) and hydrochlorothiazide (HCT) as a template was electrodeposited on it [27]. During the design of MIP, computational studies were carried out to select pyrrole and ethanol as the best monomer and solvent, respectively, for HCT template. The sensor is used to detect HCT (a drug used to

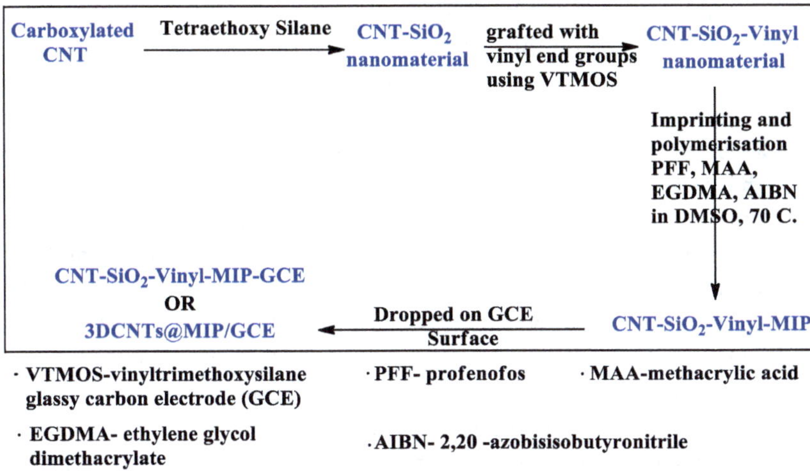

Figure 4.2: A diagrammatic representation of various steps for the synthesis of 3DCNTs@MIP/GCE.

lower higher blood pressure) in the presence of other drugs such as methimazole, benzimidazole, acetaminophen, and aspirin that interfere during HCT detection. The performance of the sensor was checked in serum and pharmaceutical samples and gave a detection limit of 1.0×10^{-10} M.

A nano film of MIP was electropolymerized on MWCTN-modified PGE for the enantio-selective separation of aspartic acid (D- and L-forms) [28]. The MIP is prepared by the polymerization of indole-3-acetic acid in the presence of acetic acid as a template molecule. Template recognition is done by the method of doping and de-doping where template was lost due to over oxidation.

Zhou et al. [29] constructed a super hydrophilic magnet surface imprinted nanocomposite by one-pot condensation of formaldehyde, resorcinol, and melamine on MWNTs@Fe_3O_4. It was selective for the determination of triazines, a potent herbicide in water samples.

A completely reusable, highly stable, and ultrasensitive electrochemical sensor for identifying ultra-trace level of antibiotic metronidazole (MNZ) (in low-detection limit 4.92×10^{-5} mg L^{-1}) was prepared by coupling MIPs and MWCNTs [30]. MNZ is a nitroimidazole class of antibiotic that gradually accumulates in human body leading to health issues like seizures and peripheral neuropathy. To prepare the sensor, the suspension of COOH-MWCNT in distilled water containing sodium dodecyl sulfate was drop coated on the surface of GCE. Polymerization of dopamine (monomer in this case) was carried on the surface of GCE in the presence of MNZ template, and finally MNZ templates were extracted from the polymer using dilute sulfuric acid. SEM and TEM images reveal deposition of the thin layer of polymer of about 10–15 nm on the surface of GCE. Stepwise confirmation of above synthetic procedure as done by cyclic voltammetric studies leads to the following results: (a) MWCNT-decorated GCE shows enhanced current signal; (b) the conductivity of the GCE increased considerably due to its coupling with MWCNTs; (c) the current signals disappear after applying the layer of MIP showing non-conducting nature of MIP which blocks the redox channels to reach the electrode surface; (d) removal of MNZ templates in the last step caused the reappearance of cyclo-voltammetric current signals due to cavities created on the polymeric film that exposed the surface of GCE; (e) finally, during the analysis of MNZ as an analyte in the test solution, the cavities created were again re-filled that blocked the electron transfer channels and considerably decreased the current signal. Polymerization in this case is carried in aqueous medium which makes all experimental measurements to be easily done in water. This MWCNT-MIP-fabricated electrochemical sensor is highly effective and has outstanding sensitivity owing to its extremely low detection limit, high stability at room temperature, and reusability which is otherwise superior to other MNZ sensors reported earlier.

Integration of molecular imprinted polymer with a magnetic material brings magnetic susceptibility and separation properties together that provides faster and efficient sensor giving satisfactory separation results. Such surface-imprinting biosensor using chitosan (CS)-coated magnetic nanoparticle (MNP)-modified MNPs/CS-MWNTs was prepared for detecting bovine serum albumin (BSA) by Chen et al. [31]. Pyrrole (monomer in this case) along with template BSA was electrodeposited on MNPs/CS-MWNT. Differential pulse voltammetry showed a wide detection linear range of 1.0×10^{-4} to 1.0×10^{-10} g mL^{-1} with a low detection limit of 2.8×10^{-11} g mL^{-1} for BSA. It addition the sensor has fast balance response time of 30 s.

MWCNT-based MIP using MAA as a monomer was developed for the detection of herbicide 2,4-dichlorophenoxy acetic acid [32], an explosive trinitroperhydro 1,3,5-triazin (called RDX) [25], and herbicide 3-(3,4-dichlorophenyl)-1,1-dimethylurea (diuron) [33] in water samples. The detection limits are 1.20×10^{-9} M, 20.0 pmol L^{-1}, and 9.0×10^{-9} mol L^{-1}, respectively. The pesticide 2,4-dichlorophenoxyacetic acid was determined potentiometrically on a sensor based on ion-imprinted polymer inclusion membrane prepared by modifying MWCNT-based MIP. MWCNT-based MIP used in this case is tailored using MAA as monomer, trimethylol propane trimethacrylate as cross-linker, and 2,4-dichlorophenoxy acetic acid as a template molecule. The imprinted polymer gave signal response in the range of 1×10^{-9} to 1×10^{-5} M and could be reused a couple of times without losing its sensitivity and stability up to a period of about 3 months. For the detection of RDX, nano-MIP-MWCNT-based voltammetric sensor was prepared by copolymerized of methacrylic acid (monomer) and EGDMA (cross-linker) by precipitation polymerization method. MIP for the detection of diuron in water samples was developed by bulk polymerization that along with carboxylated-MWCNT was immobilized on a carbon paste electrode. MAA, selected by computational simulation studies, was a preferred choice to be used as a monomer for the determination of diuron. The interaction occurred between amino groups of diuron (template) and –COOH groups of MAA (monomer). This interaction and fabrication amazingly increased the sensitivity of carbon paste electrode by approximately eight-folds. A related work by Fernandez et al. showed that sensitivity for diuron herbicide is increased by fourfolds (detection limit = 0.58×10^{-5} mol L^{-1}) when the GCE is modified with SWCNT [34]. The MIP for this sensor was prepared by free radical polymerization process having MAA as a monomer and EGDMA as a cross-linker.

A carboxyl-functionalized MWCNT-modified GCE was reported by Karimian et al. for the determination of cefixime by using acrylamide as functional monomer and *N*, *N*-methylenebisacrylamide as a cross-linker [35]. This voltammetric sensor gave signals which were linearly related to cefixime concentration between 0.01 and 600.0 μM and a detection limit of 0.001 μM.

Various polymerization processes are tested to overcome the limitations associated to bulk polymerization technique. Sol-gel technique is one such method. Conductivity of such sol-gel-based MIP systems can be increased by using conducting polymers and nanomaterials.

Among various conducting polymers, pyrrole, due to its high stability, biocompatibility, and easy synthesis is one of the best considered choices to be used as a monomer on the CNT-based MIPs for the detection of a variety of analytes [36, 37]. Deiminiat et al. [38] developed a new electrochemical imprinted sensors based on polypyrrole and functional MWCNT for the determination of tramadol. Detection limit of 0.03 nM shows excellent sensitivity and selectivity of this sensor. The sensitivity for tramadol was also analyzed in real samples.

Electropolymerization of MIP with SWCNTs was carried out for the determination of brucine in human serum samples by Liu et al. [39]. Brucine is an analgesic and anti-inflammatory drug whose overdose can be harmful leading to death in extreme cases. A voltammetric SWCNT-MIP-based sensor on GCE was prepared by using poly-*o*-phenylenediamine as monomer and brucine as template molecule. Compared to the majority of work focusing on the use of MWCNTs with MIPs, lesser reports have been found on the use of SWCNT to modify MIPs.

4.2.2 Graphene-based molecular imprint polymers

Graphene (G) is a one-atom-thick 2D carbon sheet providing large surface area almost twice as that of SWCNTs. It has hexagonally arranged sp^2 hybridized carbon atoms in a honeycomb like fashion, commonly referred to as “wonder material” in consistence with its fascinating properties. Mostly it is used in its modified form such as graphene oxide, reduced graphene oxide, and graphene nanoribbons:

(i) Graphene oxide (GO): it is the oxygenated form of graphene, where G is oxidized with oxidizing agents like $KMnO_4$ and $NaNO_3$. The carbon atoms are decorated with oxygen atoms as functional groups.
(ii) Reduced GO (rGO): it is the reduced form of graphene oxide formed by reducing the GO with various reducing agents like hydrazine hydrate.

(iii) Graphene nanoribbons (GNRs): In this form, G sheets are in the form of thin nano-sized ribbons.

Graphene and its derivatives owing to their extraordinary thermal and electrical conductivity, large surface area, fast electron transfer rate, and unique electronic properties are used to improve the basic properties of traditional MIPs. It has high potential to be used in future electrochemical sensing [40–42] and detection of antibiotics, pesticides, herbicides, biological molecules, and other agricultural, food, and environmental pollutants along with contaminants produced by the charge transfer processes. G can be easily made functional via both covalent ad non covalent interactions. During covalent functionalization the electronic arrangement around the carbon atoms change while in non-covalent interaction there is no impact on the structure of G nanosheets causing surface adsorptions only.

Deposition of graphene on MIP is however a difficult task in organic solvents due to its inert surface, hydrophilic nature, and low solubility in maximum solvents like chloroform, acetone, and toluene. In most of the work, MIPs are prepared in the organic solvents whereas the analyte extraction is done in aqueous medium. Due to different polarity of solvents used, there is differential swelling/shrinkage of polymers in different solvents, affecting the sensitivity and selectivity of MIP-based sensors. To solve the problem, the initial step of MIP synthesis can also be carried out in aqueous medium using graphene-based MIP owing to their high hydrophylicity. The G derivatives such as GO and rGO are more reactive due to active functional groups on their surface that can be easily blended with MIPs. The G-based sensors so produced are decorated with favorable properties like high sensitivity, good selectivity, stability, and reusability.

After the discovery of graphene in 2004, the first G-based MIP electrochemical sensor was prepared by Liu et al. in 2012 for the detection of 2,4-dinitrophenol [43]. They used GO-MIP on the GCE taking *o*-phenylenediamine as both functional monomer and cross-linker. Luo et al. [44] formulated a unique strategy for manufacturing MIP. They used vinyl group-functionalized graphene by immobilizing 4-vinylcarbazole on the graphene through π-π interaction. Polymerization between MAA (monomer) and EGDMA (cross-linker) and 4-nitrophenol (template) lead to the formation of MIP for the detection of 4-nitrophenol which is a potent wide spread pollutant.

In the study done by Beluomini and coworkers [6] on the carbon-based MIPs, it was found that initially the CNT were the major choice for nanomaterial selection to modifying the MIPs. But by the end of last decade, the reports published indicate that G-based MIP sensors have surpassed the publications related to CNT-MIP. A simple monolayer of MIP was deposited on the surface of unmodified G sheets by Zang and his group [45]. In this modified procedure the monomer *p*-vinylbenzoic acid was first combined with graphene and then complexed with imidacloprid (template) using EGDMA (cross-linker). MIP/G so produced was applied on GCE to give MIP/G-modified GCE sensor to detect imidacloprid pesticide. The introduction of graphene to the MIP increased the conductivity of polymer as reflected by the peak current on MIP/GCE that was lower than MIP/G-modified GCE. Due to irreversible cyclovoltammetric reduction, it could be easily removed from the binding sites during the initial scans only. The sensor had excellent reproducibility which is otherwise difficult to achieve in electrochemical sensors.

MIP based on graphene is prepared by Gupta et al. [46] by depositing G on the surface of GCE followed by electropolymerization of 4-amino-3-hydroxy-1-naphthalenesulfonic acid and melamine in the presence of melatonin as a template. Cyclovoltammetric response of the sensor showed linear variation of current signal with concentration of melatonin in 0.05–100 $\mu M\ L^{-1}$ range. The efficiency of the sensor was checked in the biological human samples with detection limit of 0.006 $\mu M\ L^{-1}$.

A nano-sized MIP film was made for the determination of testosterone steroid up to femtomolar to micromolar levels on the surface of GO sheets. The extreme selectivity and sensitivity (detection limit up to 0.4 fM) of the biosensor owe to the MIP design and large surface area provided by GO nanolayers. The authors used electrochemical impedance spectroscopy to see electrode surface interactions to detect the recognition signal of testosterone [47]. A graphene-chitosan molecularly imprinted film is developed to modify the surface of GCE for the determination of L-5-hydroxytryptophan [48]. Easy and timely determination of L-5-hydroxytryptophan is important to detect and monitor the problem posed by it in human body. The sensor exhibits superiority in sensitive detection and selective recognition of L-5-hydroxytryptophan than the classical methods and other surface-modified electrodes developed earlier. The super selectivity (detection limit of 6.0 nM) of this sensor reflects its capacity to recognize the target species among the other coexisting structural analogues and its ability to not interfere with them.

Cloxacillin (CLO) is an antibiotics used against *S. aureus* in mastitis treatment in lactating cows. The antibiotic is readily transferred from the raw milk to various dairy products. Jafari et al. [49] produced an electrochemical MIP nanosensor to selectively extract CLO from milk and other biological samples. MIP preparation fabricates MAA as monomer, 2,2′-azobis-2-methylpropionitrile as initiator and ethylene glycol dimethylacrylate as a cross-linker in the presence of CLO templates. The sensor was prepared by using MIP-GO-gold nanoparticle modified on screen-printed carbon electrode giving a detection limit of 36 nM. The optimum conditions for the removal of CLO (92%) were pH = 8.5, contact time = 89 min, and MIP = 0.79 g.

A highly efficient sensor with rGO for the determination of fluxapyroxad (a fungicide) was recently prepared [50]. Indole-6-carboxylic acid and platinum nanoparticles (Pt-NPs) were immobilized on amino-functionalized reduced graphene oxide (NH_2-r-GO) to give Pt-NPs-NH_2-r-GO to be deposited on GCE. The Pt-NPs-NH_2-r-GO/GCE sensor showed an excellent detection limit.

Picric acid was determined on a sensor (MIP/rGO/PGE) prepared by using rGO on PGE [51]. Pyrrole was polymerized on its surface using picric acid templates. To give the best possible sensitivity to the sensor, initial computational studies were put in use to study the interactions between the MIP and the template. The resulting sensor gave good detection limit of 1.4 μM, high reproducibility, and mechanical strength.

In a work done by Wang et al. [52], graphene was functionalized via π-π interaction on GCE by using diallyldimethylammonium chloride as a monomer for the determination of 4-chlorophenol. The modified sensor is highly sensitive with detection limit of 0.3 μmol L^{-1} showing excellent selectivity for analyte and non-interference with structurally related compounds. Recently GO-based MIP sensor was synthesized by Soman et al. [53] for the real-time selective determination of uric acid in blood serum samples, giving limit of detection of 0.4257 μM. In this work, GO is first changed to Si-GO form thus making it more conducting by inserting vinyltrimethoxysilane within its layers. MIP is prepared by copolymerization of itaconic acid and MMA in presence of uric acid templates. The sensor with modified GO is confirmed to be highly reproducible after repeatedly used under altering conditions of time and concentration.

Apart from most of the studies based on modification of GCE; an MIP based on gold (Au) electrode was decorated with double composite, that is, chitosan-platinum nanoparticles (CS-PtNPs) and graphene-gold nanoparticles (G-AuNPs) to give G-AuNPs/CS-PtNPs/gold electrode [54]. Mercaptonicotinic acid (MNA) and erythromycin used as monomer and template, respectively, were then deposited on G-AuNPs/CS-PtNPs/gold by Au–S bond and H-bond interaction followed by electropolymerization of $HAuCl_4$, MNA, and erythromycin. The resulting polymer was highly stable, reproducible, and selective for the determination of erythromycin in various edible products with limit of detection up to 2.3×10^{-8} mol L^{-1}.

Cannabidiol (CBD), an active ingredient present in cannabis, is an important drug used to treat seizures. Tang and group [55] developed a sensor for the detection of CBD. A zirconium-based metal organic framework (MOF), UiO-66 (prepared by mixing Zr_6O_{32} units and 2-aminoterephthalate linkers) was mixed with G in water to form G-UiO-66 which was then dropped on the working electrode of screen-printed electrode (SPE) to give G-UiO-66/SPE. Subsequently the modified electrode was mixed with magnetic MIP to generate magnetic MIP/G-UiO-66/SPE sensor. In this process 3-methacryloxypropyltrimethoxysilane-modified Fe_3O_4 magnetic nanoparticles, CBD, MAA, ethyleneglycol dimethacrylate, and 2,2-azobisisobutyronitrile were used to prepare mag-MIP.

Olaquindox (OLA) is a chemical substance having antimicrobial action and growth promoting effects in cattle. When found in food, it poses a very harmful effect to humans. Very recently, in a work published by Han et al. [56] an electrochemical sensor was developed for the determination of OLA. They used nitrogen-doped G quantum dots (N-GQDs) and a Ni-based metal-organic framework functionalized with Ag NP (Ag/Ni-MOF) to form a unique honeycomb structures that was modified on GCE. MIP was prepared by elctropolymerization on G/Ni-MOF/N-GQDs/GCE. Modification of GCE with NPs increased its surface areas and enhanced the electron transfer rate. The MIP sensor so developed had extremely low detection limit of 2.2 nmol L^{-1} for OLA.

Xing et al. and Wang et al. in two independent works prepared an imprinting sensor by coupling the advantages of carbon-based NMs, G, and MWLNT for the detection of tryptamine and vanillin, respectively [57, 58]. Tryptamine is responsible for hypertension when overconsumed in a protein-rich diet. Xing and coworkers initially elctropolymerized pyrrole and sulfonated graphene (SG) on GCE to obtain polypyrrole PPy-SG/GCE. Hyaluronic acid-MWCNTs (HA-MWCNTs) and an MIP layer (prepared by using tryptamine as the template, and *para*-aminobenzoic acid (PABA) as the monomer) were then casted on it to give MIPs/HA-MWCNTs/PPy-SG/GCE electrochemical sensor. For the ultra-sensitive estimation of vallinin, a sensor based on GO/carboxylated-MWCNs/Au-NP/MIP was prepared. This association of various nanoparticles increased the rate of electron transfer processes giving the detection limit as low as 6.23×10^{-10} mol L^{-1}.

Recently, Shao and his team [59] developed a selective and sensitive electrochemical sensor MIP/MXene/Au/GCE by dropping MIP on MXene and gold (Au) nanoparticles. Fe_3O_4 NPs were initially introduced on G and subsequently dopamine was polymerized on Fe_3O_4. MIP was prepared using 4-vinylpyridine as a monomer for the selective detection of tetrabromobisphenol A in a water sample. Differential pulse voltammetric studies surprisingly gave low detection limit of 0.0144 nM L^{-1}.

A polydopamine-rGO (PDA-rGO) and MIP-modified GCE was fabricated for highly sensitive and selective detection of 2,4-dichlorophenol (2,4-DCP) by Liu et al. [60]. Dopamine was auto-polymerized in the rGO solution to give PDA-rGO, whose surface was deposited with electropolymerized MIP having *o*-phenylenediamine (monomer) and 2,4-DCP (template). The MIP components have π-π stacking interactions with PDA-rGO forming a compact sensor with more binding sites. During the design of G-based MIPs the choice of a monomer is very crucial. Table 4.2 presents the selective work related to G-based MIP.

4.2.3 Carbon dot-based molecular imprint polymers

CDs are a newly developed zero-dimensional modern luminescent carbon nanoparticle having size below 10 nm. Accidently discovered in 2004, during the purification of SWCNT, a distinct route for their synthesis was given 2 years later in 2006 by Sun et al. where he utilized them for enhancing fluorescence emission [62]. Since then they are studied for their important properties like light harvesting, chemical inertness, biocompatibility, less cytotoxicity, high quantum yield, excellent and tunable photoluminescence, bright fluorescence, electron transfer ability, and small size [63]. Moreover they can be easily functionalized and can be prepared by pure green routes in aqueous medium due to their excellent water solubility. They are nowadays frequently prepared from the biowaste. The fluorescent properties of CD are adjusted by proper size selection and variation of the surface

Table 4.2: Molecularly imprinted polymers based on G and its derivatives such as GO and rGO.

Nanocomposite	Template/analyte	Monomer	Detection limit	Reference
MIP/GO	2,4-Dinitrophenol	*o*-Phenylenediamine	—	[43]
MIP/G modified GCE	Imidacloprid	*p*-Vinylbenzoic acid	0.10 μM	[45]
MIP/G/GCE	Melatonin	4-Amino-3-hydroxy-1-naphthalenesulfonic acid and melamine	0.006 μM L^{-1}	[46]
MIP/GO	Testosterone	—	0.4 fM	[47]
MIP/G-chitosan/GCE	L-5-hydroxytryptophan	—	6.0 nM	[48]
MIP-GO-gold nanoparticle modified on screen printed carbon electrode	Cloxacillin	MAA	36.00 nM	[49]
MIP/Pt-NPs-NH_2-r-GO/PGE	Fluxapyroxad	Indole-6-carboxylic acid	1.00×10^{-10} M L^{-1}	[50]
MIP/rGO/PGE	Picric acid	Pyrrole	1.4 μM	[51]
MIP/G/GCE	4-Chlorophenol	Diallyldimethylammonium chloride	0.3 μmol L^{-1}	[52]
MIP/Si-GO	Uric acid	Itaconic acid and MMA	0.4257 μM L^{-1}	[53]
G-AuNPs/chitosan-PtNPs/gold electrode	Erythromycin	Mercaptonicotinic acid	2.3×10^{-8} mol L^{-1}	[54]
Magnetic MIP/G-UiO-66/SPE	Cannabidiol	MAA	5.00×10^{-8} M L^{-1}	[55]
G/Ni-MOF/N-GQDs/GCE	Olaquindox	Aniline	2.2 nmol L^{-1}	[56]
MIP/hyaluronic acid-MWCNT/polypyrrole-sulfonated G/GCE	Tryptamine	*para*-Aminobenzoic acid	7.4×10^{-8} mol L^{-1}	[57]
GO/carboxylated-MWCNs/ionic liquid/Au-NP/MIPs	Vanillin	Pyrrole	6.23×10^{-10} mol L^{-1}	[58]
MIP/MaaXene/Au/GC	Tetrabromobisphenol A	4-Vinylpyridine	0.0144 nM	[59]
PDA/rGO/MIP	2,4-Dichlorophenol	*o*-Phenylenediamine	0.8 nM	[60]
MIP/graphene	4-Nitrophenol	MAA	5 nm	[44]
MIP/ Fe_3O_4/GO-modified glassy paste electrode	Patulin (in apple juice)	1-Ethyl-3-methylimidazolium tetrafluoroborate	3.33×10^{-4} nM	[61]

functional groups. The emission behavior can further be modified by doping the CDs. They find wide applications in the current hot research fields like biosensing, bioimaging, photocatalysis, biological vehicles, biological dyes, and determining hazardous substances in food and environment [64]. Most recently they are the class of nano particles that are used for fluorescent resonance energy transfer processes.

Despite various important properties and advantages related to MIP there are certain challenges that need to be resolved. Coupling the physical properties like high sensitivity and recognition specificity of MIP with fluorescence detection ability of CD is a new revolution in the field of analytical chemistry for manufacturing optical sensors. MIP/CD-based sensors being more economic, handy, fast, and convenient to operate are much preferred over the sophisticated traditional ones such as GC and HPLC. The CDs are easily synthesized and have very less toxicity so they are preferably used over the conventional semiconductor quantum dots [65].

Depending upon the fluorescence behavior of the target analytes, the CD-based MIP are synthesized by both direct and indirect methods. Direct procedure is followed for the analytes having in-built fluorescence properties while in the indirect route the analogs of fluorescent functional monomers are used. Transfer of fluorescence resonance energy between CD and MIP was used to manufacture a novel sensor by Li et al. [66] for detecting dimethoate pesticide in vegetables. Here MIP was prepared on electrode surface by electropolymerization of methyl red doped-*o*-phenylenediamine and dimethoate as a monomer and a template, respectively. Limitations associated with the earlier used fluorescent dyes in sensors owe to their high toxicity and low sensitivity which is overcome by the use of CD. The above sensor shows good stability up to a period of 1 week.

A nanopomegranate-type CD-MIP was prepared by placing CD as seeds in polymer matrix of *N*-(β-aminoethyl)-γ-aminopropyl trimethoxysilane (monomer) and quinoline (template) to detect a toxic chemical quinolone in coking waste water. Combined results of MIT and fluorescence detection resulted in high sensitivity with good photo luminescence of the sensor with detection limit of 6.7 mg/L [67]. Similarly, another fluorescent CD-MIP sensor was prepared by Xu et al. Here MIP was fabricated on CD in a one-pot reaction by a non-hydrolytic sol-gel process [68]. It is used to detect sterigmatocystin, a carcinogenic mycotoxin produced by some Aspergillus stains. The CD enhanced the signal for sterigmatocystin as detected by MIP. During MIP synthesis, 1,8-dihydroxyanthraquinone was used as an alternative template.

Vinyl functionalized CDs (V-CDs) were used to modify the MIP which is prepared by using *N*-isopropylacrylamide and 4-vinylphenylbronic acid as monomers, ammonium peroxodisulfate, and *N,N'*-methylene bisacrylamide as an initiator and a crosslinker, respectively, by Sun et al. [69]. This fluorescent nanosensor was used for the detection of alpha-fetoprotein (a protein produced by the fetus) in the biological samples such as amniotic fluid and blood of a pregnant mother as a novel method for the detection of tumor markers with a remarkable detection limit of 0.474 ng mL^{-1}.

An equipment-free paper-based, safe, and a handy visual sensor for the detection of trace-level silver ions (Ag^+) (with detection limit 10.0 mg L^{-1}) was prepared by using CD by Patra et al. [70]. It led to the detection of Ag^+ ions with naked eyes. CD was for the first time prepared from fluorescein dye as a precursor. Then fabricating with thermo-responsive MIP to elevate its selectivity, it followed a totally green synthetic route. The sensor exhibits minimum toxicity during determination of Ag^+ in living cells. The utility is designed in such a way that it could be even operated with non-technical hands giving fast results within less than 10 s. It also finds application in calligraphy and in other craft-related works.

Silane-doped CD (Si-CDs)-based MIP (MIP@Si-CDs) as an optical sensor was prepared to detect a systemic broad-spectrum insecticide, acetamiprid with 2 nM detection limit [71]. In this procedure organosilane-impregnated CD (Si-CDs) are prepared by reacting citric acid and 3-aminopropyltriethoxysilane, and finally MIPs (having 3-aminopropyltriethoxysilane as monomer and tetramethoxysilane as a cross-linker) are deposited on Si-CDs using sol-gel method.

Very recently, a CD-doped MOF-MIP complex was prepared where CD is solely derived from tryptophan and MOF. This sensor is equipped with the high adsorbing properties of MOF, high selectivity of MIP, and excellent fluorescence properties of CD for fast detection and scavenging of tetracycline; a broad spectrum antibiotic used to cure bacterial infection in humans and animal husbandry [72]. Excessive use of this antibiotic has led to global concerns due to its bioaccumulation in various animal products like milk, meat, and honey. The auto fluorescence of the honey overlaps with the fluorescence emission spectrum of the CDs; therefore CD cannot be used alone to detect tetracycline in honey samples. This limits the applications of CD. To prevent the fluorescence interference, a single-hole hollow MIP (HMIP) embedded CDs (HMIP@CD) was prepared by Li et al. by microwave-assisted sol-gel technique [73]. This could quickly adsorb tetracycline from honey samples. Solid MIP has the surface binding sites whose number could be increased by utilizing hollow molecular imprint polymers (HMIPs). Here, HMIP@CD is dispersed in dilute honey. After the adsorption process is over, it is centrifuged and dispersed in phosphate buffer solution and finally checking for fluorescence. In this method the detection limit is 3.0 ng/mL. In another work, tetracycline was determined in fish by using MIP-CD, prepared by reverse microemulsion method by Yang et al. [74]. This fluorescent sensor collaborates the benefits of synchronous fluorescence spectroscopy with MIP.

To detect tetracycline in the milk samples a simple microwave-assisted method was used for the first time by Hou et al. for preparing an eco-friendly, highly selective, efficient and accessible MIP-CD sensor [75]. This approach simplified the experimental procedure and made the polymerization reaction amazingly fast. This makes direct determination of trace levels of tetracycline in milk samples. CD are modified with 3-aminopropyltriethoxysilane by simply microwaving their ethanolic solution for about four minutes to produce silica-coated CDs (Si-CDs). The MIP-Si-CDs were prepared via a typical sol-gel polymerization by using tetracycline as template, 3-aminopropyltriethoxysilane as a functional monomer, and tetraethoxysilane as a cross-linker. This sensor gave detection limit of 5.48 nM for tetracycline in milk samples. A novel N-doped CD-MIP sensor was prepared for recognition and determination of tetracycline in milk powder [76]. N-doped CD is produced by microwave-assisted pyrolysis method by using citric acid and ethylenediamine as a carbon and nitrogen source, respectively. N-doped CDs was treated with 3-aminopropyl triethoxysilane as a precursor of silica layer (N-CDs-NH_2). MIP-coated CD (CD-MIP) was synthesized by a typical sol-gel process by using 3-aminopropyl triethoxysilane as a monomer giving the detection limit of 0.054 μg mL^{-1}. The operational conditions like effect of solvent, amount of material, temperature, and pH are optimized carefully to get the best possible results for the novel fluorescent sensor. In another work on N-doped CD, a florometric sensor N-CDs@MIP was manufactured by a reverse microemulsion technique for the selective and sensitive determination of 2,4,6-trinitrophenol in tap and lake water (detection limit of 0.15 nM). 3-Aminopropyltriethoxysilane and tetraethoxysilane were used as functional monomer and as a cross-linker, respectively [77]. Some of the CD-based MIPs are summarized in Table 4.3.

Table 4.3: Molecular imprint polymers based on carbon dots.

Nanocomposite	Template/ analyte	Special feature	Detection limit	Reference
CDs@MOF@MIP	Quercetin	Zeolitic imidazolate framework used as MOF	2.9 nM	[78]
CD@MIP	Alpha-fetoprotein	Vinyl-functionalized CD used. A thermo-responsive and pH-responsive monomers used. Detection of alpha-fetoprotein In human serum samples	0.474 ng mL^{-1}	[69]
N-CDs@MIP	2,4,6-Trinitrophenol	Nitrogen-doped carbon dots infused with a MIP. Reverse microemulsion technique used. Used for tap and lake water samples.	0.15 nM	[77]
CDs@MIP	Caffeic acid	MIP on silane-functionalized CD. Fluorescence of CDs@MIPs was quenched by electron transfer reaction caffeic acid and CDs@MIP	0.11 μM	[79]
N-doped CD-MIP sensor	Tetracycline	N-doped CD used. Fluorescent sensor for determination of tetracycline in milk powder. Ethylenediamine used as a source of N for N-doped CD.	0.054 μg mL^{-1}	[76]

Table 4.3 (continued)

Nanocomposite	Template/ analyte	Special feature	Detection limit	Reference
MIP-CD-COF	4-Ethylguaiacol	Fluorescent nanoprobe based on COF. Polymer is formed by reactions such as Schiff-base, C–C bond formation, nitrile trimerization, boronic acid-based coupling. MIP doped with COFs is grafted on carbon nanodots.	17 ng/mL	[80]
MIL-based MIP nanocomposite	Doxorubicin hydrochloride drug	Prepared by stirring auxillary encapsulation method. MIL and nitrogen-doped CDs used in MOF-based MIP. Used for recognition and drug delivery system in cancerous area.	–	[81]
CD-FMIPH	Tetracycline	CD-based fluorescent molecularly imprinted photonic crystal hydrogel (FMIPH) strip prepared. The inverse opal photonic crystals with 3D macroporous structure, which provides mass transfer channels for the diffusion and adsorption of the target.	0.067 μg mL^{-1}.	[82]
CQDs-MIPs	Promethazine hydrochloride	Less toxic CQD prepared by hydrothermal method from orange juice as a carbon source. CQD wrapped with MIP by sol-gel method.	0.5 μmol L^{-1}	[83]
MIP@CDs	Naproxen	CDs embedded in silica imprinted film through one-step sol gel polymerization technique. Anti-interference with co existing and recovery studies done.	0.03 μM	[84]

Eissa et al. [85] gave a detailed study of drug micro extraction through MIPs and/or CD, their selectivity, and recyclability, keeping in view their pharmaceutical and greener approaches. They also highlighted the advantages associated with multiple-template-based MIP that will be probably the major future analytical field to be worked upon. Kecili, Hussain, and Hussain collectively made the recent study on the design and preparation of environmental friendly MIP sensors based on green CDs [86]. They detailed the various preparatory methods for CDs and their applications as MIP@CD sensors in analysis of food and environmental samples. In a separate work, they reviewed the environmental applications, analysis of pollutants, and future prospects related to fluorescent nano sensors of green CD-based MIPs [87]. MIT is now heading its way toward development of green MIT, now called GREENIFICATION to utilize green principles to all aspects related to it [88, 89].

4.3 Conclusion and future prospects

MIP involves both easy identification and quantification of the aimed target molecule by replacing the complex, sophisticated traditional analytical approaches which were used earlier. Chi et al. [64] elaborated the use of CNM@MIP sensors for the detection of food hazards. However it is found that work related to the use of carbon-based MIP sensors in food industry such as food production, processing, storage and packing is comparatively limited. Various studies related to the functioning of multi-template MIPs (mt-MIPs) on the basis of their functions and properties associated with the use of various cross-linkers, functional monomers, and graphene oxide are done. The choice of a monomer is amazingly crucial. It will alter the binding forces and capacity of MIP during the extraction processes. Any MIP should not only fulfill the demand of being highly selective and sensitive but future developments of MIP should be aimed to be (a) being eco-friendly, following the greener routes of synthesis (b) pose minimum toxicity to the environment and human health (c) recyclable and regenerative (d) highly stable and biocompatible (e) having multi-template recognition sites that simultaneously detect and extract various analytes thus saving time and energy (f) and have greater shelf-life. In comparison to other allotropic forms of carbon, CDs are the new class of fluorescent CNM famous for their low toxic and ecofriendly nature, making MIP more efficient and convenient to use. Various reports are seen to put green CDs in use now. Prioritizing the environmental concerns and long-term human health issues greener routes for the synthesis, and use of CNM-based MIP will be the future challenge to the analysts.

Abbreviations

AIBN	2,20-Azobisisobutyronitrile
CBD	Cannabidiol
CE	Carbon electrode
CBMIP	Carbon-based molecule imprint polymer
CLO	Cloxacillin
CNP	Carbon nanoparticles
CNT	Carbon nanotubes
CS-PtNPs	Chitosan-platinum nanoparticles
CD	Carbon dots
CQD	Carbon quantum dots
2,4-DCP	2,4-Dichlorophenol
DNP	2,4-Dinitrophenol

DIURON	3-(3,4-Dichlorophenyl)-1,1-dimethylurea
2D	Two-dimensional
2,2-DCP	2,4-Dichlorophenol
EGDMA	Ethylene glycol dimethacrylate
EME	Electromembrane extraction
G	Graphene
GO	Graphene oxide
rGO	Reduced graphene oxide
GNR	Graphene nano ribbons
GC	Gas chromatography
GQD	Graphene quantum dots
GR-AuNPs	Graphene-gold nanoparticles
GCE	Glassy carbon electrode
HMIP	Hollow MIP
HCT	Hydrochlorothiazide
HA	Hyaluronic acid
MIT	Molecular imprint technology
MWCNT	Multi-walled carbon nanotubes
MIP	Molecular imprint polymer
MNP	Magnetic nanoparticles
mt-MIP	Multi-template MIPs
MAA	Methyl acrylic acid
MOF	Metal-organic framework
NP	Nanoparticles
PGE	Pencil graphite electrode
PFF	Profenofos
PDA-rGO	Polydopamine-rGO
PPy	Polypyrrole
SMIT	Surface molecular imprinting technology
SMIP	Surface molecularly imprinted polymers
SPE	Screen-printed electrode
SWCNT	Single-walled carbon nanotubes
MNZ	Metronidazole
V-CD	Vinyl-functionalized CD

References

[1] Wulff G., & Sarhan A. (1972). The use of polymers with enzyme analogous structures for the resolution of racemates, *Angewandte Chemie International Edition*. 11, 341–343.

[2] Leibl N., Haupt K., Gonzato C., & Duma L. (2021). Molecularly imprinted polymers for chemical sensing: A tutorial review, *Chemosensors*. 9, 123.

[3] Jamieson O., Soares T. C. C. C., De Faria B. A., et al. (2020). Screen printed electrode based detection systems for the antibiotic amoxicillin in aqueous samples utilising molecularly imprinted polymers as synthetic receptors, *Chemosensors*. 8, 5.

[4] Tarannum N., Hendrickson O. D., Khatoon S., Zherdev A. V., & Dzantiev B. B. (2020). Molecularly imprinted polymers as receptors for assays of antibiotics, *Critical Reviews in Analytical Chemistry*. 50, 291–310.

[5] Brauer B., Unger C., Werner M., & Lieberzeit P. A. (2021). Biomimetic sensors to detect bioanalytes in real-life samples using molecularly imprinted polymers: A review, *Sensors*. 21, 5550.

[6] Beluomini M. A., Silva J. L., Cardoso de S. A., et al. (2019). Electrochemical sensors based on molecularly imprinted polymer on nanostructured carbon materials: A review, *Journal of Electroanalytical Chemistry*. 840, 343–366.

[7] Pauling L. (1940). A theory of the structure and process of formation of antibodies, *Journal of the American Chemical Society*. 62, 2643–2657.

[8] Basak S., Venkatram R., & Singhal R. S. (2022). Recent advances in the application of molecularly imprinted polymers (MPs) in food analysis, *Food Control*. 139, 109074.

[9] Mayes A. G., & Mosbach K. (1996). Molecularly imprinted polymer beads: Suspension polymerization using a liquid perfluorocarbon as the dispersing phase, *Analytical Chemistry*. 68(21), 3769–3774.

[10] Say R., Birlik E., Ersoz A., Yılmaz F., Gedikbey T., & Denizli A. (2003). Preconcentration of copper on ion-selective imprinted polymer microbeads, *Analytica Chimica Acta*. 480(2), 251–258.

[11] Yoshimatsu K., Reimhult K., Krozer A., Mosbach K., Sode K., & Ye L. (2007). Uniform molecularly imprinted microspheres and nanoparticles prepared by precipitation polymerization: The control of particle size suitable for different analytical applications, *Analytica Chimica Acta*. 584(1), 112–121.

[12] Yáñez–Sedeño P., Campuzano, & Pingarrón J. M. (2017). Electrochemical sensors based on magnetic molecularly imprinted polymers: A review, *Analytica Chimica Acta*. 960, 1–17.

[13] Ansari S., & Karimi M. (2017). Recent configurations and progressive uses of magnetic molecularly imprinted polymers for drug analysis, *Talanta*. 167, 470–485.

[14] Dong C., Shi H., Han Y., Yang Y., Wang R., & Men J. (2021). Molecularly imprinted polymers by the surface imprinting technique, *European Polymer Journal*. 145, 11023.

[15] Dickey F. H. (1949). The preparation of specific adsorbents, *Proceedings of the National Academy of Sciences of the United States of America (PNAS)*. 35(5), 227–229.

[16] Shah N., Islam M. U., Haneef M., & Park J. K. (2012). A brief overview of molecularly imprinted polymers: From basics to applications, *Journal of Pharmaceutical Research*. 5, 3309–3317.

[17] Augusto F., Carasek E., Silva R. G. C., Rivellino S. R., Batista A. D., & Martendal E. (2010). New sorbents for extraction and microextraction techniques, *Journal of Chromatography A*. 1217, 2533–2542.

[18] Pandey H., Khare P., Singh S., & Singh S. (2020). Carbon nanomaterials integrated molecularly imprinted polymers for biological sample analysis: A critical review, *Materials Chemistry and Physics*. 239, 121966.

[19] Dai H., Xiao D., He H., Li H., Yuan D., & Zhang C. (2015). Synthesis and analytical applications of molecularly imprinted polymers on the surface of carbon nanotubes: A review, *Microchimica Acta*. 182, 893–908.

[20] Ansari S. (2017). Combination of molecularly imprinted polymers and carbon nanomaterials as a versatile biosensing tool in sample analysis: Recent applications and challenges, *TrAC Trends in Analytical Chemistry*. 93, 134–151.

[21] Bakhshandeh B., Sorboni S. G., Haghighi D. M., Ahmadi F., Dehghani Z., & Badiei A. (2022). New analytical methods using carbon-based nanomaterials for detection of Salmonella species as a major food poisoning organism in water and soil resources, *Chemosphere*. 287, 132243.

[22] Kamyab H., Chelliapan S., & Tavakkoli O. (2022). A review on carbon based molecularly-imprinted polymers (CBMIP) for detection of hazardous pollutants in aqueous solutions, *Chemosphere*. 308(3), 136471.

[23] Chen P. Y., Nien P. C., Hu C. W., & Ho K. C. (2010). Detection of uric acid based on multiwalled carbon nanotubes polymerized with a layer of molecularly imprinted PMAA, *Sensors & Actuators: B. Chemical*. 146, 466–471.

[24] Tahmasebi Z., Davarani S. S. H., & Asgharinezhad A. A. (2016). An efficient approach to selective electromembrane extraction of naproxen by means of molecularly imprinted polymer-coated multi-walled carbon nanotubesreinforced hollow fibers, *Journal of Chromatography A*. 1470, 19–26.
[25] Alizadeh T., Atashi F., & Ganjali M. R. (2019). Molecularly imprinted polymer nano-sphere/multi-walled carbon nanotube coated glassy carbon electrode as an ultra-sensitive voltammetric sensor for picomolar level determination of RDX, *Talanta*. 194, 415–421.
[26] Amatatongchai M., Sroysee W., Sodkrathok P., Kesangam N., Chairam S., & Jarujamrus P. (2019,). Novel three-dimensional molecularly imprinted polymer-coated carbon nanotubes (3D-CNTs@MIP) for selective detection of profenofos in food, *Analytica Chimica Acta*. 1076, 64–72.
[27] Nezhadali A., & Mojarrab M. (2014). Computational study and multivariate optimization of hydrochlorothiazide analysis using molecularly imprinted polymer electrochemical sensor based on carbon nanotube/polypyrrole film, *Sensors and Actuators B: Chemical*. 190, 829–837.
[28] Prasad B. B., & Pandey I. (2013). Electrochemically imprinted molecular recognition sites on multiwalled carbon-nanotubes/pencil graphite electrode surface for enantioselective detection of d- and l-aspartic acid, *Electrochimica Acta*. 88, 24–34.
[29] Zhou T., Ding J., He Z., et al. (2018). Preparation of magnetic superhydrophilic molecularly imprinted composite resin based on multi-walled carbon nanotubes to detect triazines in environmental water, *Chemical Engineering Journal*. 334, 2293–2302.
[30] Liu Y., Liu J., Tang H., et al. (2015). Fabrication of highly sensitive and selective electrochemical sensor by using optimized molecularly imprinted polymers on multi-walled carbon nanotubes for metronidazole measurement, *Sensors and Actuators B: Chemical*. 206, 647–652.
[31] Chen H. J., Zhang Z. H., Luo L. J., & Yao S. Z. (2012). Surface-imprinted chitosan-coated magnetic nanoparticles modified multi-walled carbon nanotubes biosensor for detection of bovine serum albumin, Sens, *Sensors and Actuators B: Chemical*. 163, 76–83.
[32] Anirudhan T. S., & Alexander S. (2014). Multiwalled carbon nanotube based molecular imprinted polymer for trace determination of 2, 4-dichlorophenoxyaceticacid in natural water samples using a potentiometric method, *Applied Surface Science*. 303, 180–186.
[33] Wong A., Foguel M. V., Khan S., et al. (2015). Development of an electrochemical sensor modified with MWCNT-COOH and MIP for detection of diuron, *Electrochimica Acta*. 182, 122–130.
[34] Fernandez-Cori R., Morales Gomero J. C., Huayhuas-Chipana B. C., Taboada Sotomayor M. D. P., & Ruiz-Montoya J. G. (2015). Nanostructured sensors for determination of 3- (3,4-Dichlorophenyl)-1,1-Dimethylurea based in molecularly imprinted polymers (MIPs) deposited in screen printed carbon Nanotubes, *ECS Transaction*. 66, 33–41.
[35] Karimian N., Gholivand M. B., & Malekzadeh G. (2016). Cefixime detection by a novel electrochemical sensor based on glassy carbon electrode modified with surface imprinted polymer/multiwall Carbon Nanotubes, *Journal of Electroanalytical Chemistry*. 771, 64–72.
[36] Nezhadali A., Senobari S., & Mojarrab M. (2016). 1,4-dihydroxyanthraquinone electrochemical sensor based on molecularly imprinted polymer using multi-walled carbon nanotubes and multivariate optimization method, *Talanta*. 146, 525–532.
[37] Nezhadali A., & Mojarrab M. (2016). Computational design and multivariate optimization of an electrochemical metoprolol sensor based on molecular imprinting in combination with carbon nanotubes, *Analytica Chimica Acta*. 924, 86–98.
[38] Deiminiat B., Rounaghi G. H., & Zavar M. H. A. (2017). Development of a new electrochemical imprinted sensor based on poly-pyrrole, sol–gel and multiwall carbon nanotubes for determination of tramadol, *Sensors and Actuators B: Chemical*. 238, 651–659.
[39] Liu P., Zhang X., Xu W., Guo C., & Wang S. (2012). Electrochemical sensor for the determination of brucine in human serum based on molecularly imprinted poly-ophenylenediamine/SWNTs composite film, *Sensors and Actuators B: Chemical*. 163, 84–89.

[40] Radu G. L., Lițescu S. C., Enache A., Albu C., & Eremia S. A. V. (2023). Synergism between graphene and molecularly imprinted polymers in developing electrochemical sensors for agri-food and environmental analyses, *Chemosensors*. 11(7), 380.
[41] Fu L., Mao S., Chen F., et al.. (2022) Graphene-based electrochemical sensors for antibiotic detection in water, food and soil. A scientometric analysis in cite space (2011–2021), *Chemosphere*. 297, 134127.
[42] Sridhar K., Inbaraj B. S., & Chen B. H. (2022). An improved surface enhanced raman spectroscopic method using a paper-based grape skin-gold nanoparticles/graphene oxide substrate for detection of rhodamine 6G in water and food, *Chemosphere*. 301, 134702.
[43] Liu Y., Zhu L., Zhang Y., & Tang H. (2012). Electrochemical sensoring of 2,4-dinitrophenol by using composites of graphene oxide with surface molecular imprinted polymer, *Sensors and Actuators B: Chemical*, 171–172, 1151–1158.
[44] Luo J., Cong J., Liu J., Gao Y., & Liu X. (2015). A facile approach for synthesizing molecularly imprinted graphene for ultrasensitive and selective electrochemical detecting 4-nitrophenol, *Analytica Chimica Acta*. 864, 74–84.
[45] Zhang M., Zhao H. T., T J X., et al. (2017). Molecularly imprinted polymer on graphene surface for selective and sensitive electrochemical sensing imidacloprid, *Sensors and Actuators B: Chemical*. 252, 991–1002.
[46] Gupta P., & Goyal R. N. (2015). Graphene and co-polymer composite based molecularly imprinted sensor for ultratrace determination of melatonin in human biological fluids, *RSC Advances*. 5, 40444–40454.
[47] Liu W., Ma Y., Sun G., Wang S., Deng J., & Wei H. (2017). Molecularly imprinted polymers on graphene oxide surface for EIS sensing of testosterone, *Biosensors and Bioelectronics*. 92, 305–312.
[48] Chen L., Lian H. T., Sun X. Y., & Liu B. (2017). Sensitive detection of L-5- hydroxytryptophan based on molecularly imprinted polymers with graphene amplification, *Analytical Biochemistry*. 526, 58–65.
[49] Jafari S., Dehghani M., Nasirizadeh N., Baghersad M. H., & Azimzadeh M. (2019). Label-free electrochemical detection of cloxacillin antibiotic in milk samples based on molecularly imprinted polymer and graphene oxide-gold nanocomposite, *Measurement: Journal of the International Measurement Confederation*. 145, 22–29.
[50] Shi L., Duan J., Xu S., Yang H., & Yang G. (2021). One-step synthesis of ultrasmall platinum nanoparticles supported on amino-functionalized graphene to create an electrochemical molecularly imprinted polymer sensor for fluxapyroxad, *International Journal of Electrochemical Science*. 16, (7), 210767.
[51] Karthika P., Shanmuganathan S., & Viswanathan S. (2022). Electrochemical sensor for picric acid by using molecularly imprinted polymer and reduced graphene oxide modified pencil graphite electrode, *Proceedings of the Indian National Science Academy*. 88, 263–276.
[52] Wang B., Okoth O. K., Yan K., & Zhang J. (2016). A highly selective electrochemical sensor for 4-chlorophenol determination based on molecularly imprinted polymer and PDDA-functionalized graphene, *Sensors and Actuators B: Chemical*. 236, 294–303.
[53] Soman G., Vandana M., & Hegde G. (2023). Molecularly imprinted graphene based biosensor as effective tool for electrochemical sensing of uric acid, *Sensors International*. 4, 100243.
[54] Lian W., Liu S., Yu J., et al. (2012). Electrochemical sensor based on gold nanoparticles fabricated molecularly imprinted polymer film at chitosan-platinum nanoparticles/graphene-gold nanoparticles double nanocomposites modified electrode for detection of erythromycin, *Biosensors and Bioelectronics*. 38, 163–169.
[55] Tang X., Gu Y., Tang P., & Liu L. (2022). Electrochemical sensor based on magnetic molecularly imprinted polymer and graphene-UiO-66 composite modified screen-printed electrode for cannabidiol detection, *International Journal of Electrochemical Science*. 17, 220562.

[56] Han S., Sun R., Zhao L., Yan C., & Chu H. (2023). Molecularly imprinted electrochemical sensor based on synergistic interaction of honeycomb-like Ni-MOF decorated with AgNPs and N-GQDs for ultra-sensitive detection of olaquindox in animal-origin food, *Food Chemistry*. 418, 136001.

[57] Xing X., Liu S., Yu J., Lian W., & Huang J. (2012). Electrochemical sensor based on molecularly imprinted film at polypyrrole-sulfonated graphene/hyaluronic acid-multiwalled carbon nanotubes modified electrode for determination of tryptamine, *Biosensors and Bioelectronics*. 31, 277–283.

[58] Wang X., Luo C., Li L., & Duan H. (2015). An ultrasensitive molecularly imprinted electrochemical sensor based on graphene oxide/carboxylated multiwalled carbon nanotube/ionic liquid/gold nanoparticle composites for vanillin analysis, *RSC Advances*. 5, 92932–92939.

[59] Shao Y., Zhu Y., Zheng R., Wang P., Zhao Z., & An J. (2022). Highly sensitive and selective surface molecularly imprinted polymer electrochemical sensor prepared by Au and MXene modified glassy carbon electrode for efficient detection of tetrabromobisphenol A in water, *Advanced Composites and Hybrid Materials*. 5, 3104–3116.

[60] Liu Y., Liang Y., Yang R., Li J., & Qu L. (2019). A highly sensitive and selective electrochemical sensor based on polydopamine functionalized graphene and molecularly imprinted polymer for the 2, 4-dichlorophenol recognition and detection, *Talanta*. 195, 691–698.

[61] Afzali Z., Mohadesi A., Karimi M. A., & Fathirad F. (2022). A highly selective and sensitive electrochemical sensor based on graphene oxide and molecularly imprinted polymer magnetic nanocomposite for patulin determination, *Microchemical Journal*. 177, 107215.

[62] Sun Y. P., Zhou B., Lin Y., et al. (2006). Quantum-sized carbon dots for bright and colorful photoluminescence, *Journal of the American Chemical Society*. 128, 7756.

[63] Cui L., Ren X., Sun M., Liu H., & Xia L. (2021). Carbon dots: Synthesis, properties and applications, *Nanomaterials (Basel)*. 11(12), 3419.

[64] Chi H., & Liu G. (2023). Carbon nanomaterial-based molecularly imprinted polymer sensors for detection of hazardous substances in food: Recent progress and future trends, *Food Chemistry*. 420, 136100.

[65] Mazzier D., Favaro M., Agnoli S., et al. (2014). Synthesis of luminescent 3D microstructures formed by carbon quantum dots and their self-assembly properties, *Chemical Communications*. 50, 6592–6595.

[66] Li S., Luo J., Yin G., et al. (2015). Selective determination of dimethoate via fluorescence resonance energy transfer between carbon dots and a dye-doped molecularly imprinted polymer, *Sensors and Actuators B: Chemical*. 206, 14–21.

[67] Sun W., Li R., Liu W., & Liu X. (2023). Carbon dot-based molecularly imprinted fluorescent nanopomegranate for selective detection of quinoline in coking wastewater, *Spectrochimica Acta Part A: Molecular and Biomolecular Spectroscopy*. 284, 121770.

[68] Xu L., Fang G., Pan M., & Wang X. (2015). One-pot synthesis of carbon dots-embedded molecularly imprinted polymer for specific recognition of sterigmatocystin in grains, *Biosensors and Bioelectronics*. 77, 950–956.

[69] Sun C., Pan L., Zhang L., et al. (2019). A biomimetic fluorescent nanosensor based on imprinted polymers modified with carbon dots for sensitive detection of alpha-fetoprotein in clinical samples, *Analyst*. 144, 6760–6772.

[70] Patra S., Choudhary R., Roy E., Madhuri R., & Sharma P. K. (2017). Triple signalling mode carbon dots-based biodegradable molecularly imprinted polymer as a multi-tasking visual sensor for rapid and "on-site" monitoring of silver ions, *Journal of Materials Chemistry C*. 5, 11965–11976.

[71] Shirani M. P., Rezaei B., & Ensafi A. A. (2019). A novel optical sensor based on carbon dots embedded molecularly imprinted silica for selective acetamiprid detection, *Spectrochimica Acta Part A: Molecular and Biomolecular Spectroscopy*. 210, 36–43.

[72] Bai Q., Wang H., Xu Y., Wang H., Guan K., & Gong B. (2023). Dual-functional molecularly imprinted doped carbon dot based on metal-organic frameworks for tetracycline adsorption and determination, *Microchimica Acta*. 190(12), 463.

[73] Li H., Zhao L., Xu Y., et al. (2018). Single-hole hollow molecularly imprinted polymer embedded carbon dot for fast detection of tetracycline in honey, *Talanta*. 185, 542–549.
[74] Yang J., Lin Z. Z., Nur A. Z., et al. (2018). Detection of trace tetracycline in fish via synchronous fluorescence quenching 22 with carbon quantum dots coated with molecularly imprinted silica, *Spectrochimica Acta Part A*. 190, 450–456.
[75] Hou J., Li H., Wang L., et al. (2016). Rapid microwave-assisted synthesis of molecularly imprinted polymers on carbon quantum dots for fluorescent sensing of tetracycline in milk, *Talanta*. 146, 34–40.
[76] Wang Q., Wu Y., Bao X et al. (2022). Novel fluorescence sensor for the selective recognition of tetracycline based on molecularly imprinted polymer-capped N-doped carbon dots, *RSC Advances*. 12(38), 24778–24785.
[77] Murugan K., Jothi V. K., Rajaram A., & Natarajan A. (2022). Novel metal-free Fluorescent sensor based on molecularly imprinted polymer N-CDs@MIP for highly selective detection of TNP, *ACS Omega*. 7(1), 1368–1379.
[78] Xu L., Pan M., Fang G., & Wang S. (2019). Carbon dots embedded metal-organic framework@molecularly imprinted nanoparticles for highly sensitive and selective detection of quercetin, *Sensors and Actuators B*. 286, 321–327.
[79] Xu X., Xu G., Wei F., et al. (2018). Carbon dots coated with molecularly imprinted polymers: A facile bioprobe for fluorescent determination of caffeic acid, *Journal of Colloid and Interface Science*. 529, 568–574.
[80] Liu H., Zhang Y., Zhang D., et al. (2019). A fluorescent nanoprobe for 4-ethylguaiacol based on the use of a molecularly imprinted polymer doped with a covalent organic framework grafted onto carbon nanodots, *Microchimica Acta*. 186, 182–192.
[81] Shi Y., Wang Y., Zhu J., Liu W., Khan M. Z. H., & Liu X. (2020). Molecularly Imprinting Polymers (MIP) based on nitrogen doped carbon dots and MIL-101(Fe) for Doxorubicin Hydrochloride delivery, *Nanomaterials*. 10(9), 1655.
[82] Pan M., Wang Y., Yang J., Li H., Han X., & Wang S. (2024). Carbon dots-based fluorescent molecularly imprinted photonic crystal hydrogel strip: Portable and efficient strategy for selective detection of tetracycline in foods of animal origin, *Food Chemistry*. 433, 137407.
[83] Ensafi A. A., Nasr-Esfahani P., & Rezaei B. (2018). Synthesis of molecularly imprinted polymer on carbon quantum dots as an optical sensor for selective fluorescent determination of promethazine hydrochloride, *Sensors and Actuators B: Chemical*. Volume 257, 889–896.
[84] Li K., Zhang M., Ye X., et al. (2021). Highly sensitive and selective detection of naproxen *via* molecularly imprinted carbon dots as a fluorescent sensor, *RSC Advances*. 11, 29073–29079.
[85] Eissa M. S., Imam M. S., Elrahman M. A., et al. (2024). Magnetic molecularly imprinted polymers and carbon dots molecularly imprinted polymers for green micro-extraction and analysis of pharmaceuticals in a variety of matrices, *Microchemical Journal*. 205, 111235.
[86] Kecili R., Hussain C. G., & Hussain C. M. (2024). Chapter 14, Molecularly imprinted polymer (MIP)-coated green carbon dots-based sensing platforms, *Green Imprinted Materials from Design to Environmental and Food Applications*. First edition, Elsivier, 435–452.
[87] Kecili R., Hussain C. G., & Hussain C. M. (2023). Fluorescent nanosensors based on green carbon dots (CDs) and molecularly imprinted polymers (MIPs) for environmental pollutants: Emerging trends and future prospects, *Trends in Environmental Analytical Chemistry*. 40, e00213.
[88] Arabi M., Ostovan A., Li J., et al. (2021). Molecular imprinting: Green perspectives and strategies, *Advanced Materials*. 33(30), 2100543.
[89] Kanwal A., Bibi N., Hyder S., et al. (2022). Recent advances in green carbon dots (2015–2022): Synthesis, metal ion sensing, and biological applications, *Beilstein Journal of Nanotechnology*. 13, 1068–1107.

Part 2: Performance evaluation of molecular imprinting-based smart carbon nanomaterials

Ziyin Mei, Martins Lateef Yusuf, Jiamei Liu, Shaoguang Li,
Hui Li*, and Fan Xia

Chapter 5
Characterization methods of carbon-based molecularly imprinted polymers

Abstract: Carbon-based molecularly imprinted polymers (MIPs) have emerged as highly promising materials for applications across sensor fabrication as biorecognition elements, environmental remediation as adsorbents, and biotechnology for drug delivery. The performance of these MIPs is largely dictated by the properties and optimization of the polymer cavities, making accurate and effective characterization essential. This chapter reviews the key characterization techniques applied to various properties of carbon-based MIPs, including structural, morphological, physical, surface area, pore size, and spectroscopic properties. We highlight studies in which carbon-based MIPs were synthesized, modified, and characterized using these techniques to demonstrate the role in evaluating and optimizing specific MIP attributes. Finally, we discuss future prospects including the potential for in situ and real-time characterization methods.

Ziyin Mei and Martins Lateef Yusuf, These authors contributed equally to this chapter.

Acknowledgments: This work was supported by the National Natural Science Foundation of China (22474130, 22274144, 22090050), the National Key Research and Development Program of China (2021YFA1200403), and the Joint NSFC-ISF Research Grant Program (grant no.: 22161142020).

***Corresponding author: Hui Li,** State Key Laboratory of Geomicrobiology and Environmental Changes, Engineering Research Center of Nano-Geomaterials of the Ministry of Education, Faculty of Materials Science and Chemistry, China University of Geosciences, Wuhan 430074, China, e-mail: lihui-chem@cug.edu.cn
Ziyin Mei, State Key Laboratory of Geomicrobiology and Environmental Changes, Engineering Research Center of Nano-Geomaterials of the Ministry of Education, Faculty of Materials Science and Chemistry, China University of Geosciences, Wuhan 430074, China, e-mail: meiziyin@cug.edu.cn
Martins Lateef Yusuf, State Key Laboratory of Geomicrobiology and Environmental Changes, Engineering Research Center of Nano-Geomaterials of the Ministry of Education, Faculty of Materials Science and Chemistry, China University of Geosciences, Wuhan 430074, China, e-mail: martins_yusuf@cug.edu.cn
Jiamei Liu, State Key Laboratory of Geomicrobiology and Environmental Changes, Engineering Research Center of Nano-Geomaterials of the Ministry of Education, Faculty of Materials Science and Chemistry, China University of Geosciences, Wuhan 430074, China, e-mail: 1202320155@cug.edu.cn
Shaoguang Li, State Key Laboratory of Geomicrobiology and Environmental Changes, Engineering Research Center of Nano-Geomaterials of the Ministry of Education, Faculty of Materials Science and Chemistry, China University of Geosciences, Wuhan 430074, China, e-mail: lisg@cug.edu.cn
Fan Xia, State Key Laboratory of Geomicrobiology and Environmental Changes, Engineering Research Center of Nano-Geomaterials of the Ministry of Education, Faculty of Materials Science and Chemistry, China University of Geosciences, Wuhan 430074, China, e-mail: xiafan@cug.edu.cn

https://doi.org/10.1515/9783111503325-005

Keywords: Carbon-based, molecularly imprinted polymers, characterization methods, surface morphology

5.1 Introduction

In the field of chemical diagnostics, biosensor design has taken center stage due to its precision and versatility in detecting specific biomolecules [1]. The optimization of biosensor performance for target-specific sensing in biological samples has significantly advanced the field [2]. At the core of these biosensors are biorecognition elements, which are essential for selective biomolecule detection [3]. Various biorecognition elements, such as nucleic acids [4, 5], antibodies [6], enzymes [7], proteins [8], and molecularly imprinted polymers (MIPs) [9, 10], have been utilized for this purpose. Despite demonstrating significant potential, many of these biorecognition elements face challenges in terms of stability and lifespan, especially in complex biological matrices. MIPs, often referred to as synthetic antibodies, have garnered considerable attention due to their specificity, durability, and cost-effectiveness [11, 12]. MIPs contain molecularly imprinted binding sites that can be tailored to specific targets, allowing for precise biorecognition. While MIPs are highly efficient and less costly than natural antibodies, they often suffer from limited detection sensitivity due to slow electron transfer, which can impede their use in highly sensitive applications [13, 14]. Researchers have responded to this limitation by enhancing MIP design to create active sites that facilitate faster electron transfer. These modifications are achieved through strategic choices in material composition and synthesis methods, with a particular focus on the incorporation of nanomaterials to improve electron mobility and target affinity [15, 16].

Carbon-based nanomaterials such as graphene, carbon nanotubes (CNTs), and MXene have shown great promise in MIP synthesis due to their unique properties [17, 18]. Graphene, known for its large surface area and excellent conductivity, has been widely used to improve the sensitivity of MIPs. MXene, a two-dimensional nanomaterial with high electrical conductivity, stability, and ease of functionalization, is another valuable addition to carbon-based MIPs. For example, recent research by Liu et al. [19] demonstrated a graphene-MXene composite MIP that exhibited outstanding electrochemical sensitivity for cortisol detection, highlighting the potential of these carbon-based materials in biosensor applications.

The effective characterization of carbon-based MIPs is critical, ensures that their performance aligns with the intended biorecognition objectives. Characterization techniques help verify that these materials possess the necessary structural, chemical, and morphological attributes for effective biomolecule detection [20]. Key characterization methods include structural analysis, spectroscopy, and morphological assessment (Figure 5.1), each offering specific insights essential to tailoring MIPs for high specificity and sensitivity.

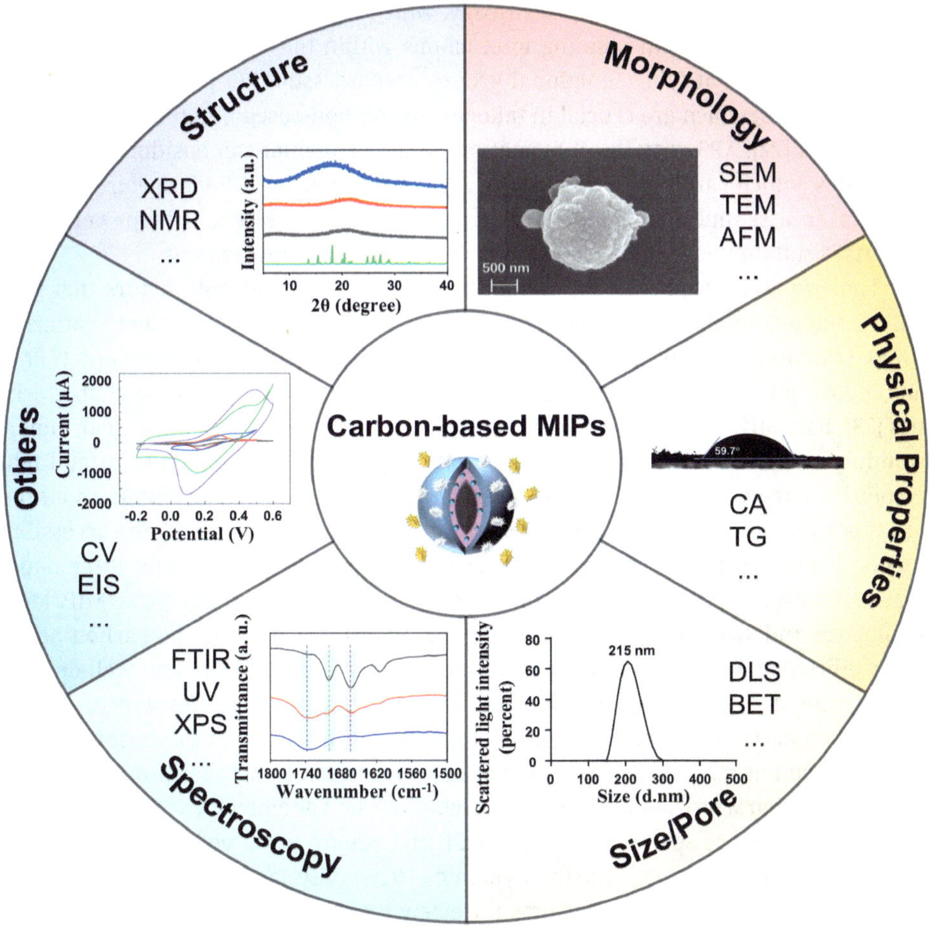

Figure 5.1: Characterization methods of carbon-based molecularly imprinted polymers.

Structural analysis methods such as nuclear magnetic resonance (NMR) and X-ray diffraction (XRD) are utilized for understanding the composition and arrangement of atoms within carbon-based MIP. NMR, on the one hand, provides detailed information on the molecular structure and composition of carbon-based MIPs. This allows researchers to evaluate the functional groups present and determine if they are optimized for the target binding site. XRD, on the other hand, is used to identify the crystalline structure of the carbon-based MIP and assess its degree of order. The details of the data of XRD based on the crystallinity and phase of the carbon-based MIP can indicate the efficiency of the synthetic process and the structural stability under various conditions [19].

Another important characterization is the spectroscopy analysis, including Fourier transform infrared spectroscopy (FTIR), X-ray photoelectron spectroscopy (XPS), and ultraviolet-visible (UV-vis) spectroscopy, which is essential for characterizing the chemical composition and bonding interactions within the MIP [21]. FTIR is particularly useful for identifying functional groups and assessing the presence of specific chemical bonds, which are crucial in tailoring the carbon-based MIP for specific target interactions [22]. XPS provides information on the elemental composition and oxidation states, which can impact its electron transfer properties. UV-vis spectroscopy allows for an assessment of the electronic properties of the MIP and helps verify the material's suitability for optoelectronic or electrochemical applications.

The size and morphology of carbon-based MIPs are important factors that provides valuable details on the property. Techniques such as dynamic light scattering (DLS), scanning electron microscopy (SEM), transmission electron microscopy (TEM), and atomic force microscopy (AFM) are commonly used to analyze these characteristics [23]. Basically, DLS measures particle size distribution by providing a clear understanding of the MIP's particle stability and dispersity. SEM and TEM offer high-resolution images of the surface and internal morphology, revealing details about the MIP's porosity and surface structure, which are important for assessing the accessibility of the binding sites. AFM provides three-dimensional surface profiling at the nanoscale, giving a comprehensive view of the topography and roughness of the MIP. Morphological and size data are particularly valuable for evaluating the carbon-based MIP's efficiency in capturing target molecules, as increased porosity and well-defined binding sites are often associated with enhanced biorecognition performance.

Carbon-based MIPs are frequently applied in electrochemical biosensors [19, 22] and contaminant removal [23] due to their favorable electron transfer properties. Electrochemical characterization techniques such as cyclic voltammetry (CV) [24], electrochemical impedance spectroscopy (EIS) [25], and square wave voltammetry [26] are used to evaluate electron transfer dynamics, sensitivity, and stability of the MIP-modified electrodes. This chapter aims to review the array of characterization techniques utilized for carbon-based MIPs, offering researchers in the field a comprehensive guide for selecting appropriate methods based on their specific research objectives. We will also discuss limitations inherent in each technique and present perspectives on future approaches that could further enhance the characterization of carbon-based MIPs.

There are several reviews on MIP materials and their applications as recognition probe [27], water analysis [28], chemical sensing, and catalysis [29]. This chapter aims to review the array of characterization techniques utilized for carbon-based MIPs, offering researchers in the field a comprehensive guide for selecting appropriate methods based on their specific research objectives. We will also discuss limitations inherent in each technique and present perspectives on future approaches that could further enhance the characterization of carbon-based MIPs.

5.2 Structural characterization

When modifying carbon-based MIPs to enhance properties for electrochemical sensing, precise structural characterization is essential. Techniques such as NMR and XRD are invaluable for confirming the effectiveness of optimization strategies, identifying limitations in the modification process, and assessing the structural integrity of the material. These methods provide critical insights, allowing researchers to evaluate whether the modified MIP meets performance criteria for specific applications. Through this structural analysis, scientists can make informed decisions about the MIP's suitability and potential efficacy in real-world electrochemical sensing applications.

5.2.1 Nuclear magnetic resonance

Since MIPs are typically insoluble, solid-state NMR techniques, which are ideal for analyzing insoluble materials, are frequently employed for characterizing carbon-based MIPs [30]. Solid-state NMR provides detailed insights into the types of chemical bonds present and the degree of chemical interactions within the MIP, which are crucial for confirming the material's functionality [21]. For electrochemical sensing, structural information obtained from NMR is essential to verify that the materials used in modification effectively enhance the MIP's performance for target detection.

In a recent study by Qu et al. [31], solid-state NMR was used in characterizing an MIP membrane based on an ordered mesoporous carbon nanosphere modified with methacrylic acid (MAA) for the detection of dibenzothiophene (DBT). NMR analysis revealed valuable information on the interaction mechanism between MAA and DBT. Specifically, the NMR spectrum showed that the hydrogen proton chemical shift of the –COOH group in MAA shifted from 12.71 to 12.78 ppm, indicating bond formation between the sulfur atom in DBT and the –COOH group of MAA via hydrogen bonding (Figure 5.2A-**i**). This chemical shift confirmed the modified MAA-MIP's enhanced adsorption capacity and selectivity for DBT, validating the role of NMR in assessing the MIP's specificity.

Solid-state NMR have also been employed to analyze a MIP synthesized for the selective removal of diclofenac (DCF) from solution [32]. DCF was used as the template molecule to create a composite (MIP-DCF) with a thiourea functional group. The NMR analysis confirmed interactions between DCF and the monomer within the pre-polymerization complex, verifying the success of the imprinting process and the binding capacity of the MIP for diclofenac. This study further highlighted NMR's utility in confirming the integrity of the polymer-template interaction within the MIP.

More recent studies have demonstrated the versatility of NMR for MIP characterization, employing both ^{129}Xe NMR and ^{1}H time-domain NMR (TD-NMR) to investigate the hierarchical porosity and cross-linking density within MIPs [33]. These advanced

NMR techniques provided insights into the porosity and structural organization of the MIP, key factors that influence the MIP's accessibility and performance in target capture.

Collectively, these studies underscore the importance of NMR as a characterization tool for carbon-based MIPs, offering researchers a nondestructive, in-depth method to confirm the structural and chemical properties critical to effective molecular recognition. Solid-state NMR, with its ability to characterize insoluble materials and reveal structural details, remains invaluable in optimizing MIPs for various applications, particularly in electrochemical sensing and environmental monitoring.

5.2.2 X-ray diffraction

Like NMR, XRD is a vital tool for the structural characterization of carbon-based MIPs. XRD is primarily used to identify the crystalline structure of the MIP and assess its degree of order or amorphous nature, which is crucial for performance, especially in applications such as electrochemical sensing. The crystallinity or amorphous structure can impact electron transfer rates and stability, thereby affecting the MIP's sensing efficiency.

In a study by Wang et al. [34], XRD was instrumental in addressing sensing limitations of carbon-based MIPs by characterizing a novel ferrocene (Fc)/PEDOT composite for amino acid detection. The structural composition of this ferrocene-MIP composite was compared with noncomposite and non-imprinted molecular polymer (NIP) samples. While all samples exhibited similar structural features before polymerization, XRD revealed differences in crystallinity after the process. The MIP composite showed lower crystallinity than the sharp peaks seen in the noncomposite material, likely due to the formation of template cavities within the MIP, which disrupted the crystalline structure (Figure 5.2B-**i**). This reduction in crystallinity may contribute to improved electrochemical sensing properties, as the amorphous structure provided lower oxidation potential with minimized interference.

Another study by Qu et al. [31], previously discussed in Section 2.2.1, also utilized XRD to verify the impact of MAA modification on MIP structure. The introduction of MAA into the MIP synthesis process led to a loss of crystallinity, resulting in increased cavity formation within the MIP (Figure 5.2A-**ii**). These cavities enhanced the MIP's surface area and accessibility, potentially improving the material's adsorption capacity and target specificity.

These studies show the promising role of XRD in providing essential structural insights into carbon-based MIPs. Distinguishing between crystalline and amorphous regions and assessing the effects of modifications on structural order, XRD allows researchers to tailor MIPs more effectively for specific applications. Understanding these structural characteristics enables the precise engineering of MIPs for improved performance in electrochemical sensing and other molecular detection tasks.

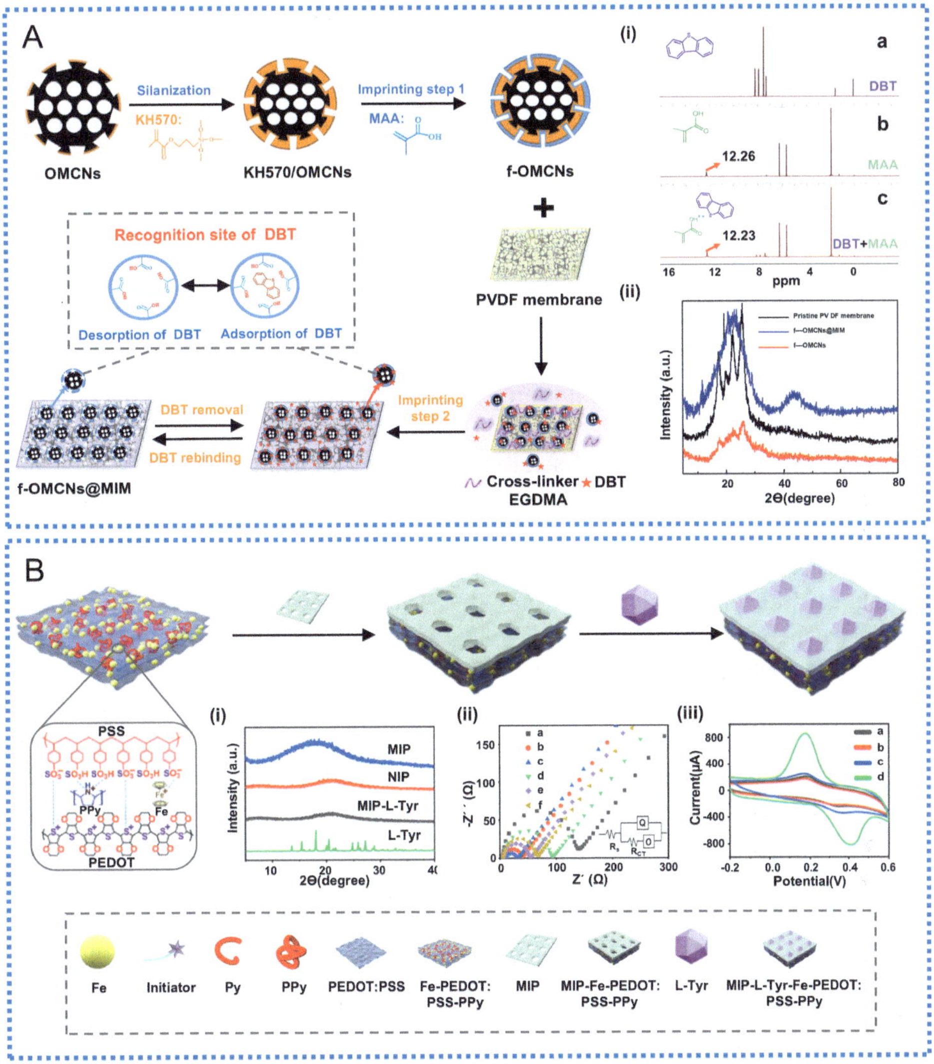

Figure 5.2: (A) Schematic illustration of MIP synthesis and structural characterization: (i) NMR patterns of modified, unmodified, and non-MIP; (ii) XRD patterns of modified, un-modified, and non-MIP. Copyright 2024 with permission from Elsevier. (B) Schematic illustration of MIP synthesis: (i) XRD patterns of modified, un-modified, and non-MIP; (ii) EIS of different MIP-modified electrodes; and (iii) CV of different MIP-modified electrodes. Copyright 2024 with permission from American Chemical Society.

5.3 Morphological characterization

Characterizing the surface morphology of carbon-based MIPs can aid in understanding their microstructures and compositions, thereby establishing correlations between structure and performance. This section introduces three commonly used morphological characterization methods, namely SEM, TEM, and AFM, along with their specific applications.

5.3.1 Scanning electron microscopy

SEM is frequently utilized to acquire information on the surface microtopography or particle size of samples. Its operating principle involves utilizing an accelerated electron beam to scan the surface of solid samples point by point, collecting the various signals generated thereby, and subsequently obtaining microscopic images of the samples [20, 35].

In research endeavors concerning carbon-based MIPs, SEM is often employed to distinguish imprinted polymers from NIPs. For instance, Liu et al. [19] reported an electrochemical biosensor based on graphene composite materials for MIPs (Figure 5.3A), which can be used for ultrasensitive detection of cortisol. Specifically, they combined Ti_3C_2Tx-MXene sheets with graphene (Gr) to prepare a porous composite (MXG). On electrodes modified with the MXG composite, they synthesized molecularly imprinted polymers (CS/MXG/GCE-MIP) using cortisol as a template through electropolymerization. SEM was used to characterize the surface morphology of the CS/MXG/GCE-MIP sensor before and after elution of cortisol. Prior to elution, the electrode surface was covered with a wrinkled thin film (Figure 5.3A-**i**). Upon removal of the cortisol template using 95% ethanol, the electrode surface revealed pores of varying sizes and became rough (Figure 5.3A-**ii**). In contrast, the surface morphology of the NIP sensor, serving as a control group, exhibited minimal changes before and after template elution, indicating the successful preparation of MIPs with cortisol as the template on the electrode surface. In the study by Chen et al. [36], due to the removal of template molecules from the MIP film, the SEM images of the MIP film exhibited a porous structure, appearing rougher than that of the NIP film.

In another work, Budak et al. [37] also utilized SEM to characterize the surface morphology of MIP sensors and NIP sensors. Compared to NIP sensors, the polymer film on MIP sensors exhibited a higher porosity, which facilitated the entry of template molecules into recognition sites, effectively enhancing the sensitivity of the sensors. Unlike the aforementioned works, Maria et al. [38], in addition to SEM characterization of morphology, combined SEM with energy-dispersive X-ray spectroscopy (EDX) to further elucidate the chemical composition of MIPs. Specifically, SEM images revealed the presence of larger white particles on the surface of MIP electrodes with uneluted template molecules (2,4-dichlorophenol (2,4-DCP)). However, following the removal of template molecules, the MIP electrode surface became devoid of these white particles, and the EDX peak for chlorine significantly decreased. This phenomenon further confirmed the successful synthesis of MIPs and the effective removal of 2,4-DCP [38].

5.3.2 Transmission electron microscopy

TEM can provide high-resolution images of the internal structure of samples by accelerating an electron beam to penetrate thin film samples and collecting transmitted electrons as imaging signals [39]. As shown in Figure 5.3B, Tan et al. [22] synthesized imprinted nanoparticles on graphene oxide modified with Fe_3O_4 (GO/Fe_3O_4) using dopamine self-polymerization with sarafloxacin as the template. These nanoparticles (PDA@GO/Fe_3O_4) exhibit advantages of rapid binding and high capacity, enabling their application in the removal of fluoroquinolone antibiotics from seawater with a removal efficiency as high as 95%. During the synthesis process, they characterized both GO/Fe_3O_4 and PDA@GO/Fe_3O_4 using TEM. As illustrated in Figure 5.3B(i), the Fe_3O_4 particles attached to the surface of graphene oxide have an average diameter of 80–100 nm. After dopamine self-polymerization, a uniform layer of polydopamine (PDA) adhered to the surface of GO/Fe_3O_4 (Figure 5.3B-**ii** and **iv**), with an average thickness of the PDA film being approximately 10–20 nm (Figure 5.3B-**iii**).

TEM is also commonly used to characterize the shape and size of MIPs particles. For instance, in the work of Ozcelikay et al. [40], they first verified the uniform distribution of the synthesized imprinted polymer beads on the surface of a glassy carbon electrode (GCE) using SEM and then obtained the particle size of 10 nm through TEM imaging. In another study, Zhang et al. [41] also employed TEM to characterize the size of the core-shell-imprinted polymers (C-MIPs) they synthesized on carbon quantum dots. By analyzing more than 200 particles in the TEM images, they found that the diameter of the core of C-MIPs (CDs@SiO_2) was approximately 55 nm, and the diameter of C-MIPs was approximately 63.3 nm, indicating the successful synthesis of the imprinted polymer layer on the surface of CDs@SiO_2.

In some studies, researchers use both TEM and SEM to reveal the internal structural features and external morphological details of the samples, respectively, and the two sets of data from these two techniques provide strong evidence that the imprinted polymers have been successfully introduced onto the surface of the carbon substrate [42]. For example, in the work of Yao et al., SEM images showed that the MIPs consisted of tubular and spherical structures adhered to each other, while TEM images revealed that the spherical structures had a diameter of 200 nm and that both the tubular and spherical structures were covered with a thin film. Together, these data from both scanning electron microscopes confirmed the successful synthesis of the MIPs [43].

5.3.3 Atomic force microscopy

AFM is a type of scanning probe microscopy, which typically uses probes with sharp tips to scan surfaces to obtain information about the surface topography or other features of the sample [39]. The operating principle of AFM involves using a cantilever with a tip to scan the surface of the sample for imaging. Specifically, when the probe comes into contact with the sample surface, due to the interactive forces between the tip and the sample surface, the position of the laser beam shining on the cantilever changes. By collecting the movement variations of the laser beam reflected from the cantilever, a corresponding surface morphology image can be constructed [20].

In the study of MIPs, AFM can be utilized to understand the uniformity and roughness of the polymer film surface. As shown in Figure 5.3C, Zidarič et al. [23] developed a portable MIPs sensor that can be used for the detection of a single drop of insulin test solution (50 μL). Specifically, they used insulin as a template and employed CV to polymerize pyrrole onto a screen-printed carbon electrode (SPCE). After removing the insulin from the polypyrrole, imprinted cavities were formed. Subsequently, they used AFM to obtain surface morphology images of the bare SPCE electrode, the SPCE electrode polymerized with MIPs (MIPs-SPCE), and the MIP-SPCE after template molecule removal at different scales. Here, we mainly introduce the AFM images of the MIPs-SPCE before and after template removal, as shown in Figure 5.3C-**i** and **ii**. MIPs can form a uniform thin film on the SPCE surface. After removing the template molecules (Figure 5.3C-**ii**), the surface of the MIPs-SPCE becomes rough, with a significant increase in the surface roughness (S_a).

Amatatongchai et al. [44], on the other hand, compared the AFM images of MIP-coated CNTs (CNTs@MIP) and CNTs@NIP and found that the surface of CNTs@MIP exhibited a more porous structure than CNTs@NIP, with greater roughness and depth than the NIP film. These results indicate that the removal of target molecules creates imprinted cavities on the polymer surface.

5.4 Characterization of physical properties

After understanding the structural and morphological characteristics of MIPs, researchers typically turn their attention to the physical properties of MIPs, which often influence their performance in practical applications. In this section, we primarily introduce two common analytical methods and case studies for assessing properties: contact angle (CA) and thermogravimetric analysis (TGA).

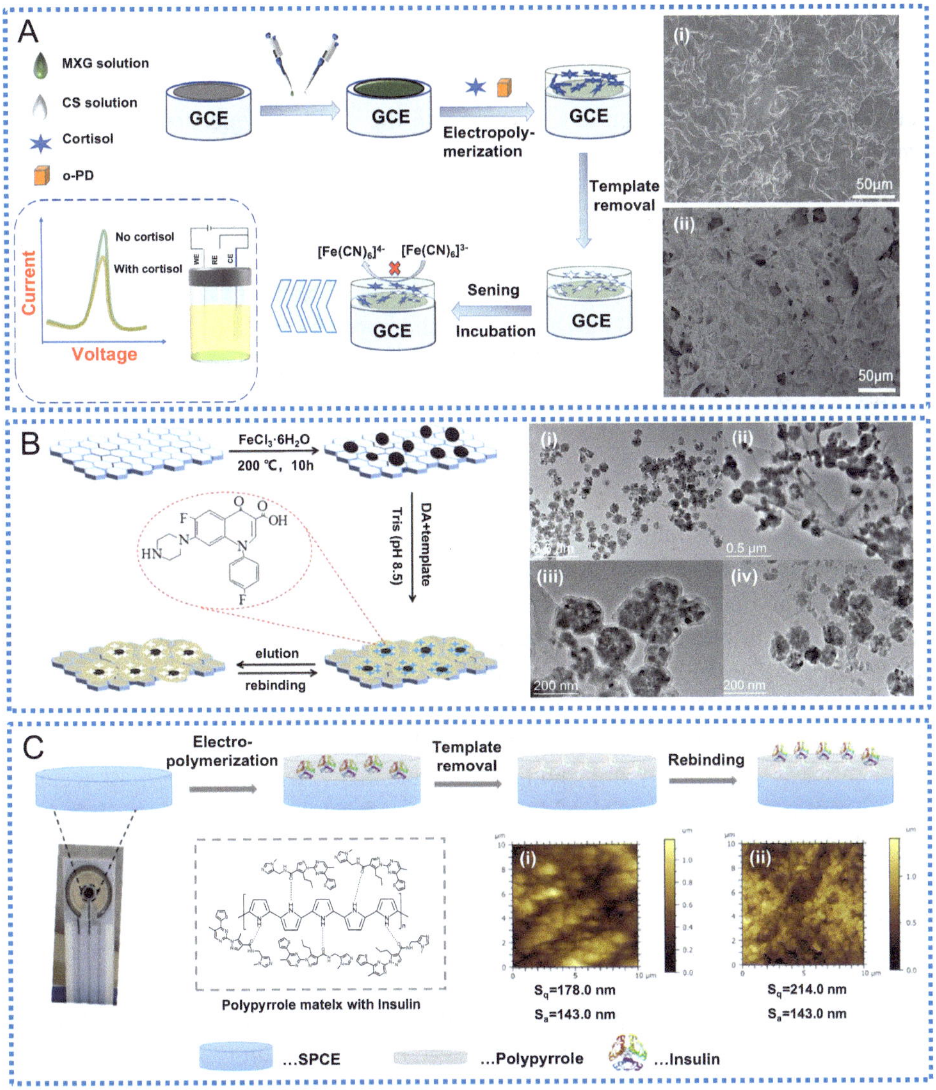

Figure 5.3: (A) Schematic illustration of MIP synthesis and morphological characterization, SEM pattern of (i) MIP before elution of template and (ii) MIP after elution of template. Copyright 2024 with permission from American Chemical Society. (B) Schematic illustration of MIP Synthesis, TEM pattern of (i) GO/Fe_3O_4, (ii) PDA@GO/Fe_3O_4, (iii) PDA@GO/Fe_3O_4, and (iv) NPDA@GO/Fe_3O_4. Copyright 2024 with permission from Springer Nature. (C) Schematic illustration of MIP synthesis, AFM pattern of (i) MIPs-SPCE before elution of template and (ii) MIPs-SPCE after elution of template. Copyright 2024 with permission from The Royal Society of Chemistry.

5.4.1 Contact angle

CA is a vital parameter for characterizing the wettability of solid surfaces, typically related to the surface tension of the liquid and the surface properties of the solid. A smaller CA indicates a greater attraction of the material surface to water and hence better wettability. For example, in the work of Li et al. [45], they reported a sensor utilizing a molecularly imprinted polymer membrane (PCA-MIMs) for detecting microcystin-LR (MC-LR) in water. As shown in Figure 5.4A, they introduced Ag nanoparticles and magnetized CNTs to modify the polyvinylidene fluoride (PVDF) membrane, followed by synthesizing MIMs on PVDF via photopolymerization. By observing the water contact angles of both the PVDF membrane (Figure 5.4A-**i**) and PCA-MIMs (Figure 5.4A-**ii**), they found that the CA of PCA-MIMs was significantly smaller than that of the PVDF membrane, indicating that the MIMs could enhance the membrane's hydrophilicity. This improvement in property favors the adsorption and detection of MC-LR by the sensor in aquatic environments.

Given the broad application prospects of MIMs in wastewater treatment, Qu et al. [46] designed a multilayered molecularly imprinted composite membrane (mMIM) based on carbon nanospheres and PDA. In this study, they enhanced the membrane's selective recognition ability by increasing the hydrophilicity and imprinting area of mMIM. Due to the polymerization of dopamine on the surface of the substrate membrane (polyethersulfone, PES), introducing numerous –OH and $-NH_2$ groups, the water contact angle of the membrane decreased from 46° (for PES) to 32° after modifying with mMIM, effectively improving its hydrophilicity, which makes mass transfer more convenient. Erdoğan et al. [47] similarly compared the water contact angles of a bare electrode (128.9°) and an modified electrode by MIP (13.1°), demonstrating that the polymerization of MIP can effectively enhance the electrode's hydrophilicity.

CA can also help us understand the differences between MIP and NIP. In Figure 5.4B, Hasabnis et al. [48] developed a nanocomposite-based biosensor for detecting cardiac troponin I. They prepared MIP using AuNPs, graphene quantum dots (GQDs, blue luminescent), and 2-aminophenol (2-AP) as raw materials and analyzed the contact angles of this MIP film (Figure 5.4C-**i**) and NIP film (Figure 5.4C-**ii**). They found that the former (CA = 30.1°) exhibited better hydrophilicity than the latter (CA = 40.4°). Additionally, through fluorescence microscopy, they observed distinct patchy fluorescence on the MIP film, unlike the NIP film, which may represent cavities formed after template molecule elution.

5.4.2 Thermogravimetric analysis

TGA can be employed to evaluate the thermal stability of polymers. As depicted in Figure 5.4C, Mariappan et al. [49] reported an electrochemical sensor platform based on electroactive MIPs (E-MIPs), which can be utilized for the detection of progesterone

(P4) in clinical samples. Specifically, they synthesized E-MIPs through copolymerization of acrylic acid (AA) and ferrocenylmethyl methacrylate using P4 as the template molecule. The incorporation of the ferrocene moieties endowed the MIPs with electroactivity, thereby enabling electrochemical detection of P4. Figure 5.4C-**i** illustrates the thermal stability performance of E-MIPs and electro-NIPs (E-NIPs) in this work. Both E-MIPs and E-NIPs exhibited good thermal stability up to 120 °C. As the temperature further increased, the weight of both materials began to decrease rapidly, reaching approximately 75% weight loss at 430 °C. These data indicate that the polymers undergo decomposition above 120 °C, but the minimal weight loss at room temperature suggests their good stability for practical applications.

MIPs synthesized based on multiwalled carbon nanotubes (MWNTs-MIPs) can remain stable at higher temperatures. For instance, Zhang et al. [50] reported MWNTs-MIPs for the adsorption of furan in serum, which exhibited minimal weight change up to 280 °C. Cao et al. [51] reported MWNTs-MIPs for the adsorption of perfluorooctanoic acid, with a weight loss of only 2.89% at 300 °C.

For MIPs with more complex compositions, changes in TGA can explain their decomposition processes. For example, Mostafa et al. [52] synthesized a novel fluorescent MIP membrane (FMIM) for the detection of pepsin. The differential scanning calorimetric assay thermogram of this FMIM membrane showed a strong endothermic peak around 100 °C. By analyzing the TGA curve of FMIM, they found that with increasing temperature, FMIM first underwent moisture loss around 100 °C, while membrane decomposition occurred around 450 °C. Abubakar et al. [53] synthesized an MIP based on nanoporous carbon (MIP/NC) for the absorption of Hg(II) in aqueous solutions. In TGA studies on MIP/NC, the weight change of this material mainly exhibited three stages: the first change occurred around 100 °C due to water evaporation; the second weight reduction occurred between 200 and 300 °C, attributed to the cysteine complex in the matrix; and the final decomposition ranged from 400 to 450 °C, primarily due to the degradation of the lignin complex.

5.5 Characterization of surface area and pore size

The core property of carbon-based MIPs that supports their applications in sensor design, drug delivery, pollutant removal, and other fields lies in their porosity or the presence of cavities within the polymer. This porosity is essential for molecular recognition and target binding. During synthesis, characterizing these pores is crucial to understanding and optimizing the MIP's functionality. Brunauer-Emmett-Teller (BET) and DLS are primary techniques used to evaluate the pore size and distribution of MIPs in solution, especially before they are immobilized on electrode surfaces for sensor applications. After surface modification, SEM, TEM, and AFM are employed as

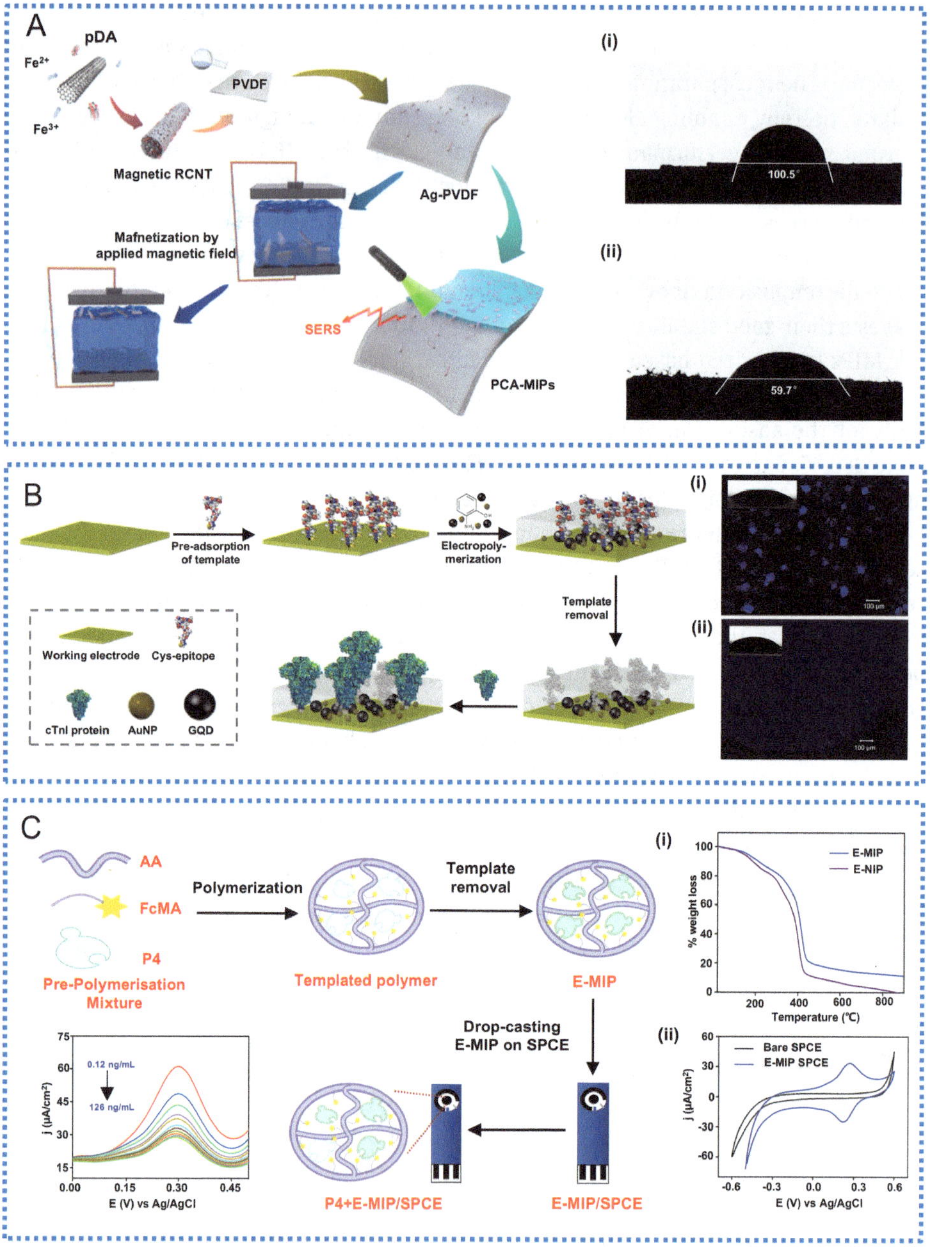

Figure 5.4: (A) Schematic illustration of MIP synthesis and CA pattern of (i) PVDF and (ii)PCA-MIMs. Copyright 2024 with permission from American Chemical Society. (B) Schematic illustration of MIP synthesis and contact angle pattern of (i) MIP film and (ii) NIP film. Copyright 2024 with permission from American Chemical Society. (C)Schematic illustration of MIP synthesis: (i) TGA pattern of E-MIP and E-NIP and (ii) CV of bare SPCE and E-MIP SPCE. Copyright 2024 with permission from American Chemical Society.

post-characterization tools to examine the MIP's surface morphology and porosity in greater detail, ensuring suitability for the intended application.

5.5.1 Brunauer-Emmett-Teller

The specific surface area of inorganic materials, especially carbon-based MIPs, is a key factor influencing their performance in applications like sensing, drug delivery, and pollutant removal. BET method is commonly employed to measure the surface area of porous materials accurately, although its limitations are noted when applied to materials with very large surface areas [54, 55]. Nevertheless, BET analysis remains highly useful for characterizing the surface area of carbon-based nanomaterials, offering insights into pore size, distribution, and the interaction capabilities of these materials [56].

For carbon-based MIPs, BET analysis can reveal crucial details about cavity structure and target interaction. When MIPs are used as biorecognition elements in sensors, their performance, particularly the limit of detection (LOD), can be influenced by binding affinity and the tendency of the MIP to retain or release its target. In one study, Nofar et al. [11] addressed the limitations of high LOD in MIPs by implementing an "on-off" detection approach that relies on the target's uptake into, and release from, the MIP's cavities through positive potential repulsion. BET was utilized in this study to optimize the pore size of the MIP, allowing the cavities to hold the target strongly enough for high detection sensitivity while permitting release as needed. BET analysis revealed a substantial difference in pore characteristics between the MIP and the NIP, with the MIP displaying significantly more cavities. The specific surface area for the MIP was found to be 282.9 m^2/g, in contrast to just 18.4 m^2/g for the NIP, indicating a high cavity presence conducive to target uptake (Figure 5.5A-**i**).

Another study by Utkarsh et al. [57] focused on improving MIP porosity to enhance target binding efficiency. BET analysis indicated a slightly higher surface area for the MIP compared to the NIP, suggesting an increased number of accessible cavities for target interaction. The study highlighted BET's role in understanding the porosity and pore distribution that enhance the binding capacity of MIPs for specific applications (Figure 5.5B).

The choice of cross-linking agents during MIP synthesis also significantly influences pore structure and target interaction. Seyed et al. [58] used BET to investigate the effect of cross-linking on pore formation and binding behavior in MIPs. They found that excessive cross-linking led to a hardened structure that restricted the flexibility of the MIP, impeding effective target binding. BET analysis showed that excessive cross-linking reduced pore accessibility, highlighting the importance of fine-tuning cross-linking levels to avoid compromising the MIP's functionality.

These studies illustrate BET's importance in understanding and controlling pore characteristics in carbon-based MIPs. When optimizing the pore structure and specific surface area of carbon-based MIP, BET analysis supports the possibility of fine-tuning carbon-based MIPs. This, in turn, allows improved performance of carbon-based MIP, particularly in applications that demand high sensitivity and precise target recognition such as electrochemical sensing.

5.5.2 Dynamic light scattering

DLS is a widely used technique for determining nanoparticle size by measuring the scattering of light as it interacts with particles in suspension. DLS is especially valuable in assessing the hydrodynamic diameter of particles and provides insights into their size distribution, aggregation state, and stability in solution [59]. DLS is commonly applied in the characterization of MIPs, especially those involving carbon-based matrices, due to its ability to yield important information regarding the porosity and size distribution of these particles.

In studies focusing on MIPs, such as the work conducted by Nofar et al. [11], DLS played a critical role in analyzing both MIP and NIP counterparts. The analysis provided an understanding of the variations in particle size distributions in different environments such as methanol. In their study, NIP particles exhibited a hydrodynamic size of around 215 ± 12.96 nm (Figure 5.5A-ii). This larger size is attributed to the lack of specific binding sites or cavities in the NIP structure, resulting in a denser, less porous morphology. In contrast, the MIP, designed with specific molecular cavities tailored to the target molecule, showed a smaller size distribution. This reduction in size is indicative of the MIP's porous nature, as the presence of cavities generally reduces the effective particle diameter measured by DLS.

The complementary use of DLS with other surface characterization techniques, like BET analysis, allows researchers to draw correlations between particle size, porosity, and surface area. BET measures surface area through nitrogen adsorption, whereas DLS offers a solution-based perspective on particle stability and size distribution. Together, these methods contribute to a comprehensive understanding of the material properties of MIPs and NIPs, especially for applications where surface functionality and porosity directly impact sensor or sorption performance.

5.6 Spectroscopic characterization

5.6.1 Fluorescence spectroscopy (FL) and UV-vis

In practical applications of MIPs, particularly for target-specific recognition, post-modifications are often introduced to enhance target binding [60]. One effective strategy is incorporating external probes into the MIP's binding cavities to increase speci-

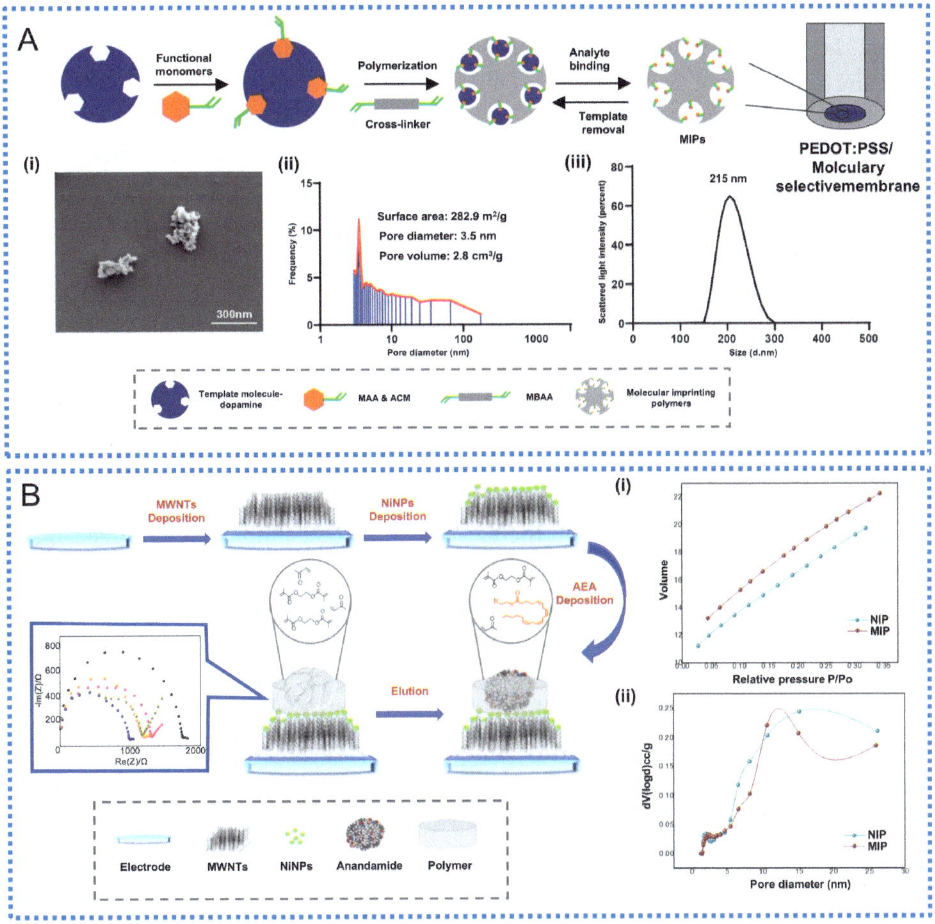

Figure 5.5: (A) Schematic illustration of MIP synthesis and characterization of surface area/pore size: (i) SEM image of the MIP particles; (ii) BET analysis of modified MIP, showing pore diameter; (iii) DLS pattern of selective MIP composite. Copyright 2024 with permission from American Chemical Society. (B) Schematic illustration of MIP synthesis: (i) BET analysis of MIP and NIP; (ii) pore diameter of MIP and NIP. Copyright 2024 with permission from American Chemical Society.

ficity and affinity. In this regard, an aptamer modified with a metal-organic framework (MOF) was introduced into the MIP's cavities to specifically bind a target virus [61]. This aptamer served as a recognition element that interacts with the target, producing a measurable fluorescent signal. Specifically, Fe_3O_4@quantum dot complexes were used as carriers for the aptamer, allowing efficient fluorescence-based detection.

The successful loading of the aptamer-MOF probe into the MIP cavities was confirmed using FL. The MIP initially showed a high fluorescence intensity, which decreased significantly upon exposure to the target virus (Figure 5.6A-**i**). This decrease in intensity was attributed to the interaction between the aptamer and the target within the MIP cavities. The binding of the target to the aptamer causes quenching of the fluorescent signal, as the target molecule occupies the cavity sites that would otherwise contribute to fluorescence. In contrast, when the target was introduced to an NIP, no significant change in fluorescence intensity was observed. This was expected, as the NIP lacks specific binding cavities, preventing effective target-aptamer interactions. Thus, fluorescence spectroscopy provided a reliable way to distinguish between MIP and NIP, emphasizing the specificity of the MIP toward the target.

Further characterization was carried out using UV-vis spectroscopy to substantiate the interaction and ensure that no fluorescence energy transfer contributed to the observed fluorescence changes. The target molecule's absorbance peak was identified at 320 nm, while the MIP particle's absorbance occurred at 530 nm when excited at 450 nm. These distinct absorbance peaks indicate that the target and MIP do not overlap spectrally, confirming that fluorescence quenching was not due to fluorescence resonance energy transfer (FRET). In FRET, energy transfer between a donor and an acceptor fluorophore occurs when their emission and excitation spectra overlap, allowing nonradiative energy transfer from the donor to the acceptor. Since there is no spectral overlap between the target and the MIP, in this case, the decrease in fluorescence intensity cannot be attributed to FRET. Instead, it is the direct result of target-aptamer binding within the MIP cavities.

Together, fluorescence and UV-vis spectroscopies highlight the specificity and efficiency of the aptamer-modified MIP for virus detection. The MIP's initial high fluorescence, coupled with significant quenching upon target binding, validates the effective integration of the aptamer probe.

5.6.2 Fourier transform infrared spectroscopy

FTIR spectroscopy is one of the main techniques frequently employed to characterize imprinted polymers by providing detailed information on their chemical composition and bond interactions. FTIR is used to examining the specific IR absorption peaks, identifying the functional groups present in different components of a MIP, including functional monomers, cross-linkers, and post-modification elements. This approach is especially valuable when distinct chemical environments in the polymer matrix lead to well-resolved, diagnostic signals. For instance, FTIR can probe noncovalent interactions, such as hydrogen bonding, although for precise identification of bond formation, it is often essential to pair FTIR with complementary techniques.

In the study described, FTIR spectroscopy was crucial for confirming the presence of key components in the sensor, specifically the cavities, quantum dot nanocarrier, and MOF. The characteristic Fe-O bond in the Fe_3O_4 precursor for the nanocarrier was observed at 574 cm^1, indicating the presence of the iron oxide structure. Peaks at 1,095 and 1,627 cm^{-1} were attributed to Si–O–Si and C=O bonds, respectively, which were present in the nanocarrier, MIP, and NIP (Figure 5.6A-**ii**). This confirmed that these structural bonds are integral to the overall material, supporting the interaction among the components.

Additional FTIR analysis by Yu et al. [62] utilized similar quantum dot fluorescence with an MIP modified by MOF to create a "dual-nanoenzyme" cascade catalysis fluorescence sensor. In their study, FTIR was used to compare the MOF-MIP with the MOF-NIP, revealing unique peak patterns that confirmed the successful integration of MOF in the MIP structure. Specifically, the peaks at 544584, and 620 cm^{-1} corresponded to Fe–O stretching, C–O stretching, and Bi–O vibrations, respectively. The MOF-MIP spectrum further displayed new peaks at 474, 797, and 1,038 cm^{-1}, attributed to Si–O and Si–O–Si bond vibrations. These additional peaks verified the formation of the MOF-MIP composite, as the distinct Si–O–Si vibrations are indicative of a structured, stable MOF-MIP matrix.

Thus, FTIR spectroscopy is invaluable for confirming bond interactions and the presence of functional groups in modified MIPs, providing essential insights into composite formation and material properties that support targeted sensing capabilities.

5.6.3 X-ray photoelectron spectroscopy

XPS is a surface-sensitive technique that measures the elemental composition and chemical state of materials by detecting photoelectrons ejected from the sample's surface upon X-ray irradiation. This technique is highly informative for analyzing the bond formation and electronic environment in MIPs. It provides valuable insight into the elemental surface distribution. Although XPS is less commonly used for characterizing MIPs compared to other techniques like NMR, FTIR, XRD, and morphological analysis, it can be instrumental when used to confirm bonding patterns and chemical composition, especially in cases involving complex MIP modifications.

As research in MIP optimization continues to expand due to the versatility and wide-ranging applications of these materials, various modification techniques are being explored to enhance the performance of MIPs. To ensure the success of these modifications, it is essential to employ complementary characterization techniques, as each offers unique information about the MIP's structural and chemical properties. XPS, for instance, provides detailed data on the chemical composition and bonding states, making it a valuable tool for assessing the success of MIP modifications, which ultimately affects the overall performance and functionality of the final product.

In the study by Yu et al. [62], XPS was used to characterize an MIP composite modified with MOFs and further coated with Fe_3O_4 nanoparticles. This MOF-MIP composite, designed for enhanced binding and selectivity, was analyzed by XPS to confirm the incorporation of MOF precursors and Fe_3O_4 nanoparticles. The XPS spectrum of the composite was compared with spectra from the MIP alone and the unmodified MOF. The MIP spectrum displayed a characteristic Si 2p peak, indicating the formation of an Si–O–Si layer within the MIP, confirming that the imprinted layer of silicon was successfully incorporated (Figure 5.6B-**i**).

For the MOF, which was synthesized from cobalt (Co) and bismuth (Bi) precursors, XPS spectra revealed distinctive peaks corresponding to Bi at 158, 164, 158, and 163 eV. These peaks were attributed to the Bi 4f5/2 and Bi 4f7/2 orbitals of Bi^0 as well as the Bi 4f5/2 and Bi 4f7/2 orbitals of Bi^{3+}, respectively. This confirmed the presence of bismuth in its intended chemical states. Furthermore, the XPS analysis detected characteristic Fe peaks in the Fe_3O_4-coated MOF-MIP composite at binding energies of 725, 712, 729, and 716 eV. These peaks correspond to the Fe 2p1/2 and Fe 2p3/2 orbitals of Fe^{2+} as well as the Fe 2p1/2 and Fe 2p3/2 orbitals of Fe^{3+}, confirming the successful coating of the composite surface with Fe_3O_4 (Figure 5.6B-**ii–iv**).

In addition to Yu et al.'s work, other studies have applied XPS to assess the chemical composition of MIPs modified with various functional groups. For instance, Qu and colleagues [31] used XPS to evaluate a modified MIP based on ordered mesoporous carbon nanospheres functionalized with MMA. The XPS spectra provided valuable information by revealing a distinct Si 2p peak, attributed to the imprinted layer in the composite. More so, increased oxygen content and the presence of C=O bonds are characteristic of MMA, indicating a successful incorporation of the functional acid into the MIP. This additional data helped confirm the formation of a modified MIP with optimized binding properties.

Another example is the synthesis of a magnetized dummy-template MIP using CNTs. In this study, the absorbent properties of CNTs were enhanced by magnetization, creating a composite with improved separation and detection capabilities [63]. XPS data analysis of the magnetized MIP composite showed a 2p peak at a binding energy of 709.25 eV, a signature of magnetization in the composite. Additional binding energies at 284.8, 286.36, and 288.67 eV in the CNT-MIP were assigned to the 3p3 hybridization of carbon atoms bonded to functional groups and monomers within the composite. This comprehensive characterization confirmed the successful magnetization and functionalization of the CNT-based MIP.

The application of XPS in these studies shows the importance of this technique in verifying the chemical composition and bonding in modified MIPs. XPS ensures that the modifications contribute to the MIP's intended properties through the identification and confirmation of the intended elements and bond formations. Whether to detect elemental incorporation, analyze bonding states, or confirm surface modifications, XPS provides a level of detail essential for advancing MIP design and optimization, proving invaluable for applications requiring precise molecular recognition and selectivity.

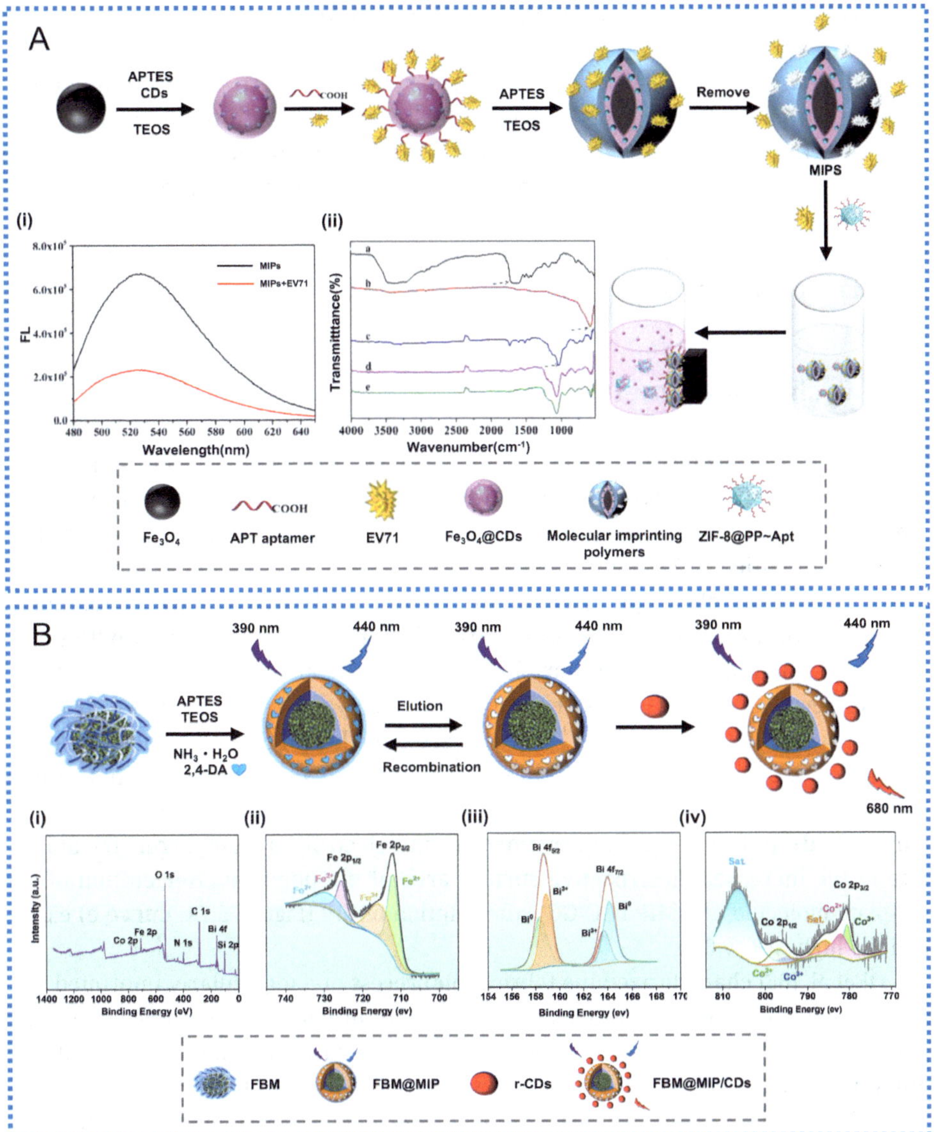

Figure 5.6: (A) Schematic illustration of MIP synthesis and spectroscopy characterization: (i) fluorescence spectroscopy of the MIP particles; (ii) FTIR of the MIP particles. Copyright 2024 with permission from American Chemical Society. (B) Schematic illustration of MIP synthesis and spectroscopy characterization: (i–iv) XPS spectra of modified MIP composite. Copyright 2024 with permission from American Chemical Society.

5.7 Other characterization methods

Depending on the practical application scenarios of polymers, there are numerous methods for characterizing carbon-based MIPs. For instance, in the previously mentioned work (Figures 5.1A-**ii, iii**, and 5.4B-**ii**), when carbon-based MIPs are utilized in sensors, researchers often compare the peak current and charge transfer resistance (R_{ct}) of the electrode before and after modifying with MIPs. Therefore, in this section, we primarily supplement and introduce electrochemical methods used for the characterization of MIP sensors.

5.7.1 Cyclic voltammetry

CV allows for the acquisition of information such as the rate of electron transfer, diffusion coefficient, redox potential, and the quantity of adsorbate in an adsorption system [64]. As illustrated in Figure 5.7A, in the work conducted by Chuiprasert et al. [65], they modified GCEs with MIPs synthesized based on reduced graphene oxide (rGO), creating an electropolymerized MIP sensor (MIP/rGO/GCE) with recognition sites for ciprofloxacin (CIP). This sensor exhibited a 1.5-fold higher sensitivity to CIP compared to antibiotics such as enrofloxacin, ofloxacin, and sulfamethoxazole. To assess the electrode's charge-transfer ability, they employed CV to study the electroactivity of the electrode in $[Fe(CN)_6]^{3-/4-}$ and extracted the height of the anodic peak as the response (I_p). Compared to the bare GCE (Figure 5.7A, **curve a**), the GCE with the introduction of rGO and polymers showed significantly enhanced I_p values (Figure 5.7A, **curves b–d**), indicating an improvement in the electrode's charge-transfer ability. Due to the increased electroactive surface area of the polymer after elution of the template molecule, the MIP/rGO/GCE after elution of CIP (Figure 5.7A, **curve e**) exhibited the highest I_p value.

Qi et al. [66] characterized the fabrication process of a molecularly imprinted sensor for tebuconazole recognition using CV. Unlike previous work, their sensor employed Prussian blue (PB) as a probe and tested the current signals of electrodes at different stages in a KNO_3 electrolyte. The results showed that the electrode modified with PB exhibited a significant decrease in current signal after polymerizing the MIP, indicating that the polymer hindered signal transmission. After removing the template molecule, the redox peak reappeared, suggesting the formation of cavities on the polymer. Upon rebinding tebuconazole, the current decreased again, demonstrating the sensor's effectiveness for tebuconazole detection.

Conclusions drawn from CV can often be verified using other electrochemical methods. For instance, Khosropour et al. [67] reported a molecularly imprinted sensor based on multiwalled CNTs, also using CV to characterize the I_p of electrodes at different stages of fabrication, with the results further validated by differential pulse voltammetry (DPV) and EIS.

In Figure 5.7B, Ma et al. [68] prepared a porous carbon nanomaterial (ZC) and constructed a molecularly imprinted electrochemical sensor for detecting *tert*-butyl hydroquinone (TBHQ) in edible oils. Since TBHQ is electroactive, they chose to conduct electrochemical tests in a Britton-Robinson (BR) buffer solution. As shown in Figure 5.7B-**i**, TBHQ generated a redox peak at the electrode surface during testing (Figure 5.7B-**i**, **curve a**), while no significant peak current was detected in the BR buffer solution after elution of TBHQ (Figure 5.7B-**i**, **curve b**). When the electrode was placed back into a solution containing TBHQ, the redox peak was again detected (Figure 5.7B-**i**, **curve c**). The same trend was also observed in the DPV images (Figure 5.7B-**ii**). These two tests collectively demonstrated the successful construction of the MIP sensor and its excellent adsorption capacity.

5.7.2 Electrochemical impedance spectroscopy

EIS can be utilized to characterize the resistance properties of different electrodes, thereby aiding in the derivation of information regarding the electrode's structure or reaction process. In the Nyquist impedance spectrum, the semicircular segment represents the charge transfer resistance (R_{ct}), while the linear portion signifies the diffusion process [69, 70].

For instance, in Figure 5.7C, Wang et al. [71] designed a sensor capable of rapid and ultrasensitive detection of metronidazole (MNZ). Specifically, they first modified nitrogen-doped CNTs with $CuCo_2O_4$ nanoparticles ($CuCo_2O_4$/N-CNTs) to enhance the material's surface area and conductivity. Using MNZ as a template and aniline as a functional monomer, they then synthesized MIPs via electropolymerization on the $CuCo_2O_4$/N-CNTs-modified GCE, constructing a sensor capable of detecting MNZ in samples such as human serum and urine. During electrode fabrication, they characterized the bare electrode and electrodes with various modification layers using EIS. As shown in Figure 5.7C-**i**, due to the excellent conductivity of both $CuCo_2O_4$ and N-CNTs, the R_{ct} of the $CuCo_2O_4$-modified GCE (**curve b**) and the $CuCo_2O_4$/N-CNTs-modified GCE (**curve c**) were lower compared to the bare electrode (**curve a**). The $CuCo_2O_4$/N-CNTs/GCE (**curve d**) exhibited an even lower R_{ct}, indicating the combination of $CuCo_2O_4$ and N-CNTs provided a larger surface area and superior conductivity. Due to the low conductivity of polyaniline, the R_{ct} significantly increased after assembling MIP on the $CuCo_2O_4$/N-CNTs/GCE (**curve e**). Compared to NIP/$CuCo_2O_4$/N-CNTs/GCE (**curve f**), MIP/$CuCo_2O_4$/N-CNTs/GCE (**curve e**), after template removal, possessed cavities that facilitated the redox reaction of potassium ferricyanide, resulting in a smaller R_{ct}. Correspondingly, when these cavities were occupied by template molecules, that is, before elution of MIP/$CuCo_2O_4$/N-CNTs/GCE, the R_{ct} reached its maximum (Figure 5.7C-**ii**, **curve g**).

Such applications are not uncommon. Sulym et al. [72] fabricated a GCE modified with L-histidine functionalized multiwalled CNTs-MIP for tetracycline detection in human serum and other samples. Akyıldırım et al. [73] reported a molecularly imprinted sensor based on 5-(4-hydroxybenzylidenamino)-2-mercaptobenzimidazole functionalized GQDs for citrinin analysis. Beytur et al. [74] designed a GCE-based molecularly imprinted sensor for ceftizoxime detection. These studies all employed EIS to analyze the Nyquist plots and R_{ct} of bare electrodes, MIP electrodes, MIP electrodes after template removal, and electrodes upon rebinding of target molecules. They arrived at a generally consistent conclusion: the bare electrode exhibited the smallest R_{ct} due to easier electron transfer on its surface. Polymerization of MIP on the electrode surface increased the R_{ct} by inhibiting electron transfer. Removal of template molecules increased polymer porosity, resulting in a decreased R_{ct}. When MIP was rebound with template molecules, the R_{ct} increased again.

5.8 Summary and future perspectives

In this chapter, we summarize commonly used techniques for the characterization of carbon-based MIPs. We organized them by specific characteristics to guide readers according to the intended application of the carbon-based MIP. These characterization methods are essential for assessing and optimizing carbon-based MIPs, especially as modifications are continuously made to tailor their properties for specific applications. Broadly, the characterizations can be categorized into structural, morphological, physical, surface and pore, spectroscopic, and other analyses.

Structural characterizations, including NMR and XRD, are employed for analyzing chemical bonding and the crystallinity of carbon-based MIP, respectively. For morphological insights, techniques such as SEM, TEM, and AFM are used. These methods allow scientists to examine the size, internal structure, and surface topography of carbon-based MIP, which is especially valuable when MIPs are immobilized on electrodes for sensor applications. In physical characterizations, CA and TG are used to determine the wettability and thermal stability, essential properties for MIP stability, and longevity in various environments.

Owing to the fact that the pore structure and surface area are imperative for MIPs, especially when used in sensing applications where target interaction within the polymer cavities. For this, BET analysis and DLS provide precise measurements of pore size and surface area. These data are used for understanding how modifications affect the MIP's capacity to interact with target molecules. Spectroscopic techniques such as UV-visible and fluorescence spectroscopy offer insights into the MIP's response to light, which can be useful in optical sensing applications, while FTIR and XPS help identify bond interactions and chemical compositions, confirming successful modifications.

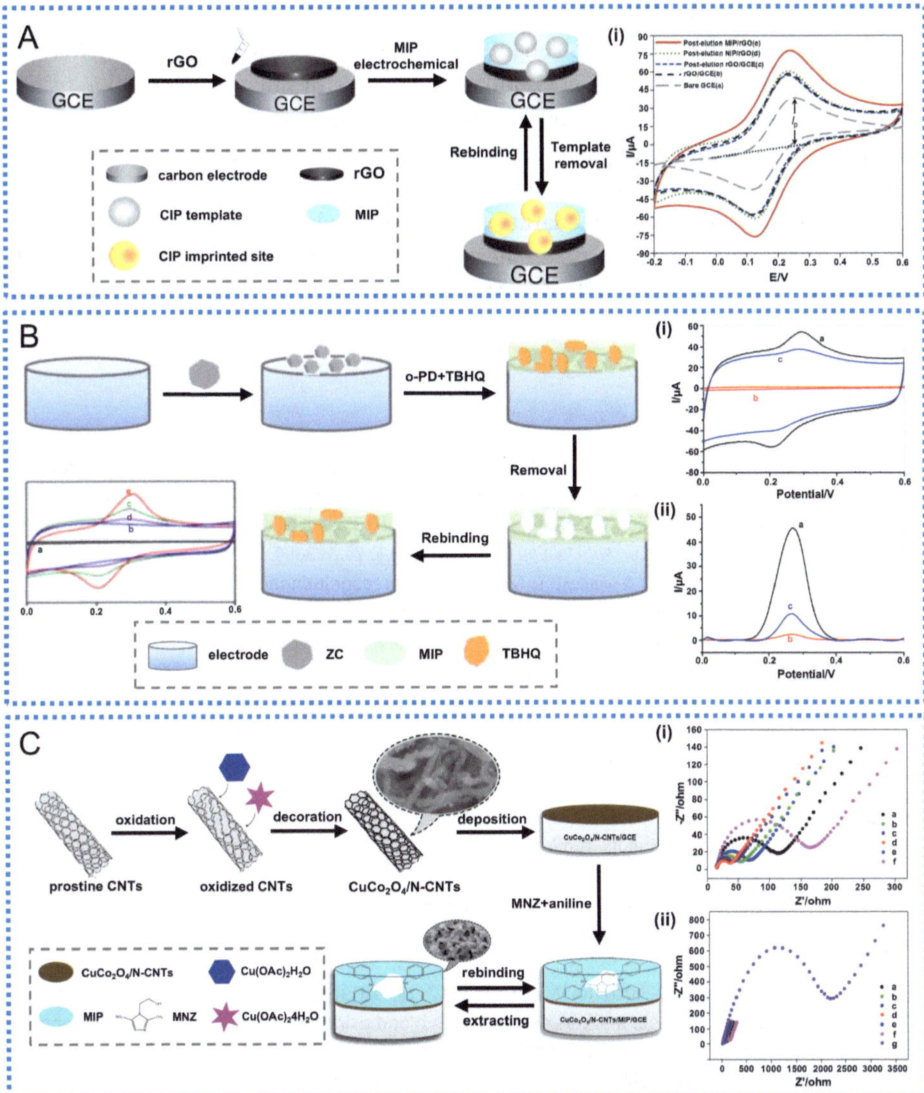

Figure 5.7: (A) Schematic illustration of MIP synthesis: (i) CV curves of different electrodes. Copyright 2024 with permission from American Chemical Society. (B) Schematic illustration of MIP Synthesis: (i) CV curves of different electrodes and (ii) DPV curves of different electrodes. Copyright 2024 with permission from Elsevier. (C)Schematic illustration of MIP synthesis and (i) and (ii) EIS of different electrodes. Copyright 2024 with permission from Elsevier.

Given that carbon-based MIPs are increasingly used to replace natural antibodies in biosensing applications, there is a strong need for enhanced characterization techniques to capture intricate internal and cross-linked structures. In situ characterization techniques, which can reveal real-time insights into cross-linking and cavity structure, would be particularly valuable for tuning MIP properties more precisely. Furthermore, MIPs face performance limitations from factors such as solvent interference and side reactions, which are less commonly characterized but can significantly impact their functionality. Techniques that can assess these effects, while not widely reported, would greatly benefit the field by identifying and addressing these limitations.

References

[1] Purohit B., Vernekar P. R., Shetti N. P., & Chandra P. (2020). Biosensor nanoengineering: Design, operation, and implementation for biomolecular analysis, *Sensors International*. 1, 100040.
[2] Teng Y., Zhang J., Jiang T., Zou Y., Gong X., & Yan Y. (2022). Biosensor-enabled pathway optimization in metabolic engineering, *Current Opinion in Biotechnology*. 75, 102696.
[3] Bazin I., Tria S. A., Hayat A., & Marty J. L. (2017). New biorecognition molecules in biosensors for the detection of toxins, *Biosensors and Bioelectronics*. 87, 285–298.
[4] Zhang Z., Sen P., Adhikari B. R., Li Y., & Soleymani L. (2022). Development of Nucleic-Acid-Based electrochemical biosensors for clinical applications, *Angewandte Chemie International Edition*. 61(50), e202212496.
[5] Li H., Dauphin-Ducharme P., Ortega G., & Plaxco K. W. (2017). Calibration-free electrochemical biosensors supporting accurate molecular measurements directly in undiluted whole blood, *Journal of the American Chemical Society*. 139(32), 11207–11213.
[6] Dkhar D., Kumari S., Mahapatra R., Divya S., Kumar R., Tripathi T., & Chandra P. (2022). Antibody-receptor bioengineering and its implications in designing bioelectronic devices, *International Journal of Biological Macromolecules*. 218, 225–242.
[7] Shahbaz A., Hussain N., Intisar A., Bilal M., & Iqbal H. M. N. (2022). Immobilized enzymes-based biosensing cues for strengthening biocatalysis and biorecognition, *Catalysis Letters*. 152(9), 2637–2649.
[8] Sitkov N. O., Karasev V. A., Luchinin V. V., & Zimina T. M. (2019). Development of biosensors for express-detection of protein markers of diseases in blood using peptide biorecognition elements, *AIP Conference Proceedings*. 2140(1), 020072.
[9] Cieplak M., & Kutner W. (2016). Artificial biosensors: How can molecular imprinting mimic biorecognition?, *Trends in Biotechnology*. 34(11), 922–941.
[10] Huang X., Xia L., & Li G. (2023). Recent progress of molecularly imprinted optical sensors, *Chemosensors*. 11(3), 168.
[11] Mintz Hemed N., Leal-Ortiz S., Zhao E. T., & Melosh N. A. (2023). On-demand, reversible, ultrasensitive polymer membrane based on molecular imprinting polymer, *ACS Nano*. 17(6), 5632–5643.
[12] Çorman M. E., Cetinkaya A., Ozcelikay G., Özgür E., Atici E. B., Uzun L., & Ozkan S. A. (2021). A porous molecularly imprinted nanofilm for selective and sensitive sensing of an anticancer drug ruxolitinib, *Analytica Chimica Acta*. 1187, 339143.
[13] Mehmandoust M., Erk N., Karaman C., & Karaman O. (2022). An electrochemical molecularly imprinted sensor based on $CuBi_2O4/rGO@MoS_2$ nanocomposite and its utilization for highly selective and sensitive for linagliptin assay, *Chemosphere*. 291, 132807.

[14] Nurdin M., Arham Z., Irna W. O., Maulidiyah M., Kurniawan K., Salim L. O. A., Irwan I., & Umar A. A. (2022). Enhanced-charge transfer over molecularly imprinted polyaniline modified graphene/ TiO_2 nanocomposite electrode for highly selective detection of fipronil insecticide, *Materials Science in Semiconductor Processing*. 151, 106994.
[15] He S., Zhang L., Bai S., Yang H., Cui Z., Zhang X., & Li Y. (2021). Advances of molecularly imprinted polymers (MIP) and the application in drug delivery, *European Polymer Journal*. 143, 110179.
[16] Mohsenzadeh E., Ratautaite V., Brazys E., Ramanavicius S., Zukauskas S., Plausinaitis D., & Ramanavicius A. (2024). Design of molecularly imprinted polymers (MIP) using computational methods: A review of strategies and approaches, *Wiley Interdisciplinary Reviews: Computational Molecular Science*. 14(3), e1713.
[17] Kamyab H., Chelliapan S., Tavakkoli O., Mesbah M., Bhutto J. K., Khademi T., Kirpichnikova I., Ahmad A., & Aljohani A. A. (2022). A review on carbon-based molecularly-imprinted polymers (CBMIP) for detection of hazardous pollutants in aqueous solutions, *Chemosphere*. 308, 136471.
[18] Gui R., Guo H., & Jin H. (2019). Preparation and applications of electrochemical chemosensors based on carbon-nanomaterial-modified molecularly imprinted polymers, *Nanoscale Advances*. 1(9), 3325–3363.
[19] Liu H., Qin W., Li X., Feng L., Gu C., Chen J., Tian Z., Chen J., Yang M., Qiao H., Guo X., Zhang Y., Zhao B., & Yin S. (2023). Molecularly imprinted electrochemical sensors based on Ti_3C_2Tx-MXene and graphene composite modifications for ultrasensitive cortisol detection, *Analytical Chemistry*. 95(44), 16079–16088.
[20] De Middeleer G., Dubruel P., & De Saeger S. (2016). Characterization of MIP and MIP functionalized surfaces: Current state-of-the-art, *TrAC - Trends in Analytical Chemistry*. 76, 71–85.
[21] Hasanah A. N., Safitri N., Zulfa A., Neli N., & Rahayu D. (2021). Factors affecting preparation of molecularly imprinted polymer and methods on finding template-monomer interaction as the key of selective properties of the materials, *Molecules*. 26, 18.
[22] Tan F., Liu M., & Ren S. (2017). Preparation of polydopamine-coated graphene oxide/Fe3O4 imprinted nanoparticles for selective removal of fluoroquinolone antibiotics in water, *Scientific Reports*. 7, 1.
[23] Zidarič T., Majer D., Maver T., Finšgar M., & Maver U. (2023). The development of an electropolymerized, molecularly imprinted polymer (MIP) sensor for insulin determination using single-drop analysis, *The Analyst*. 148(5), 1102–1115.
[24] Li J. P., Zhao J., & Wei X. P. (2009). A sensitive and selective sensor for dopamine determination based on a molecularly imprinted electropolymer of o-aminophenol, *Sensors & Actuators, B: Chemical*. 140(2), 663–669.
[25] Yarman A., Kurbanoglu S., Jetzschmann K. J., Ozkan S. A., Wollenberger U., & Scheller F. W. (2018). Electrochemical MIP-Sensors for Drugs, *Current Medicinal Chemistry*. 25(33), 4007–4019.
[26] Wong A., Foguel M. V., Khan S., De Oliveira F. M., Tarley C. R. T., & Sotomayor M. (2015). Development of an electrochemical sensor modified with MWCNT-COOH and MIP for detection of diuron, *Electrochimica Acta*. 182, 122–130.
[27] Liu Y., Wang L., Li H., Zhao L., Ma Y., Zhang Y., Liu J., & Wei Y. (2024). Rigorous recognition mode analysis of molecularly imprinted polymers-Rational design, challenges, and opportunities, *Progress in Polymer Science*. 150, 101790.
[28] Azizi A., & Bottaro C. S. (2020). A critical review of molecularly imprinted polymers for the analysis of organic pollutants in environmental water samples, *Journal of Chromatography A*. 1614, 460603.
[29] Vasapollo G., Sole R. D., Mergola L., Lazzoi M. R., Scardino A., Scorrano S., & Mele G. (2011). Molecularly Imprinted Polymers: Present and Future Prospective *IInt, Journal of Molecular Sciences*. 12(9), 5908–5945.
[30] Cormack P. A. G., & Elorza A. Z. (2004). Molecularly imprinted polymers: Synthesis and characterisation, *Journal of Chromatography B-Analytical Technologies in the Biomedical and Life Sciences*. 804(1), 173–182.

[31] Qu Y., Qin L., Yang Y., & Liu X. (2021). Porous carbon nanosphere-based imprinted composite membrane for selective and effective separation of dibenzothiophene, *Separation and Purification Technology*. 265, 118530.
[32] Samah N. A., Sánchez-Martín M.-J., Sebastián R. M., Valiente M., & López-Mesas M. (2018). Molecularly imprinted polymer for the removal of diclofenac from water: Synthesis and characterization, *Science of the Total Environment*. 631–632, 1534–1543.
[33] Boventi M., Mauri M., Golker K., Wiklander J. G., Nicholls I. A., & Simonutti R. (2022). Porosity of molecularly imprinted polymers investigated by 129Xe NMR spectroscopy, *ACS Applied Polymer Materials*. 4(12), 8740–8749.
[34] Wang Z., Chen Z., Ma Z., & Han H. (2024). Label-free mode based on Ferrocene/PEDOT:PSS–PPy for molecularly imprinted electrochemically ultrasensitive detection of amino acids, *Analytical Chemistry*. 96(35), 14298–14305.
[35] Alsafi S. R., Rostamabadi H., Assadpour E., & Jafari S. M. (2020). Morphology and microstructural analysis of bioactive-loaded micro/nanocarriers via microscopy techniques; CLSM/SEM/TEM/AFM, *Advances in Colloid and Interface Science*. 280, 102166.
[36] Chen B., Zhang Y., Lin L., Chen H., & Zhao M. (2020). Au nanoparticles @metal organic framework/polythionine loaded with molecularly imprinted polymer sensor: Preparation, characterization, and electrochemical detection of tyrosine, *Journal of Electroanalytical Chemistry*. 863, 114052.
[37] Budak F., Cetinkaya A., Kaya S. I., Bellur Atici E., & Ozkan S. A. (2023). Investigations on the electrochemical behavior of sunitinib and metabolites N-desethyl-sunitinib and sunitinib-N-oxide and its selective determination using molecularly imprinted polymer-based sensor, *Electrochimica Acta*. 472, 143434.
[38] Ann Maria C. G., Varghese A., & Nidhin M. (2024). Advanced electrochemical detection of 2,4-dichlorophenol in water with molecularly imprinted chitosan stabilized gold nanoparticles, *Journal of the Electrochemical Society*. 171, 10.
[39] Kim H., Murata M. M., Chang H., Lee S. H., Kim J., Lee J. H., Rho W.-Y., & Jun B.-H. (2021). Optical and electron microscopy for analysis of nanomaterials, *Nanotechnology for Bioapplications*. 1309, 277–287.
[40] Ozcelikay G., Karadas-Bakirhan N., Taskin-Tok T., & Ozkan S. A. (2020). A selective and molecular imaging approach for anticancer drug: Pemetrexed by nanoparticle accelerated molecularly imprinting polymer, *Electrochimica Acta*. 354, 136665.
[41] Zhang Y., Li S., Ma X.-T., He X.-W., Li W.-Y., & Zhang Y.-K. (2020). Carbon dots-embedded epitope imprinted polymer for targeted fluorescence imaging of cervical cancer via recognition of epidermal growth factor receptor, *Microchimica Acta*. 187, 4.
[42] Xu X., Guo P., Luo Z., Ge Y., Zhou Y., Chang R., Du W., Chang C., & Fu Q. (2017). Preparation and characterization of surface molecularly imprinted films coated on multiwall carbon nanotubes for recognition and separation of lysozyme with high binding capacity and selectivity, *RSC Advances*. 7(30), 18765–18774.
[43] Yao B., Gu L., Huang L., Li R., Fan Z., Chen Z., Qin D., & Gao L. (2024). Using magnetic molecularly imprinted polymer technology for determination of fish serum glucose levels, *Polymers*. 16(11), 1538.
[44] Amatatongchai M., Sroysee W., Sodkrathok P., Kesangam N., Chairam S., & Jarujamrus P. (2019). Novel three-Dimensional molecularly imprinted polymer-coated carbon nanotubes (3D-CNTs@MIP) for selective detection of profenofos in food, *Analytica Chimica Acta*. 1076, 64–72.
[45] Li H., Wang D., Zhang D., Zhou J., Yang W., Su Z., Sun W., & Li C. (2024). Light-initiated imprinted membrane-based biomimetic SERS sensor toward selective detection of trace MC-LR, *Analytical Chemistry*. 96(15), 5887–5896.
[46] Qu Y., Qin L., Guo M., Liu X., & Yang Y. (2022). Multilayered molecularly imprinted composite membrane based on porous carbon nanospheres/pDA cooperative structure for selective adsorption and separation of phenol, *Separation and Purification Technology*. 280, 119915.

[47] Erdoğan N. Ö., Uslu B., & Aydoğdu Tığ G. (2023). Development of an electrochemical biosensor utilizing a combined aptamer and MIP strategy for the detection of the food allergen lysozyme, *Microchimica Acta*. 190, 12.

[48] Hasabnis G. K., & Altintas Z. (2024). Cardiac Troponin I-responsive nanocomposite materials for voltammetric monitoring of acute myocardial infarction, *ACS Omega*. 9(28), 30737–30750.

[49] Mariappan S. A., & Manickam P. (2024). Electro-responsive Molecularly Imprinted Polymers (E-MIPs) for rapid detection of progesterone in human serum samples, *Analytical Chemistry* 96(38), 15222–15230

[50] Zhang W.-J., Li D., Xu Y., Jiang Z., Chen Y., & Wang P. (2019). Synthesis and application of novel molecularly imprinted solid phase extraction materials based on carbon nanotubes for determination of carbofuran in human serum by high performance liquid chromatography, *Journal of Agricultural and Food Chemistry*. 67(18), 5105–5112.

[51] Cao F., Wang L., Yao Y., Wu F., Sun H., & Lu S. (2018). Synthesis and application of a highly selective molecularly imprinted adsorbent based on multi-walled carbon nanotubes for selective removal of perfluorooctanoic acid, *Environmental Science: Water Research & Technology*. 4(5), 689–700.

[52] Mostafa A. M., Barton S. J., Wren S. P., & Barker J. (2024). Design and preparation of a fluorescent molecularly imprinted membrane for the selective detection of pepsin enzyme as a biomarker for gastroesophageal reflux disease, *Talanta Open*. 10, 100351.

[53] Abubakar L., Yusof N. A., Abdullah A. H., Mohammad F., Wahid M. H., Ismail S., Abdul Rahim Z., Al-Lohedan H. A., & Soleiman A. A. (2023). Molecularly imprinted polymer-based nanoporous carbon nanocomposite for effective adsorption of Hg(II) ions from aqueous suspensions, *Separations*. 10, 8.

[54] Sinha P., Datar A., Jeong C., Deng X., Chung Y. G., & Lin L.-C. (2019). Surface area determination of porous materials using the Brunauer–Emmett–Teller (BET) method: Limitations and improvements, *The Journal of Physical Chemistry C*. 123(33), 20195–20209.

[55] Irwansyah F. S., Amal A. I., Diyanthi E. W., Hadisantoso E. P., Noviyanti A. R., Eddy D. R., & Risdiana R. (2024). How to read and determine the specific surface area of inorganic materials using the brunauer-emmett-teller (bet) method. *ASEAN Journal of Science and Engineering*. 4(1), 61–70.

[56] Zou J., Fan C., Jiang Y., Liu X., Zhou W., Xu H., & Huang L. (2021). A preliminary study on assessing the Brunauer-Emmett-Teller analysis for disordered carbonaceous materials, *Microporous Mesoporous Mater*. 327, 111411.

[57] Jain U., Soni S., Balhara Y. P. S., Khanuja M., & Chauhan N. (2020). Dual-layered nanomaterial-based molecular pattering on polymer surface biomimetic impedimetric sensing of a bliss molecule, anandamide neurotransmitter, *ACS Omega*. 5(19), 10750–10758.

[58] Torabi S. J., Mohebali A., Abdouss M., Shakiba M., Abdouss H., Ramakrishna S., Teo Y. S., Jafari I., & Ghomi E. R. (2021). Synthesis and characterization of a novel molecularly imprinted polymer for the controlled release of rivastigmine tartrate, *Materials Science & Engineering C-Materials for Biological Applications*. 128, 112273.

[59] Farkas N., & Kramar J. A. (2021). Dynamic light scattering distributions by any means, *Journal of Nanoparticle Research*. 23(5), 11.

[60] Henry O. Y. F., Cullen D. C., & Piletsky S. A. (2005). Optical interrogation of molecularly imprinted polymers and development of MIP sensors: A review, *Analytical and Bioanalytical Chemistry*. 382(4), 947–956.

[61] Tang L., Liang K., Wang L., Chen C., Cai C., & Gong H. (2022). Construction of an ultrasensitive molecularly imprinted virus sensor based on an “explosive” secondary amplification strategy for the visual detection of viruses, *Analytical Chemistry*. 94(40), 13879–13888.

[62] Chen Y., Tang K., Zhou Q., Wang X., Wang R., & Zhang Z. (2023). Integrating intelligent logic gate dual-nanozyme cascade fluorescence capillary imprinted sensors for ultrasensitive simultaneous detection of 2,4-Dichlorophenoxyacetic Acid and 2,4-Dichlorophenol, *Analytical Chemistry*. 95(49), 18139–18148.

[63] Yu M., Wang S., Wang N., Wu Y., Li L., & Lu X. (2024). Magnetic dummy-template molecularly imprinted polymer with an ultrathin shell for selective enrichment of geosmin in water sample, *Journal of Industrial and Engineering Chemistry*. 136, 349–358.
[64] Yamada H., Yoshii K., Asahi M., Chiku M., & Kitazumi Y. (2022). Cyclic voltammetry part 1: Fundamentals, *Electrochemistry*. 90(10), 102005–102005.
[65] Chuiprasert J., Srinives S., Boontanon N., Polprasert C., Ramungul N., Lertthanaphol N., Karawek A., & Boontanon S. K. (2023). Electrochemical sensor based on a composite of reduced graphene oxide and molecularly imprinted copolymer of Polyaniline–Poly(o-phenylenediamine) for ciprofloxacin determination: fabrication, characterization, and performance evaluation, *ACS Omega*. 8(2), 2564–2574.
[66] Qi P., Wang J., Wang Z., Wang X., Wang X., Xu X., Xu H., Di S., Zhang H., Wang Q., & Wang X. (2018). Construction of a probe-immobilized molecularly imprinted electrochemical sensor with dual signal amplification of thiol graphene and gold nanoparticles for selective detection of tebuconazole in vegetable and fruit samples, *Electrochimica Acta*. 274, 406–414.
[67] Khosropour H., Saboohi M., Keramat M., Rezaei B., & Ensafi A. A. (2023). Electrochemical molecularly imprinted polymer sensor for ultrasensitive indoxacarb detection by tin disulfide quantum dots/ carbon nitride/multiwalled carbon nanotubes as a nanocomposite, *Sensors & Actuators, B: Chemical*. 385, 133652.
[68] Ma Y., Li J., & Wang L. (2021). Porous carbon derived from ZIF-8 modified molecularly imprinted electrochemical sensor for the detection of tert-butyl hydroquinone (TBHQ) in edible oil, *Food Chemistry*. 365, 130462.
[69] Szekeres K. J., Vesztergom S., Ujvári M., & Láng G. G. (2021). Methods for the determination of valid impedance spectra in non-stationary electrochemical systems: Concepts and techniques of practical importance, *ChemElectroChem*. 8(7), 1233–1250.
[70] Arman S., Tilley R. D., & Gooding J. J. (2024). A review of electrochemical impedance as a tool for examining cell biology and subcellular mechanisms: Merits, limits, and future prospects, *The Analyst*. 149(2), 269–289.
[71] Wang Y., Yao L., Liu X., Cheng J., Liu W., Liu T., Sun M., Zhao L., Ding F., Lu Z., Zou P., Wang X., Zhao Q., & Rao H. (2019). $CuCo_2O_4$/N-Doped CNTs loaded with molecularly imprinted polymer for electrochemical sensor: Preparation, characterization and detection of metronidazole, *Biosensors & Bioelectronics*. 142, 111483.
[72] Sulym I., Cetinkaya A., Yence M., Çorman M. E., Uzun L., & Ozkan S. A. (2022). Novel electrochemical sensor based on molecularly imprinted polymer combined with L-His-MWCNTs@PDMS-5 nanocomposite for selective and sensitive assay of tetracycline, *Electrochimica Acta*. 430, 141102.
[73] Akyıldırım O., Kardaş F., Beytur M., Yüksek H., Atar N., & Yola M. L. (2017). Palladium nanoparticles functionalized graphene quantum dots with molecularly imprinted polymer for electrochemical analysis of citrinin, *Journal of Molecular Liquids*. 243, 677–681.
[74] Beytur M., Kardaş F., Akyıldırım O., Özkan A., Bankoğlu B., Yüksek H., Yola M. L., & Atar N. (2018). A highly selective and sensitive voltammetric sensor with molecularly imprinted polymer based silver@gold nanoparticles/ionic liquid modified glassy carbon electrode for determination of ceftizoxime, *Journal of Molecular Liquids*. 251, 212–217.

Part 3: **Applications of molecular imprinting-based smart carbon nanomaterials**

Khasan Berdimuradov*, Akbarali Rasulov, Husan Yaxshinorov, Javokhir Abdisattorov, Elyor Berdimurodov, Omar Dagdag, and Mohamed Rbaa

Chapter 6
Environmental applications of carbon-based molecularly imprinted polymers

Abstract: Carbon-based molecularly imprinted polymers (MIPs) represent a significant advancement in the field of environmental remediation, offering enhanced selectivity, stability, and efficiency compared to traditional MIPs. This chapter provides a comprehensive overview of the development, application, and future prospects of carbon-based MIPs in environmental protection. We explore the integration of carbon materials such as graphene, carbon nanotubes, and activated carbon into MIPs, highlighting their role in improving adsorption capacity and mechanical properties. The chapter also discusses the application of these advanced materials in the removal of heavy metals, organic pollutants, and volatile organic compounds from water, air, and soil. Challenges related to scalability, cost, and practical implementation are addressed, along with emerging trends and future directions in the field. The potential of carbon-based MIPs to provide sustainable and effective solutions for environmental protection is emphasized, underscoring their importance in the future of environmental science.

Keywords: Carbon-based MIPs, environmental remediation, graphene, carbon nanotubes, activated carbon, heavy metal removal, organic pollutant removal, VOCs, soil remediation

***Corresponding author: Khasan Berdimuradov,** Karshi State Technical University, Shahrisabz Faculty of Food Engineering, 19, Khanabad Branch Street, Karshi 180119, Kashkadarya
Akbarali Rasulov, Husan Yaxshinorov, Javokhir Abdisattorov, Department of Physical Chemistry, National University of Uzbekistan, Tashkent 100034, Uzbekistan
Elyor Berdimurodov, Faculty of Chemistry, National University of Uzbekistan, Tashkent 100034, Uzbekistan
Omar Dagdag, Department of Mechanical Engineering, Gachon University, Seongnam 13120, Republic of Korea
Mohamed Rbaa, The Higher Institute of Nursing Professions and Health Techniques of Casablanca, P.O. Box 20250 Casablanca, Morocco; Laboratory of Materials, Nanotechnology, and Environment, Faculty of Sciences, Mohammed V University in Rabat, P.O. Box 1014, Agdal-Rabat, Morocco

https://doi.org/10.1515/9783111503325-006

6.1 Introduction

Molecularly imprinted polymers (MIPs) represent a highly versatile class of synthetic materials designed to exhibit selective recognition for specific target molecules. These polymers are synthesized by polymerizing functional and cross-linking monomers in the presence of a template molecule, which is later removed, leaving behind cavities that are complementary in size, shape, and functional groups to the target molecule. This unique characteristic allows MIPs to mimic the behavior of natural receptors, making them highly suitable for various applications including sensing, catalysis, and separation processes [1–3]. The significance of MIPs lies in their ability to provide tailored recognition properties combined with the robustness, reusability, and stability of synthetic polymers. Unlike natural recognition systems, such as antibodies or enzymes, MIPs can withstand harsh chemical environments, high temperatures, and prolonged use without losing their functionality. This durability makes them particularly appealing for industrial applications where extreme conditions are often encountered. Moreover, MIPs offer a cost-effective alternative to biological recognition elements, as they can be synthesized in large quantities with relatively simple and scalable procedures [4–6]. In recent years, the integration of carbon-based materials into MIPs has emerged as a powerful strategy to enhance their performance. Carbon materials, such as carbon nanotubes (CNTs), graphene, and activated carbon (AC), are known for their exceptional mechanical, electrical, and thermal properties. When combined with MIPs, these materials can significantly improve the binding capacity, selectivity, and sensitivity of the polymers. For instance, the high surface area and unique electronic properties of graphene can facilitate more efficient interaction between the polymer and the target molecule, leading to faster recognition processes and higher binding affinities [7–9]. Additionally, the incorporation of carbon-based materials can enhance the mechanical strength and conductivity of MIPs, making them more suitable for applications in harsh environments or in the development of electrochemical sensors.

The importance of environmental applications for MIPs cannot be overstated, given the growing global concerns over pollution and the need for sustainable solutions. Traditional methods of pollution control, such as adsorption using AC, chemical precipitation, and membrane filtration, often suffer from limitations such as low selectivity, high operational costs, and the generation of secondary waste. MIPs, with their ability to selectively bind to and remove specific pollutants from complex mixtures, offer a promising alternative. For instance, carbon-based MIPs have been successfully applied in the removal of heavy metals, pesticides, and pharmaceuticals from water, providing a more targeted approach to water purification compared to conventional methods. Their selective nature not only improves the efficiency of pollutant removal but also minimizes the consumption of resources and the production of waste [10–12]. Furthermore, MIPs hold significant potential for the detection and monitoring of environmental contaminants. The high selectivity and sensitivity of

MIPs make them ideal candidates for the development of sensors that can detect trace levels of pollutants in air, water, or soil. These sensors could play a crucial role in environmental monitoring, enabling the early detection of pollution events and facilitating timely intervention measures. In this context, carbon-based MIPs, with their enhanced properties, are particularly well-suited for the fabrication of advanced environmental sensors. In conclusion, the combination of MIPs with carbon-based materials represents a significant advancement in the development of materials for environmental applications [13–15]. These hybrid materials offer the selectivity and robustness of MIPs, coupled with the superior properties of carbon materials, making them highly effective for pollution control and environmental monitoring. As global environmental challenges continue to escalate, the development and application of carbon-based MIPs are likely to become increasingly important in the pursuit of sustainable and efficient solutions for environmental protection.

6.2 Environmental applications

6.2.1 Water treatment

Carbon-based MIPs have emerged as powerful tools in water treatment, offering enhanced selectivity and adsorption capacities for the removal of heavy metals and organic pollutants (Table 6.1). The integration of carbon materials such as graphene oxide (GO), AC, and CNTs with MIPs has significantly improved their performance, particularly in the selective removal of contaminants like lead (Pb), mercury (Hg), cadmium (Cd), arsenic (As), chromium (Cr), copper (Cu), zinc (Zn), cobalt (Co), and manganese (Mn) from water sources [1, 5, 16]. For instance, GO-based MIPs have achieved over 95% removal efficiency of Pb(II) ions, while AC/GO hybrid MIPs have demonstrated up to 90% removal efficiency for Hg(II) ions in wastewater, even in the presence of competing ions. CNT-based MIPs have shown remarkable results as well, with 92% removal efficiency for Cd(II) and 88% for Cr(VI) in industrial effluents, highlighting their stability and rapid adsorption kinetics in complex matrices. The effectiveness of these MIPs extends to the removal of organic pollutants, such as pesticides and volatile organic compounds (VOCs), with GO-CNT hybrids achieving high reusability and fast adsorption rates for Cu(II) ions, and GO-AC composites reaching 90% efficiency in Mn(II) removal under varying pH conditions [2, 17, 18]. The integration of these advanced materials into water treatment processes underscores their potential to offer sustainable and efficient solutions for environmental remediation.

The study [28] presents an advanced approach to tackling the environmental challenge of mercury (Hg(II)) pollution. The researchers synthesized a MIP combined with nanoporous carbon (NC) to create a nanocomposite aimed at selectively adsorbing Hg(II) ions from water (Figure 6.1). The fabrication involved bulk polymerization,

Table 6.1: Removal of additional heavy metals using carbon-based MIPs in water treatment.

Metal type	Carbon-based MIPs types and names	Main results	References
Lead (Pb)	Graphene oxide (GO)-based MIPs	Achieved over 95% removal efficiency in contaminated water samples, with high selectivity for Pb(II) ions.	[1, 5, 16]
	GO-MIP composite	Enhanced binding affinity due to high surface area and effective imprinting of Pb(II) ions.	
Mercury (Hg)	Activated carbon (AC)-MIPs	Demonstrated up to 90% removal efficiency of Hg(II) ions in wastewater, with minimal interference from other ions.	[2, 17, 18]
	AC/GO hybrid MIPs	Improved adsorption capacity and selectivity, particularly in the presence of competitive metal ions.	
Cadmium (Cd)	Carbon nanotubes (CNTs)-MIPs	Achieved 92% removal efficiency of Cd(II) ions from industrial effluents, with high reusability over multiple cycles.	[3, 19, 20]
	CNT/graphene composite MIPs	Enhanced mechanical strength and conductivity, leading to rapid and efficient Cd(II) ion removal.	
Arsenic (As)	Graphene oxide (GO)-MIPs	Achieved selective removal of As(III) with over 85% efficiency in groundwater samples.	[13, 21]
	GO/activated carbon Composite MIPs	Improved binding affinity and selectivity, particularly under varying pH conditions.	
Chromium (Cr)	Carbon nanotube (CNT)-MIPs	Demonstrated 88% removal efficiency for Cr(VI) in industrial wastewater, with high selectivity for Cr species.	[12, 22, 23]
	CNT/GO hybrid MIPs	Achieved high adsorption capacity and fast removal rates for Cr(VI) in complex matrices.	
Copper (Cu)	Graphene oxide (GO)-MIPs	Achieved 93% removal efficiency of Cu(II) ions in contaminated water, with high selectivity over other cations.	[6, 23–25]
	GO-CNT hybrid MIPs	Demonstrated rapid adsorption kinetics and high reusability in multiple cycles.	
Zinc (Zn)	Carbon nanotubes (CNTs)-MIPs	Achieved 89% removal efficiency of Zn(II) ions from industrial effluents, with high stability under varying pH conditions.	[26]
	CNT/graphene composite MIPs	Enhanced selectivity and faster adsorption rates compared to traditional adsorbents.	

Table 6.1 (continued)

Metal type	Carbon-based MIPs types and names	Main results	References
Cobalt (Co)	Activated carbon (AC)-MIPs	Demonstrated 85% removal efficiency of Co(II) ions in mining wastewater, with minimal interference from other metal ions.	[26]
	AC/graphene oxide hybrid MIPs	Improved adsorption efficiency and reusability across multiple regeneration cycles.	
Manganese (Mn)	Graphene oxide (GO)-MIPs	Achieved 90% removal efficiency of Mn(II) ions from groundwater, with high selectivity and stability under different pH conditions.	[27]
	GO-activated carbon composite MIPs	Enhanced adsorption capacity and reduced processing time for Mn(II) removal.	

leading to a nanocomposite with a high-specific surface area of 884.9 m^2/g, optimized for effective adsorption. The study found that the nanocomposite achieved a maximum adsorption capacity of 116 mg/g at pH 4, with the process best described by the Freundlich isotherm model and pseudo-second-order kinetics, indicating a favorable, spontaneous, and endothermic adsorption reaction. Further tests demonstrated the nanocomposite's effectiveness in real-world conditions, particularly in recovering 87.4% of Hg(II) from condensate samples from the oil and gas industry. These results suggest that the MIP/NC nanocomposite is a highly promising material for selective mercury removal in polluted environments. The study underscores the potential for combining different absorbents to enhance the removal of heavy metals and pharmaceuticals, positioning the MIP/NC nanocomposite as a valuable tool in environmental remediation efforts.

Carbon-based MIPs have demonstrated exceptional efficiency in removing various organic pollutants from water, providing a targeted and effective solution for water purification. For pesticides, GO-based MIPs have achieved a 92% removal efficiency of atrazine, with high selectivity and stability across varying pH conditions, while GO-AC composite MIPs have shown rapid adsorption kinetics and enhanced binding affinity for chlorpyrifos, making them ideal for treating agricultural runoff [10, 29, 30] (Table 6.2). In pharmaceutical removal, AC-based MIPs have displayed 90% removal efficiency for ibuprofen, even in the presence of structurally similar compounds, and AC/graphene composite MIPs have effectively removed antibiotics like ciprofloxacin despite competing pollutants. Endocrine-disrupting compounds (EDCs), such as bisphenol A (BPA), have been efficiently removed using CNTs-based MIPs, with an 88% removal efficiency and high reusability, while CNT/GO hybrid MIPs have rapidly adsorbed estradiol in complex wastewater matrices [8, 31, 32]. For dye removal, GO-based MIPs achieved a 93% removal efficiency for methylene blue, with

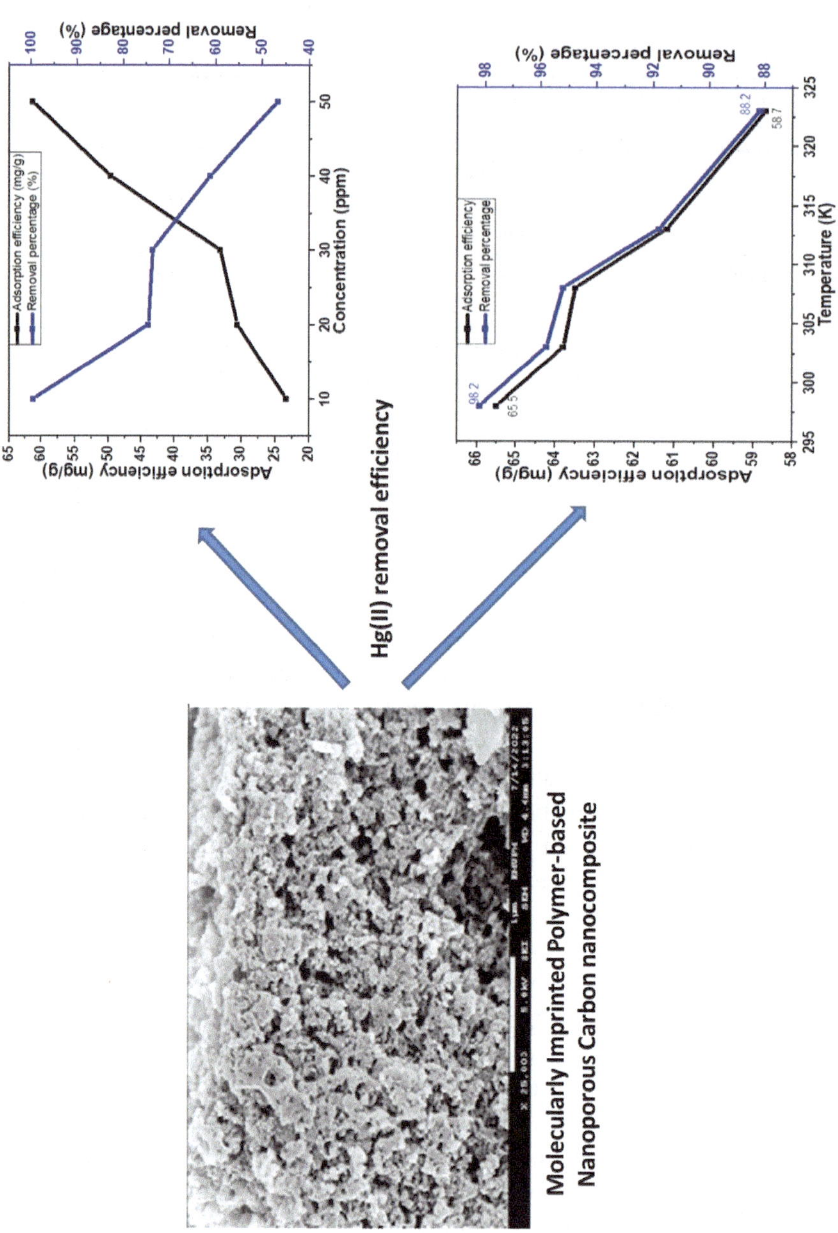

Figure 6.1: Molecularly imprinted polymer-based nanoporous carbon nanocomposite for effective adsorption of Hg(II) ions from aqueous suspensions [28].

high binding affinity and fast adsorption rates, and GO-CNT hybrid MIPs effectively removed azo dyes like Congo red, demonstrating enhanced stability under different pH conditions. In addressing polychlorinated biphenyls (PCBs), AC-based MIPs showed 85% removal efficiency, with high selectivity and minimal interference from other organic compounds, and AC/graphene composite MIPs further improved PCB removal speed and capacity [4, 33]. Lastly, phenols, including chlorophenols, were efficiently removed using CNT-based MIPs, achieving 90% removal efficiency with high selectivity and reusability, while CNT/GO hybrid MIPs demonstrated rapid adsorption and stability. These advancements highlight the significant potential of carbon-based MIPs as robust, efficient, and sustainable solutions for treating a broad spectrum of organic pollutants in water.

Table 6.2: Removal of various organic pollutants using carbon-based MIPs in water treatment.

Organic pollutant	Carbon-based MIPs types and names	Main results	References
Pesticides	Graphene oxide (GO)-MIPs	Achieved 92% removal efficiency of atrazine from water samples, with high selectivity and stability under varying pH conditions.	[10, 29, 30]
	GO-activated carbon composite MIPs	Demonstrated rapid adsorption kinetics and enhanced binding affinity for chlorpyrifos in agricultural runoff.	
Pharmaceuticals	Activated carbon (AC)-MIPs	Achieved 90% removal efficiency of ibuprofen from wastewater, with high selectivity over structurally similar compounds.	[8, 31, 32]
	AC/graphene composite MIPs	Improved adsorption capacity for antibiotics like ciprofloxacin, with effective removal in the presence of competing substances.	
Endocrine-disrupting compounds (EDCs)	Carbon nanotubes (CNTs)-MIPs	Demonstrated 88% removal efficiency of bisphenol A (BPA) from industrial effluents, with high reusability across multiple cycles.	[4, 33]
	CNT/graphene oxide hybrid MIPs	Achieved high selectivity and rapid adsorption of estradiol in complex wastewater matrices.	
Dyes	Graphene oxide (GO)-MIPs	Achieved 93% removal efficiency of methylene blue from dye-contaminated water, with high binding affinity and fast adsorption rates.	[26, 34]
	GO-CNT hybrid MIPs	Demonstrated effective removal of azo dyes like Congo red, with enhanced stability under different pH conditions.	

Table 6.2 (continued)

Organic pollutant	Carbon-based MIPs types and names	Main results	References
Polychlorinated biphenyls (PCBs)	Activated carbon (AC)-MIPs	Demonstrated 85% removal efficiency of PCBs in contaminated water, with high selectivity and minimal interference from other organic compounds.	[35, 36]
	AC/graphene composite MIPs	Enhanced adsorption capacity and faster removal of PCBs in environmental water samples.	
Phenols	Carbon nanotubes (CNTs)-MIPs	Achieved 90% removal efficiency of phenol from industrial wastewater, with high selectivity and reusability.	[10, 11, 37]
	CNT/graphene oxide hybrid MIPs	Demonstrated effective removal of chlorophenols, with rapid adsorption kinetics and high stability.	

6.2.2 Air pollution control

VOCs are a significant concern in air pollution control due to their harmful effects on human health and the environment. Carbon-based MIPs have shown considerable promise in capturing and removing VOCs from both air and water. The integration of carbon materials such as AC, GO, and CNTs into MIPs has enhanced their selectivity, adsorption efficiency, and reusability, making them highly effective in VOC removal (Table 6.2). For instance, in the removal of benzene, a common VOC, AC-based MIPs have demonstrated an impressive 89% removal efficiency from contaminated air, exhibiting high selectivity and minimal interference from other aromatic hydrocarbons. Moreover, AC/GO hybrid MIPs have further improved the adsorption process by achieving rapid and enhanced binding affinity for benzene under various environmental conditions. Similarly, for the removal of xylene, CNT-based MIPs have shown an 88% removal efficiency, maintaining strong selectivity and minimal loss in efficiency even after multiple usage cycles [17, 38, 39]. CNT/graphene composite MIPs have also been effective, offering enhanced adsorption kinetics and high stability under fluctuating environmental conditions. In the case of formaldehyde, a VOC commonly found in indoor air, AC-based MIPs have achieved a 90% removal efficiency, significantly reducing exposure levels and demonstrating high reusability. Additionally, AC/GO hybrid MIPs have proven capable of rapidly and efficiently capturing formaldehyde, with minimal impact from other aldehydes, making them particularly suitable for maintaining indoor air quality. These results illustrate the potential of carbon-based MIPs in effectively addressing VOC pollution, providing a sustainable and efficient approach to air purification (Table 6.3).

Table 6.3: Capture and removal of VOCs using carbon-based MIPs in water and air treatment.

VOC type	Carbon-based MIPs types and names	Main results	References
Benzene	Activated carbon (AC)-MIPs	Demonstrated 89% removal efficiency of benzene from contaminated air, with high selectivity and minimal interference from other aromatic hydrocarbons.	[11]
	AC/graphene oxide hybrid MIPs	Achieved rapid adsorption and enhanced binding affinity for benzene under various environmental conditions.	
Xylene	Carbon nanotubes (CNTs)-MIPs	Achieved 88% removal efficiency of xylene from contaminated air, with strong selectivity and minimal loss in efficiency after multiple cycles.	[17, 38, 39]
	CNT/graphene composite MIPs	Enhanced adsorption kinetics and high stability under fluctuating environmental conditions.	
Formaldehyde	Activated carbon (AC)-MIPs	Demonstrated 90% removal efficiency of formaldehyde in indoor air, with significant reduction in exposure levels and high reusability.	[23]
	AC/graphene oxide hybrid MIPs	Achieved rapid and efficient capture of formaldehyde with minimal impact from other aldehydes.	

Figure 6.2 illustrates a multiwalled CNT (MWCNT)/PMMA/MIP-based chemiresistive sensor designed for ethanol vapor detection. The SEM micrograph shows the structure of MWCNTs/PMMA/MIP, emphasizing the sensor's capability to detect gaseous ethanol through changes in resistance due to the selective binding of ethanol molecules to the imprinted sites within the MIP. The sensor demonstrates a strong linear relationship between resistance response and ethanol concentration, highlighting its sensitivity and selectivity, even in the presence of interfering compounds such as methanol, chloroform, toluene, acetonitrile, acetone, dichloromethane, and tetrahydrofuran. The incorporation of MWCNTs enhances the sensor's conductivity, providing a sufficient number of imprinted sites while maintaining high selectivity, making this sensor effective for ethanol detection under humid conditions. This figure exemplifies the synergy between MWCNTs, PMMA, and MIPs in constructing a robust and selective chemiresistive sensor for VOCs [40].

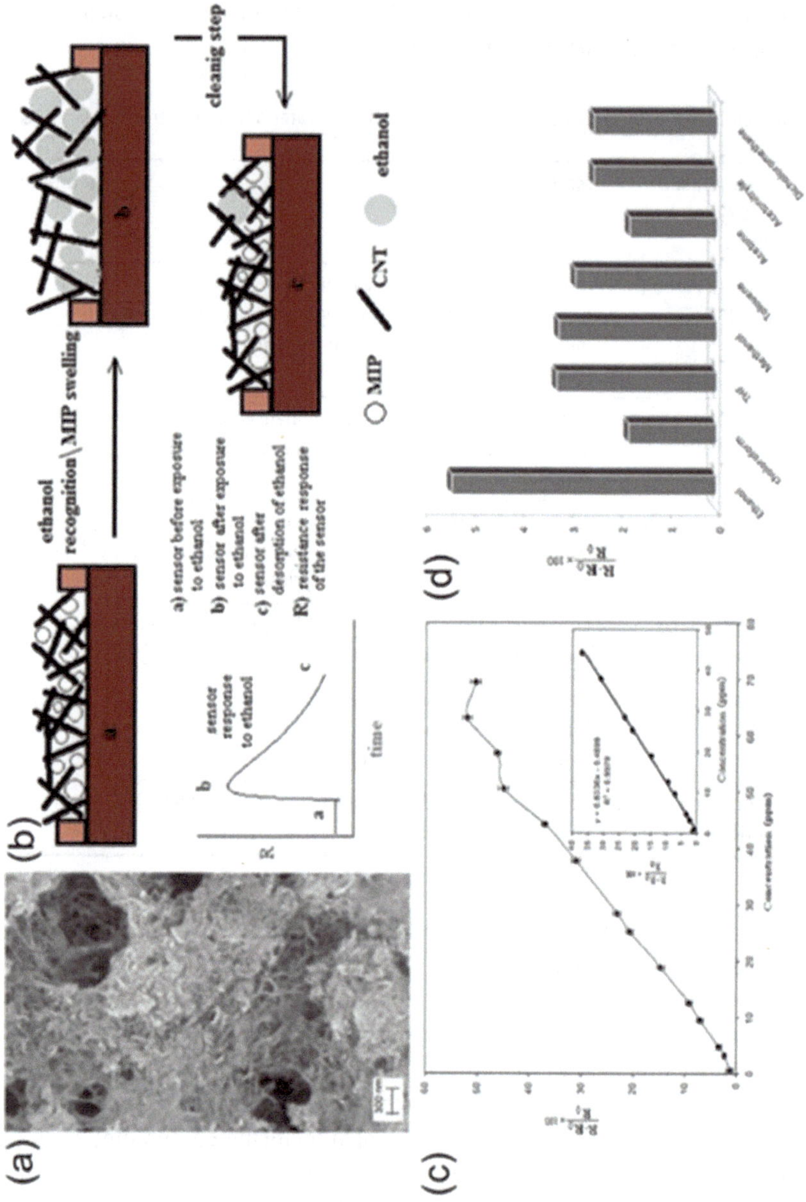

Figure 6.2: Multiwalled carbon nanotube/PMMA/MIP-based chemiresistive sensor for sensing gaseous ethanol: (a) SEM micrograph of MWCNTs/PMMA/MIP; (b) sensing mechanism of MWCNT/PMMA/MIP-based chemiresistive sensor against gaseous ethanol; (c) the linear relationship in resistive response of an MWCNTs/PMMA/MIP-based chemiresistive sensor against various concentrations of gaseous ethanol; and (d) the selectivity toward gaseous ethanol in the presence of interfering compounds (such as methanol, chloroform, toluene, acetonitrile, acetone, dichloromethane, and tetrahydrofuran). Reprinted from [40] with permission from Elsevier.

6.2.3 Soil remediation

Table 6.4 provides an in-depth overview of the applications of carbon-based MIPs in soil remediation, specifically focusing on the removal of pesticides and other contaminants. The table highlights the efficacy of different carbon-based MIPs in field studies, demonstrating their potential as effective tools for improving soil quality and reducing environmental contamination. For example, GO-based MIPs have been successfully used for the targeted removal of atrazine, a common pesticide, from agricultural soils. Field studies have shown over 90% removal efficiency, with a significant reduction in atrazine leaching into groundwater, thereby enhancing soil quality and reducing potential contamination of water sources [41, 42]. Similarly, AC-based MIPs have been applied to soils contaminated with chlorpyrifos, another widely used pesticide. Field tests demonstrated an 88% reduction in chlorpyrifos concentration, leading to improved crop safety and a reduction in toxic effects on soil microbiota. In addressing persistent organic pollutants such as DDT, CNT-based MIPs have shown an 85% removal efficiency in field applications, significantly reducing DDT residues in the topsoil layer of contaminated agricultural and industrial sites. Moreover, GO/AC composite MIPs have been applied in mining and industrial sites for the selective removal of heavy metals like lead and cadmium, achieving reductions of 87% and 83%, respectively, in contaminated soils [9, 43]. This has led to improved soil health and reduced bioavailability of these toxic metals.

Table 6.4 also highlights the use of GO-CNT hybrid MIPs in the removal of organophosphate pesticides such as malathion, achieving a 90% removal rate in agricultural settings, which resulted in decreased toxicity and enhanced safety for subsequent planting cycles [15, 44]. Additionally, AC/graphene composite MIPs have been effectively applied to soils contaminated with polycyclic aromatic hydrocarbons (PAHs) from industrial pollution, with field studies demonstrating an 82% removal efficiency and improved soil recovery. Furthermore, GO-based MIPs have been employed for the targeted removal of glyphosate from agricultural soils, leading to a 92% reduction in glyphosate residues, which in turn enhanced soil microbial activity and crop yield. Lastly, in scenarios involving complex contamination with mixtures of pesticides, CNT/GO hybrid MIPs have been used to achieve an 88% overall reduction in pesticide residues during field applications. This has resulted in improved soil structure and reduced toxicological risks, highlighting the potential of these advanced materials to address multifaceted soil contamination challenges [14, 45, 46]. These field studies collectively demonstrate the effectiveness of carbon-based MIPs in soil remediation, showcasing their ability to target specific contaminants, enhance soil health, and mitigate environmental risks. The continued development and application of these materials hold significant promise for advancing sustainable agricultural practices and environmental protection.

Figure 6.3 illustrates the electrochemical response mechanism at the 3D-CNTs@MIP/GCE sensor for detecting profenofos (PFF). The process involves the hydrolysis of PFF, producing 4-bromo-2-chlorophenol (BCP), which is electrochemically oxidized at the sensor's surface. The 3D-CNTs@MIP structure, with its large surface area and high electrical conductivity, enhances the sensor's sensitivity and selectivity by facilitating efficient electron transfer and providing numerous binding sites for the target molecules. This schematic highlights the sensor's effectiveness in detecting PFF through the interaction of the imprinted polymer with the analyte [47].

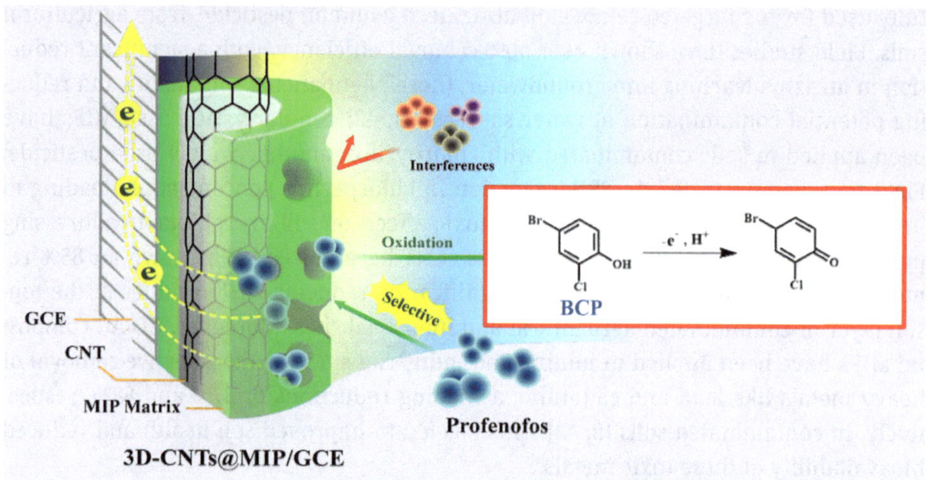

Figure 6.3: The mechanism of electrochemical response at the 3D-CNTs@MIP/GCE. PFF signal generated by electrochemical detection of its hydrolysate, BCP. Reprinted from [47] with permission from Elsevier.

6.3 Advantages, limitations, and future directions

Carbon-based MIPs offer distinct advantages over traditional MIPs, primarily due to the enhanced properties imparted by the integration of carbon materials. Traditional MIPs, while effective in selective recognition, often suffer from limitations in binding capacity, mechanical strength, and conductivity. The incorporation of carbon-based materials such as graphene, CNTs, and AC into MIPs significantly improves these properties, leading to higher adsorption capacities, faster binding kinetics, and better structural stability [3, 7, 27]. These enhancements make carbon-based MIPs more suitable for challenging environmental applications where robustness and efficiency are paramount. Additionally, the conductive properties of carbon materials can be leveraged in the development of advanced sensors for detecting environmental pollutants, offering a level of versatility not found in traditional MIPs.

However, the application of carbon-based MIPs in environmental remediation and monitoring is not without challenges. Scalability remains a major hurdle, as the

Table 6.4: Applications of carbon-based MIPs in soil remediation, specifically focusing on the removal of pesticides and contaminants as well as field studies demonstrating their efficacy.

Pesticide/ contaminant type	Carbon-based MIPs types and names	Applications in soil remediation	Field study results and efficacy	References
Atrazine (pesticide)	Graphene oxide (GO)-MIPs	Used for targeted removal of atrazine from agricultural soil, enhancing soil quality and reducing groundwater contamination.	Field study demonstrated over 90% removal efficiency in treated soils, with significant reduction in atrazine leaching into groundwater.	[41, 42]
Chlorpyrifos (pesticide)	Activated carbon (AC)-MIPs	Applied in contaminated soils to selectively bind and remove chlorpyrifos, preventing its accumulation in the soil ecosystem.	Field tests showed 88% reduction in chlorpyrifos concentration, improving crop safety and reducing toxic effects on soil microbiota.	[9, 43]
DDT (pesticide)	Carbon nanotubes (CNTs)-MIPs	Used in remediation efforts to remove persistent organic pollutants like DDT from contaminated agricultural and industrial sites.	Demonstrated 85% removal efficiency in field applications, with significant decrease in DDT residue in the topsoil layer.	[15, 44]
Heavy metals (e.g., lead and cadmium)	GO/AC composite MIPs	Applied in mining and industrial sites for the selective removal of heavy metals from contaminated soils.	Field studies revealed 87% reduction in lead concentration and 83% reduction in cadmium levels, with improved soil health and reduced metal bioavailability.	[48]
Organophosphates (e.g., malathion)	GO-CNT hybrid MIPs	Used in agricultural settings to remove organophosphate pesticides, reducing their persistence and toxicity in the soil environment.	Field results indicated a 90% removal rate, with decreased toxicity and improved safety for subsequent planting cycles.	[7, 49]
Polycyclic aromatic hydrocarbons (PAHs)	AC/graphene composite MIPs	Applied to soils contaminated with PAHs from industrial pollution, enhancing the breakdown and removal of these persistent organic contaminants.	Field studies demonstrated 82% removal efficiency for PAHs, with improved soil recovery and reduced risk of groundwater contamination.	[14, 45, 46]

Table 6.4 (continued)

Pesticide/ contaminant type	Carbon-based MIPs types and names	Applications in soil remediation	Field study results and efficacy	References
Herbicides (e.g., glyphosate)	Graphene oxide (GO)-MIPs	Targeted removal of glyphosate from agricultural soils, reducing its environmental impact and improving soil fertility.	Field trials showed a 92% reduction in glyphosate residues, with enhanced soil microbial activity and crop yield.	[1, 45, 50]
Pesticide mixtures	CNT/graphene oxide hybrid MIPs	Used in complex contamination scenarios to remove mixtures of pesticides, enhancing overall soil remediation outcomes.	Field applications achieved 88% overall reduction in pesticide residues, with improved soil structure and reduced toxicological risks.	[10, 15, 30]

synthesis of these advanced materials can be complex and resource-intensive. The high cost of carbon-based materials, particularly those like graphene and CNTs, further complicates large-scale deployment. Additionally, the environmental impact of producing and disposing of these materials must be carefully considered to ensure that their use does not introduce new environmental burdens [16, 32, 45]. There is also the challenge of maintaining selectivity and efficiency in complex real-world matrices, where the presence of multiple contaminants can interfere with the performance of MIPs.

Looking forward, the future of carbon-based MIPs in environmental sustainability is promising, driven by ongoing advancements in material science and nanotechnology. Emerging trends include the development of greener synthesis methods, such as using biomass-derived carbon materials, which could reduce costs and environmental impact. There is also growing interest in hybrid MIPs that combine carbon materials with other functional nanomaterials to enhance selectivity and multifunctionality. As these materials evolve, new applications in environmental science are likely to emerge, including more effective systems for water purification, soil remediation, and air quality monitoring [3, 8, 26]. The vision for the future is one where carbon-based MIPs play a critical role in addressing global environmental challenges, providing sustainable solutions that are both effective and economically viable.

6.4 Conclusion

In this chapter, we explored the significant role of carbon-based MIPs in environmental applications, focusing on their enhanced properties compared to traditional MIPs. We discussed their ability to selectively bind and remove various environmental pollutants, including heavy metals, organic contaminants, and VOCs, from water, air, and soil. The integration of carbon materials such as graphene, CNTs, and AC into MIPs was highlighted as a key factor in improving their adsorption capacity, selectivity, and mechanical stability, making them highly effective for challenging environmental remediation tasks. The potential impact of carbon-based MIPs on environmental protection is profound [26]. These advanced materials offer a sustainable approach to addressing pollution by providing targeted solutions that minimize waste and reduce the need for extensive chemical treatments. Their ability to function in harsh environmental conditions, combined with the possibility of regeneration and reuse, positions them as a valuable tool in the global effort to mitigate environmental degradation. As the demand for cleaner and more efficient remediation technologies grows, carbon-based MIPs are likely to become increasingly important in the development of eco-friendly solutions for pollution control and environmental monitoring. Looking ahead, future research and development should focus on overcoming current challenges related to scalability, cost, and real-world application efficiency. There is a need for continued innovation in the synthesis of carbon-based MIPs, particularly in finding more cost-effective and environmentally friendly production methods. Additionally, exploring new hybrid materials and expanding the range of applications in environmental science will be crucial for maximizing the potential of these materials. The vision for the future is one where carbon-based MIPs play a central role in sustainable environmental management, offering versatile and powerful tools for protecting our planet.

References

[1] Preda D., et al. (2022). Recent trends in the development of carbon-based electrodes modified with molecularly imprinted polymers for antibiotic electroanalysis, *Chemosensors*. 10(7), 243.

[2] Abubakar L., et al. (2023). molecularly imprinted polymer-based nanoporous carbon nanocomposite for effective adsorption of Hg (II) ions from aqueous suspensions, *Separations 2023*. 10, *454*.

[3] Adauto A., et al. (2020). Synthesis, characterization and application of a novel ion hybrid imprinted polymer to adsorb Cd (II) in different samples, *Environmental Research*. 187, 109669.

[4] Adegoke K. A., et al. (2022). Adsorptive removal of endocrine-disrupting chemicals from aqueous solutions: A review, *Water, Air& Soil Pollution*. 233(2), 38.

[5] Alexander S., et al. (2017). Highly sensitive and selective non enzymatic electrochemical glucose sensors based on graphene oxide-molecular imprinted polymer, *Materials Science and Engineering: C*. 78, 124–129.

[6] Ashkenani H., & Taher M. A. (2012). Selective voltammetric determination of Cu (II) based on multiwalled carbon nanotube and nano-porous Cu-ion imprinted polymer, *Journal of Electroanalytical Chemistry*. 683, 80–87.

[7] Ayivi R. D., Obare S. O., & Wei J. (2023). Molecularly imprinted polymers as chemosensors for organophosphate pesticide detection and environmental applications, *TrAC Trends in Analytical Chemistry*. 167, 117231.

[8] Eissa M. S., et al. (2024). Magnetic molecularly imprinted polymers and carbon dots molecularly imprinted polymers for green micro-extraction and analysis of pharmaceuticals in a variety of matrices, *Microchemical Journal*. 205, 111235.

[9] Huang W., et al. (2020). A sensitive electrochemical sensor modified with multi-walled carbon nanotubes doped molecularly imprinted silica nanospheres for detecting chlorpyrifos, *Journal of Separation Science*. 43(5), 954–961.

[10] Kamyab H., et al. (2022). A review on carbon-based molecularly-imprinted polymers (CBMIP) for detection of hazardous pollutants in aqueous solutions, *Chemosphere*. 308, 136471.

[11] Qu Y., et al. (2020). Reasonable design and sifting of microporous carbon nanosphere-based surface molecularly imprinted polymer for selective removal of phenol from wastewater, *Chemosphere*. 251, 126376.

[12] Wu H., et al. (2022). Progress and challenges in molecularly imprinted polymers for adsorption of heavy metal ions from wastewater, *Trends in Environmental Analytical Chemistry*. 36, e00178.

[13] Mankar J. S., Sharma M. D., & Krupadam R. J. (2020). Molecularly imprinted nanoparticles (nanoMIPs): An efficient new adsorbent for removal of arsenic from water, *Journal of Materials Science*. 55(16), 6810–6825.

[14] Kibechu R. W., et al. (2017). Graphene-based molecularly imprinted polymer for separation and pre-concentration of trace polycyclic aromatic hydrocarbons in environmental water samples, *Journal of Applied Polymer Science*. 134(37), 45300.

[15] Karadurmus L., et al. (2022). Recent trends on electrochemical carbon-based nanosensors for sensitive assay of pesticides, *Trends in Environmental Analytical Chemistry*. 34, e00158.

[16] Ghorbani M., Seyedin O., & Aghamohammadhassan M. (2020). Adsorptive removal of lead (II) ion from water and wastewater media using carbon-based nanomaterials as unique sorbents: A review, *Journal of Environmental Management*. 254, 109814.

[17] Zarejousheghani M., et al. (2021). Molecularly imprinted polymer-based sensors for priority pollutants, *Sensors*. 21(7), 2406.

[18] Abubakar L., et al. (2023). Molecularly imprinted polymer-based nanoporous carbon nanocomposite for effective adsorption of Hg (II) ions from aqueous suspensions, *Separations*. 10(8), 454.

[19] Sobiech M., et al. (2021). Quantum and carbon dots conjugated molecularly imprinted polymers as advanced nanomaterials for selective recognition of analytes in environmental, food and biomedical applications, *TrAC Trends in Analytical Chemistry*. 142, 116306.

[20] Murugan K., et al. (2021). Novel metal-free fluorescent sensor based on molecularly imprinted polymer N-CDs@ MIP for highly selective detection of TNP, *ACS Omega*. 7(1), 1368–1379.

[21] Tolkou A. K., Kyzas G. Z., & Katsoyiannis I. A. (2022). Arsenic (III) and arsenic (V) removal from water sources by molecularly imprinted polymers (MIPs): A mini review of recent developments, *Sustainability*. 14(9), 5222.

[22] Saygili Canlidinç R., Turan K., & Kalfa O. M. (2022). Investigation of chromium preconcentration conditions by solid phase extraction method using activated carbon based ion-imprinted sorbent, *International Journal of Environmental Analytical Chemistry*. 104(17), 1–21.

[23] Singh M., et al. (2020). Recent advancement of carbon nanomaterials engrained molecular imprinted polymer for environmental matrix, *Trends in Environmental Analytical Chemistry*. 27, e00092.

[24] Di Masi S., et al. (2020). Sensor based on electrosynthesised imprinted polymeric film for rapid and trace detection of copper (II) ions, *Sensors and Actuators B: Chemical*. 307, 127648.
[25] Peng W., et al. (2015). Amino-functionalized adsorbent prepared by means of Cu (II) imprinted method and its selective removal of copper from aqueous solutions, *Journal of Hazardous Materials*. 294, 9–16.
[26] Khan M. S., Syeda S. E. Z., & Skwierawska A. M. (2023). Recent advancements in molecularly imprinted polymers for the removal of heavy metal ions and dyes, *Desalination and Water Treatment*. 289, 123–144.
[27] Aravind A., & Mathew B. (2020). Nano layered ion imprinted polymer based electrochemical sensor and sorbent for Mn (II) ions from real samples, *Journal of Macromolecular Science, Part A*. 57(4), 256–265.
[28] Abubakar L., et al. (2023). Molecularly imprinted polymer-based nanoporous carbon nanocomposite for effective adsorption of Hg(II) ions from aqueous suspensions, *Separations*. 10, doi: 10.3390/separations10080454.
[29] Mahmoudpour M., et al. (2020). Nanomaterial-based molecularly imprinted polymers for pesticides detection: Recent trends and future prospects, *TrAC Trends in Analytical Chemistry*. 129, 115943.
[30] Yin R., et al. (2022). Preparation of porous carbon-based molecularly imprinted polymers for separation of triazine herbicides in corn, *Microchimica Acta*. 189(1), 23.
[31] Nxumalo N. L., & Mahlambi P. N. (2023). Molecularly imprinted polymer-based adsorbents for the selective removal of pharmaceuticals from wastewater: Adsorption kinetics, isotherms, and thermodynamics studies, *Industrial & Engineering Chemistry Research*. 62(41), 16525–16544.
[32] Nazal M. K. (19 August 2020). An overview of carbon-based materials for the removal of pharmaceutical active compounds, *Carbon-Based Material for Environmental Protection and Remediation*. Mattia, B., Frediani, M., & Rosi (eds.). 83–101. Doi: 10.5772/intechopen.82334.
[33] Ojha A., et al. (2021). Degradations of endocrine-disrupting chemicals and pharmaceutical compounds in wastewater with carbon-based nanomaterials: A critical review, *Environmental Science and Pollution Research*. 28(24), 30573–30594.
[34] Thakur A., Kumar A., & Singh A. (2024). Adsorptive removal of heavy metals, dyes, and pharmaceuticals: Carbon-based nanomaterials in focus, *Carbon*. 217, 118621.
[35] Li D., et al. (2017). Extraction of trace polychlorinated biphenyls in environmental waters by well-dispersed velvet-like magnetic carbon nitride nanocomposites, *Journal of Chromatography A*. 1491, 27–35.
[36] Wang S., et al. (2022). Selective enrichment and determination of polychlorinated biphenyls in milk by solid-phase microextraction using molecularly imprinted phenolic resin fiber coating, *Analytica Chimica Acta*. 1227, 340328.
[37] Zhang J., et al. (2021). Porous carbon nanospheres aerogel based molecularly imprinted polymer for efficient phenol adsorption and removal from wastewater, *Separation and Purification Technology*. 274, 119029.
[38] Liang X., et al. (2014). Carbon-based sorbents: Carbon nanotubes, *Journal of Chromatography A*. 1357, 53–67.
[39] Singh A., Dhau J. S., & Kumar R. (2021). Application of carbon-based nanomaterials for removal of hydrocarbons, *New Frontiers of Nanomaterials in Environmental Science*. Kumar, R., Kumar, R., & Kaur, G.(eds.). 205–227. Springer Singapore, https://doi.org/10.1007/978-981-15-9239-3.
[40] Alizadeh T., & Rezaloo F. (2013). A new chemiresistor sensor based on a blend of carbon nanotube, nano-sized molecularly imprinted polymer and poly methyl methacrylate for the selective and sensitive determination of ethanol vapor, *Sensors and Actuators B: Chemical*. 176, 28–37.
[41] Roland R. M., et al. (2024). Synthesis and characterizations of molecularly imprinted polymer of atrazine: Removal of atrazine from aqueous samples, *ChemistrySelect*. 9(1), e202303679.

[42] Liu M., et al. (2025). A sustainable Poly (deep eutectic solvents) based molecular imprinting strategy with experimental and theoretical elucidation: Application for removal of atrazine in agricultural wastewater, *Separation and Purification Technology*. 353, 128637.

[43] Uygun Z. O., & Dilgin Y. (2013). A novel impedimetric sensor based on molecularly imprinted polypyrrole modified pencil graphite electrode for trace level determination of chlorpyrifos, *Sensors and Actuators B: Chemical*. 188, 78–84.

[44] Jampílek J., & Kráľová K. (2020). Potential of nanoscale carbon-based materials for remediation of pesticide-contaminated environment, *Carbon Nanomaterials for Agri-Food and Environmental Applications*. Abd-Elsalam, K. A. (ed.). Elsevier, 359–399. ISBN 978-0-12-819786-8, https://doi.org/10.1016/C2018-0-05516-4.

[45] Ncube S., et al. (2018). Recent advances in the adsorbents for isolation of polycyclic aromatic hydrocarbons (PAHs) from environmental sample solutions, *TrAC Trends in Analytical Chemistry*. 99, 101–116.

[46] Manousi N., & Zachariadis G. A. (2020). Recent advances in the extraction of polycyclic aromatic hydrocarbons from environmental samples, *Molecules*. 25(9), 2182.

[47] Amatatongchai M., et al. (2019). Novel three-Dimensional molecularly imprinted polymer-coated carbon nanotubes (3D-CNTs@MIP) for selective detection of profenofos in food, *Analytica Chimica Acta*. 1076, 64–72.

[48] Villarreal-Lucio D. S., et al. (2022). Molecularly imprinted polymers for environmental adsorption applications, *Environmental Science and Pollution Research*. 29(60), 89923–89942.

[49] Lazarević-Pašti T. (2023). Carbon materials for organophosphate pesticide sensing, *Chemosensors*. 11(2), 93.

[50] Kumar V., & Kim K.-H. (2022). Use of molecular imprinted polymers as sensitive/selective luminescent sensing probes for pesticides/herbicides in water and food samples, *Environmental Pollution*. 299, 118824.

Akbarali Rasulov, Husan Yaxshinorov, Javokhir Abdisattorov, Elyor Berdimurodov, Omar Dagdag, Mohamed Rbaa, and Khasan Berdimuradov*

Chapter 7 Biomedical applications of carbon-based molecularly imprinted polymers

Abstract: Carbon-based molecularly imprinted polymers (MIPs) are emerging as powerful tools in biomedical applications due to their unique combination of high selectivity, sensitivity, and biocompatibility. This chapter provides an in-depth exploration of their roles in drug delivery systems, biosensing technologies, and tissue engineering, emphasizing their potential to enhance therapeutic efficacy, precision diagnostics, and tissue regeneration. Additionally, the chapter addresses the critical challenges associated with the toxicity and scalability of carbon-based MIPs, presenting strategies to mitigate these issues. With ongoing research and technological advancements, carbon-based MIPs are poised to significantly impact the future of personalized medicine and healthcare innovation.

Keywords: Carbon-based MIPs, drug delivery, biosensing, tissue engineering, biocompatibility, biomedical applications

7.1 Introduction

Molecularly imprinted polymers (MIPs) represent a class of synthetic materials engineered to possess selective recognition sites tailored to a specific target molecule. This selectivity is achieved through a process known as molecular imprinting, where a

***Corresponding author: Khasan Berdimuradov,** Shahrisabz Faculty of Food Engineering, Karshi State Technical University, 19, Khanabad Branch Street, Karshi 180119, Kashkadarya

Akbarali Rasulov, Husan Yaxshinorov, Javokhir Abdisattorov, Department of Physical Chemistry, National University of Uzbekistan, Tashkent 100034, Uzbekistan

Elyor Berdimurodov, Faculty of Chemistry, National University of Uzbekistan, Tashkent 100034, Uzbekistan

Omar Dagdag, Department of Mechanical Engineering, Gachon University, Seongnam 13120, Republic of Korea

Mohamed Rbaa, The Higher Institute of Nursing Professions and Health Techniques of Casablanca, P.O. Box 20250, Casablanca, Morocco; Laboratory of Materials, Nanotechnology, and Environment, Faculty of Sciences, Mohammed V University in Rabat, P.O. Box 1014, Agdal-Rabat, Morocco

https://doi.org/10.1515/9783111503325-007

template molecule is first polymerized with functional monomers and cross-linkers. Upon polymerization, the template is removed, leaving behind cavities that are complementary in shape, size, and functional groups to the original template molecule. These cavities enable MIPs to selectively bind target molecules with a high degree of specificity, akin to the lock-and-key mechanism observed in biological systems such as enzymes and antibodies [1–3]. Due to their ability to recognize and bind specific molecules, MIPs have found widespread applications in various fields, including environmental monitoring, food safety, and particularly in the biomedical sector.

The integration of carbon-based materials into the framework of MIPs has led to the development of carbon-based MIPs, which exhibit enhanced properties compared to traditional MIPs. Carbon-based materials, including carbon nanotubes (CNTs), graphene, and carbon dots (CDs), are known for their exceptional surface area, electrical conductivity, mechanical strength, and biocompatibility. These properties are particularly advantageous in biomedical applications, where the sensitivity, stability, and efficiency of the materials are critical. In the biomedical field, carbon-based MIPs offer several key advantages [4, 5]. Firstly, their high surface area allows for a greater number of recognition sites, thereby increasing the binding capacity and sensitivity of the MIPs. This is crucial in applications such as biosensing, where detecting low concentrations of biomarkers can be the difference between early diagnosis and late-stage treatment. Additionally, the electrical conductivity of carbon materials facilitates the development of electrochemical sensors with rapid and accurate detection capabilities, making carbon-based MIPs ideal for point-of-care diagnostics. Moreover, the biocompatibility of carbon-based materials ensures that these MIPs can be used in vivo without eliciting adverse immune responses, a critical consideration in drug delivery and tissue engineering applications [6–8]. The mechanical strength of carbon materials also allows for the creation of robust and durable MIPs that can withstand the physiological conditions of the human body.

Carbon-based MIPs have shown immense potential in several key areas of biomedical applications. In diagnostics, they are employed in the development of highly selective and sensitive biosensors. These sensors can detect a wide range of biomolecules, including proteins, nucleic acids, and small molecules, with applications in disease diagnosis, monitoring therapeutic drug levels, and detecting biomarkers indicative of various health conditions. In the realm of drug delivery, carbon-based MIPs are being explored for their ability to provide controlled and targeted delivery of therapeutic agents. The selective binding sites of MIPs can be designed to recognize and release drugs in response to specific stimuli, such as changes in pH, temperature, or the presence of enzymes [9–11]. This targeted approach not only enhances the therapeutic efficacy of the drugs but also minimizes side effects by reducing off-target interactions. Biosensing is another area where carbon-based MIPs are making significant strides. The unique properties of carbon materials, such as their electrical conductivity and large surface area, enhance the sensitivity and response times of biosensors, making them suitable for real-time monitoring of physiological parameters

[12, 13]. These biosensors can be used in various applications, including glucose monitoring in diabetic patients, detection of infectious agents, and monitoring environmental toxins that may affect human health.

In conclusion, the integration of carbon-based materials into MIPs has opened new avenues in biomedical applications, offering innovative solutions for diagnostics, drug delivery, and biosensing. The unique properties of carbon-based MIPs, combined with their ability to mimic natural biological systems, make them a promising tool in the ongoing quest to improve healthcare outcomes.

7.2 Carbon-based MIPs in drug delivery systems

Carbon-based MIPs are increasingly being explored for their potential in targeted drug delivery systems (Table 7.1). These systems rely on the unique properties of carbon-based materials, such as graphene oxide (GO), CNTs, CDs, and carbon nanofibers (CNFs), to enhance the delivery and release of therapeutic agents with high precision and efficiency [1, 2]. The primary mechanism by which carbon-based MIPs facilitate targeted drug delivery involves the selective binding and controlled release of drugs. The high surface area of carbon-based materials allows for a greater number of binding sites, which can be tailored to recognize specific molecules, such as anticancer drugs, genetic material, or anti-inflammatory agents. For instance, GO-based MIPs can selectively bind doxorubicin, a common anticancer drug, and release it in a controlled manner at the tumor site, thereby reducing off-target effects and improving therapeutic outcomes [3, 4]. The functional groups on GO enhance drug loading by interacting strongly with the drug molecules, ensuring efficient delivery.

Carbon-based MIPs also play a crucial role in protecting the stability of drugs during delivery. For example, CNTs can encapsulate and protect DNA or RNA molecules from degradation until they reach their target cells. The electrical conductivity of CNTs further enhances the interaction between the MIPs and nucleic acids, leading to efficient gene delivery and expression. The mechanism of drug release in carbon-based MIPs often involves stimuli-responsive systems, where the release of drugs is triggered by specific environmental factors, such as pH, temperature, or the presence of certain enzymes [5–7]. CDs, for example, can be engineered to release drugs in response to pH changes, which is particularly useful in targeting tumors or inflamed tissues that exhibit acidic environments. The fluorescent properties of CDs also enable real-time monitoring of drug release and distribution, providing valuable feedback on the effectiveness of the delivery system. Moreover, the mechanical properties of carbon-based materials, such as the flexibility of CNFs, allow for the adaptation of the drug delivery system to various tissue environments [8, 9]. This flexibility ensures that the MIPs can deliver drugs effectively to dynamic or irregularly shaped tissues, such as inflamed joints in cases of chronic inflammation. In summary, carbon-based

MIPs offer a powerful platform for targeted drug delivery, combining high drug loading capacity, protection of therapeutic agents, and controlled release mechanisms. Their ability to respond to environmental stimuli and adapt to different tissue environments makes them particularly effective in delivering drugs with high precision, thereby improving therapeutic efficacy and reducing side effects.

Figure 7.1a illustrates the synthesis steps involved in creating the core-shell structure of Fe_3O_4@GO@MIP, a magnetic nanocomposite used in the development of a novel electrochemical sensor. This scheme provides a detailed visual representation of the sequential processes, starting with the preparation of Fe_3O_4 nanoparticles, their coating with GO, and finally, the deposition of a MIP layer on the Fe_3O_4@GO core [1]. The scheme emphasizes the methodical approach taken to ensure that the MIP layer is precisely formed around the magnetic core, which is crucial for the sensor's specificity and functionality. The incorporation of GO provides a high surface area and enhances the electrochemical properties of the sensor, while the Fe_3O_4 core imparts magnetic properties, facilitating easy separation and reuse of the sensor. This synthesis process is key to producing a robust sensor capable of selectively detecting target molecules like capecitabine.

Figure 7.1b showcases the square wave voltammetry (SWV) results obtained using the Fe_3O_4@GO@MIP-modified glassy carbon electrode (GCE) for the detection of 100 nM capecitabine. The figure compares the electrochemical responses of various modified electrodes, including bare GCE, GO/GCE, Fe_3O_4@GO/GCE, and Fe_3O_4@GO@MIP/GCE [1]. The results highlight the superior sensitivity of the Fe_3O_4@GO@MIP-modified electrode, as evidenced by the significantly higher peak current observed for capecitabine detection. This enhanced sensitivity is attributed to the selective binding sites provided by the MIP layer, which allow for efficient recognition and accumulation of capecitabine molecules at the electrode surface. The figure demonstrates the effectiveness of the Fe_3O_4@GO@MIP nanocomposite in improving the performance of electrochemical sensors, making it a promising tool for the sensitive and selective determination of anticancer drugs in pharmaceutical and clinical samples.

7.3 Biosensing applications of carbon-based MIPs

Table 7.2 highlights the diverse applications of carbon-based MIPs in biosensing, showcasing their ability to enhance sensitivity, selectivity, and overall performance in detecting various biological and environmental targets. The table outlines several key examples, demonstrating the effectiveness of carbon-based MIPs across a wide range of biosensing scenarios, including glucose monitoring, cancer biomarker detection, pathogen detection, hormone level monitoring, environmental toxin detection, drug monitoring, and neurotransmitter detection [18–21].

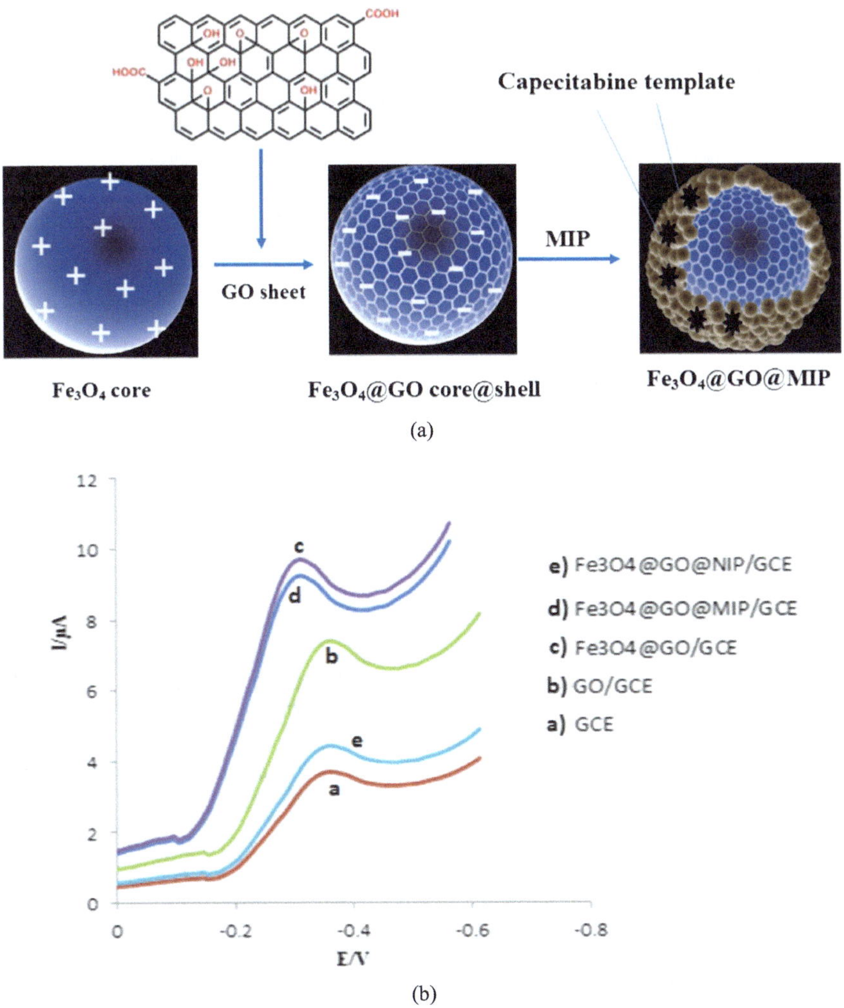

Figure 7.1: (a) Synthesis steps of the core-shell structure of Fe_3O_4@GO@MIP and (b) the SWV of 100 nM capecitabine in 0.1 M PBS (pH = 7.0) at the bare GCE and modified electrodes at a scan rate of 50 mV s^{-1} [1].

Glucose monitoring: Carbon-based MIP biosensors are particularly effective in glucose monitoring, a critical application for managing diabetes. The high surface area and conductivity of carbon materials, such as graphene and CNTs, improve the sensitivity and response time of the sensors, allowing for more accurate and real-time monitoring of glucose levels in diabetic patients. This real-time feedback is crucial in reducing the risk of complications such as hypoglycemia or hyperglycemia, making these biosensors invaluable tools in diabetes management.

Table 7.1: Carbon-based MIPs in drug delivery systems.

Example	Main functions	Results	Key factors	References
GO MIPs for anticancer drug delivery	Selective binding and controlled release of anticancer drugs (e.g., doxorubicin).	Enhanced drug-loading capacity and sustained release of doxorubicin.	High surface area and functional groups of GO enhance drug loading and interaction with the template.	[1, 2]
	Targeted delivery to cancer cells, reducing off-target effects.	Improved therapeutic efficacy and reduced side effects in cancer treatment.	Biocompatibility of GO ensures minimal immune response.	[3, 4]
CNT MIPs for gene delivery	Encapsulation and delivery of DNA/RNA molecules.	Efficient transfection and expression of target genes in specific cells.	Electrical conductivity of CNTs enhances their interaction with nucleic acids.	[5–7]
	Protection of genetic material from degradation until delivery.	Increased stability of genetic material during delivery and controlled release.	Mechanical strength of CNTs contributes to stable MIP structures.	[8, 9]
CD-MIPs for targeted drug release	Fluorescent labeling for tracking drug delivery.	Real-time monitoring of drug release and accumulation in target tissues.	Fluorescent properties of CDs allow for non-invasive tracking of drug distribution.	[10, 11]
	Stimuli-responsive release is triggered by environmental factors such as pH.	Controlled and targeted release of drugs in response to specific stimuli improves therapeutic outcomes.	Responsiveness to environmental changes (e.g., pH) ensures precise release timing.	[12, 13]
CNF MIPs for anti-inflammatory drugs	Sustained release of anti-inflammatory drugs (e.g., curcumin) at the site of inflammation.	Prolonged therapeutic effects and reduced dosing frequency in inflammatory diseases.	High surface area of CNFs provides ample binding sites for drug loading.	[14, 15]
	Targeted delivery to inflamed tissues reduces systemic exposure.	Minimized side effects and enhanced efficacy in treating chronic inflammation.	The mechanical flexibility of CNFs supports the delivery system's adaptation to tissue environments.	[16, 17]

Cancer biomarker detection: The detection of specific cancer biomarkers, such as HER2 and PSA, is essential for early diagnosis and effective treatment. Carbon-based MIP biosensors excel in this area due to their high selectivity, which allows for the accurate detection of low-abundance biomarkers. The integration of these biosensors into portable, point-of-care devices further enhances their utility, enabling quick and easy cancer screening in various settings, including remote or resource-limited areas [18, 19].

Pathogen detection: Rapid detection of pathogens, including bacteria and viruses, is critical for preventing the spread of infectious diseases. Carbon-based MIP biosensors offer quick identification of these infectious agents, enabling timely intervention and treatment. Their durability and stability in various environmental conditions make them particularly useful in low-resource settings, where minimal equipment is available, providing a cost-effective and accessible diagnostic solution [27, 28].

Hormone level monitoring: Monitoring hormone levels, such as insulin, cortisol, and estrogen, is vital for managing endocrine disorders and personalizing treatment plans. Carbon-based MIP biosensors provide accurate measurements of hormone levels, even at low concentrations, enhancing the reliability of diagnostics. These sensors' rapid response times allow for real-time monitoring and adjustments in treatment, improving patient outcomes in hormone-related conditions.

Environmental toxin detection: The detection of environmental toxins, such as heavy metals and pesticides, is crucial for protecting public health. Carbon-based MIP biosensors demonstrate high binding affinity for specific toxins, ensuring low detection limits and accurate identification of contaminants. The integration of these sensors into wearable devices allows for continuous monitoring of environmental exposures, providing on-the-go protection and early warning systems for harmful pollutants.

Drug monitoring: Ensuring that therapeutic drug levels remain within the therapeutic range is essential to avoid side effects or treatment failure. Carbon-based MIP biosensors offer high selectivity for specific drugs, enhancing the accuracy of monitoring. Their application in personalized medicine and pharmacokinetics allows for better tailoring of drug dosages to individual patient needs, supported by real-time monitoring capabilities that enable timely adjustments in drug therapy [27, 31–33].

Neurotransmitter detection: Detecting neurotransmitters such as dopamine, serotonin, and acetylcholine is crucial for understanding and managing neurological disorders. Carbon-based MIP biosensors provide high sensitivity to these neurotransmitters, allowing for the detection of minute changes in brain chemistry. This capability is particularly beneficial in the development of brain-computer interfaces and neurofeedback systems, which have therapeutic and research applications [39–41]. The biocompatibility and stability of these sensors ensure their safe use in neurobiological applications.

Overall, the table demonstrates the broad applicability and effectiveness of carbon-based MIPs in enhancing biosensing technologies. Their ability to provide highly selective, sensitive, and stable detection across a variety of targets makes them invaluable tools in both medical diagnostics and environmental monitoring. The advancements in these technologies continue to push the boundaries of what is possible in biosensing, offering promising solutions for more accurate and efficient detection in a wide range of applications.

Figure 7.2 presents a schematic of the overall synthesis process for the nanocomposite made of MIP and carbon nanoparticles (CNPs). It illustrates the step-by-step functionalization of the plastic optical fiber (POF) surface with a gold (Au) layer, followed by the deposition of the MIP + CNP nanocomposite [19]. This process is crucial for creating the sensor's active layer, which is designed to interact selectively with glucose molecules. Figure 7.3 shows a schematic diagram of the experimental setup used for glucose sensing. Part (a) depicts the overall transmission setup, where a broad-band white light source is coupled to one end of the POF, and the emerged light is collected at the other end using a spectrometer. Parts (b) and (c) illustrate the specific configurations of the unclad POF coated with MIP and an additional Au layer. These configurations demonstrate how the setup enhances the sensor's sensitivity by allowing efficient interaction between the evanescent waves and the glucose molecules in the ambient medium [19].

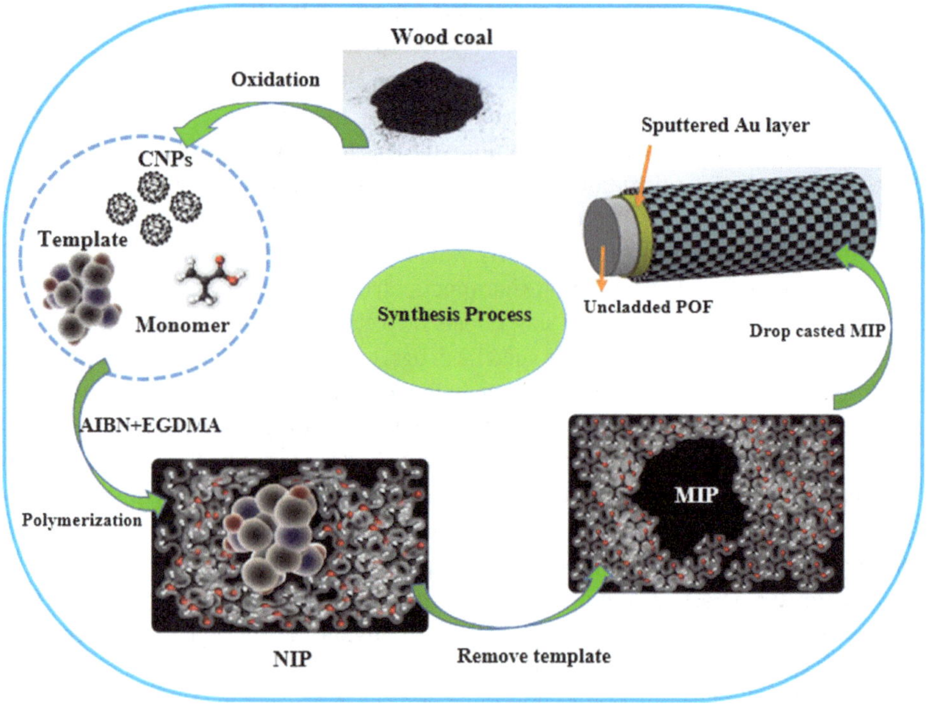

Figure 7.2: Schematic of the overall nanocomposite (MIP + CNPs) synthesis process and the POF surface functionalization with an Au layer and MIP [19].

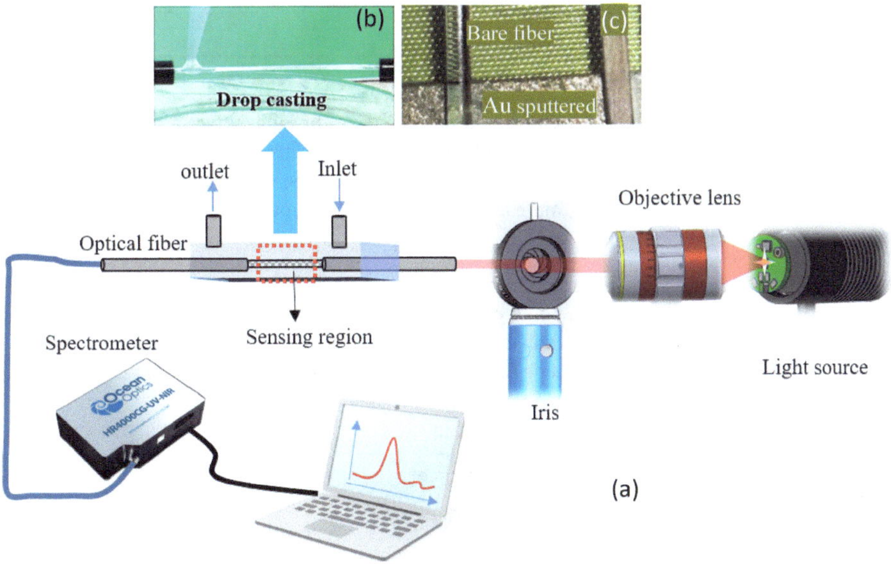

Figure 7.3: Schematic diagram of (a) glucose sensing transmission setup employing an unclad POF coated with (b) MIP, and (c) an Au layer [19].

7.4 Carbon-based MIPs in tissue engineering

Table 7.3 provides an insightful overview of the various roles that carbon-based MIPs play in tissue engineering, particularly in the design of scaffolds for tissue regeneration and bone tissue engineering. The table highlights the potential of these materials to create biomimetic scaffolds that closely resemble the extracellular matrix (ECM), which is crucial for facilitating cell attachment, proliferation, and differentiation. The high surface area and porosity of carbon-based MIPs create an environment conducive to cell growth, making them ideal for tissue regeneration applications. Furthermore, the controlled release of growth factors and signaling molecules from these scaffolds ensures a sustained and prolonged effect, which is critical for successful tissue engineering outcomes [41–43]. This controlled release mechanism is particularly important, as it allows the scaffolds to continuously support tissue growth and healing over time, mimicking the natural processes in the body.

In the specific application of bone tissue engineering, carbon-based MIPs have been shown to significantly enhance osteointegration and bone regeneration by mimicking the mineral composition of natural bone tissue. The incorporation of materials like calcium phosphate or hydroxyapatite into these MIPs further enhances bone mineralization, leading to stronger and more durable bone repair. The mechanical strength and biocompatibility of carbon-based MIPs are key factors that support the

Table 7.2: Biosensing applications of carbon-based MIPs.

Example	Main functions	Results	Key factors	References
Glucose monitoring	Selective detection of glucose in blood samples.	High sensitivity and specificity in detecting glucose levels lead to more accurate monitoring of diabetes.	The high surface area and conductivity of carbon materials improve sensor sensitivity and response time.	[18, 19]
	Real-time monitoring for diabetic patients.	Continuous glucose monitoring with real-time feedback reduces the risk of hypoglycemia or hyperglycemia.	Biocompatibility ensures safe, long-term use in glucose sensors.	[20]
Cancer biomarker detection	Detection of specific cancer biomarkers (e.g., HER2, PSA).	Early detection of cancer improves prognosis and treatment outcomes.	High selectivity of MIPs ensures accurate detection of low-abundance biomarkers.	[21–23]
	Integration into point-of-care devices for quick and easy testing.	Development of portable, user-friendly diagnostic tools for cancer screening.	Compatibility with miniaturized, portable devices for point-of-care use.	[24–26]
Pathogen detection	Rapid detection of bacteria and viruses in clinical samples.	Quick identification of infectious agents enables timely intervention and treatment.	Specific imprinting of pathogen-associated molecules enhances detection accuracy.	[11, 27, 28]
	Application in low-resource settings with minimal equipment.	Cost-effective and accessible diagnostic tools for use in resource-limited areas.	Durability and stability of carbon-based MIPs under various environmental conditions.	[29, 30]
Hormone level monitoring	Monitoring hormones such as insulin, cortisol, and estrogen.	Accurate measurement of hormone levels aids in the management of endocrine disorders.	Sensitivity to low-concentration hormones enhances the reliability of diagnostics.	[27, 31]
	Application in personalized medicine for tailored treatment plans.	Better management of hormone-related conditions improves patient outcomes.	Rapid response times allow for real-time monitoring and adjustments in treatment.	[16, 32, 33]

Table 7.2 (continued)

Example	Main functions	Results	Key factors	References
Environmental toxin detection	Detection of toxins, such as heavy metals and pesticides, in water and food samples.	Protection of public health involves identifying contaminants in the environment.	High binding affinity for specific toxins ensures low detection limits.	[34, 35]
	Integration into wearable devices for continuous monitoring.	Development of wearable biosensors for on-the-go monitoring of environmental exposures.	Flexibility and durability of carbon-based MIPs make them suitable for integration into wearable tech.	[11, 36]
Drug monitoring	Monitoring therapeutic drug levels in patients.	Ensures that drug levels remain within the therapeutic range, reducing the risk of side effects or treatment failure.	High selectivity of MIPs for specific drugs enhances the accuracy of monitoring.	[24, 27]
	Application in personalized medicine and pharmacokinetics.	Better tailoring of drug dosages to individual patients' needs.	Real-time monitoring capabilities for timely adjustments in drug therapy.	[27, 35]
Neurotransmitter detection	Detection of neurotransmitters such as dopamine, serotonin, and acetylcholine.	Better understanding and management of neurological disorders.	High sensitivity to neurotransmitters allows for the detection of minute changes in brain chemistry.	[37–39]
	Applications in brain-computer interfaces and neurofeedback systems.	Enhanced neurofeedback and brain-computer interface systems for therapeutic and research purposes.	Biocompatibility and stability ensure safe use in neurobiological applications.	[27, 40]

regeneration of bone tissue, ensuring that the scaffolds not only promote bone growth but also provide the necessary structural support during the healing process. The success of these MIP-based scaffolds in bone tissue engineering demonstrates their potential for broader applications in other areas of tissue engineering, such as cartilage and nerve regeneration [41–43]. As research continues to advance, carbon-based MIPs are likely to play an increasingly important role in developing innovative and effective solutions for complex tissue repair and regeneration challenges, offering new possibilities for regenerative medicine.

Table 7.3: Carbon-based MIPs in tissue engineering.

Example	Main functions	Results	Key factors	Reference
Scaffold design for tissue regeneration	Development of biomimetic scaffolds that mimic the extracellular matrix (ECM).	Enhanced cell attachment, proliferation, and differentiation on the scaffold lead to effective tissue regeneration.	High surface area and porosity of carbon-based MIPs provide a conducive environment for cell growth.	[41, 42]
	Incorporation of growth factors and signaling molecules into MIPs for controlled release.	Sustained release of bioactive molecules promotes tissue regeneration and healing.	Controlled release of growth factors ensures a prolonged and sustained effect, which is critical for tissue engineering.	
Bone tissue engineering	Creation of MIPs that can mimic the mineral composition of bone tissue.	Improved osteointegration and bone regeneration, leading to stronger and more durable bone repair.	Mechanical strength and biocompatibility of carbon-based MIPs support bone tissue regeneration.	[41–44]
	Incorporation of calcium phosphate or hydroxyapatite into MIPs enhances bone mineralization.	Enhanced bone mineralization and faster recovery in bone tissue engineering applications.	Mimicking the natural bone environment to encourage bone tissue growth and repair.	

7.5 Toxicity and biocompatibility of carbon-based MIPs

The use of carbon-based materials in MIPs offers numerous advantages in biomedical applications, such as high surface area, electrical conductivity, and chemical stability. However, the potential toxicity of these materials poses significant concerns that must be addressed to ensure their safe application, particularly in vivo. Carbon-based materials like CNTs, graphene, and CDs have been reported to induce cytotoxicity, oxidative stress, and inflammatory responses in various biological systems. The toxicity of these materials largely depends on their size, shape, surface chemistry, and concentration [1–3]. For example, the sharp edges of graphene or the high aspect ratio of CNTs can cause physical damage to cell membranes, leading to cell death. Moreover, the generation of reactive oxygen species by these materials can result in oxidative damage to cellular components, including DNA, proteins, and lipids.

To mitigate these toxic effects, several strategies have been developed. Surface functionalization of carbon-based materials is one of the most effective approaches to reduce their toxicity. By modifying the surface with biocompatible polymers, proteins, or small molecules, the interaction of these materials with biological systems can be modulated to minimize adverse effects. Additionally, controlling the size and shape of carbon-based materials during synthesis can help reduce their toxicity. For instance, creating spherical CDs with smooth surfaces rather than sharp-edged graphene can significantly lower cytotoxicity [4–7]. Another strategy involves the use of biodegradable carbon-based materials that can be safely metabolized and excreted by the body, reducing the risk of long-term accumulation and associated toxicities.

Assessing the biocompatibility of carbon-based MIPs is crucial for their successful application in biomedical fields. Biocompatibility refers to the ability of a material to perform its intended function without eliciting any undesirable local or systemic effects in the host organism. For carbon-based MIPs, biocompatibility is assessed through a series of in vitro and in vivo tests that evaluate their interaction with cells, tissues, and organs. In vitro tests typically involve cytotoxicity assays, such as the MTT or LDH assays, which measure cell viability after exposure to the MIPs. Hemocompatibility tests are also conducted to assess the material's effect on blood components, such as red blood cells, platelets, and plasma proteins, which is particularly important for MIPs intended for intravascular use [14–16].

In vivo assessments involve animal studies where carbon-based MIPs are implanted or injected, and their effects on the surrounding tissues and overall health are monitored over time. Parameters such as inflammation, immune response, tissue integration, and potential long-term toxicity are evaluated. Recent research findings indicate that, with appropriate surface modifications and careful design, carbon-based MIPs can exhibit excellent biocompatibility. For instance, studies have shown that PEGylated CNTs and functionalized GO can reduce cytotoxicity and improve biocompatibility, making them suitable for drug delivery and biosensing applications. Additionally, biodegradable CDs have been demonstrated to be non-toxic in animal models, showing promise for use in imaging and therapeutic applications.

Overall, while the potential toxicity of carbon-based materials is a concern, ongoing research and the development of strategies to enhance their biocompatibility are paving the way for their safe and effective use in various biomedical applications. The continuous refinement of these materials and thorough biocompatibility assessments will be critical in ensuring their successful translation from the laboratory to clinical practice.

7.6 Future perspectives and challenges

The use of carbon-based MIPs in biomedical applications is rapidly evolving, driven by advancements in nanotechnology, materials science, and biomedical engineering. Emerging trends include the integration of carbon-based MIPs with advanced sensing technologies, such as wearable and implantable biosensors, which enable real-time monitoring of physiological parameters with high precision. Additionally, the development of stimuli-responsive carbon-based MIPs is gaining traction, where the release of therapeutic agents is triggered by specific environmental changes, such as pH, temperature, or the presence of certain enzymes. This trend is particularly promising for targeted drug delivery systems, where precision and control over drug release are paramount. Moreover, the combination of carbon-based MIPs with artificial intelligence and machine learning is beginning to shape the future of personalized medicine, allowing for the design of highly specific and adaptable MIPs tailored to individual patient needs [17–21].

Despite the promising advancements, several challenges and limitations hinder the full potential of carbon-based MIPs in biomedical applications. One of the primary challenges is the complexity of synthesizing MIPs with high selectivity and binding affinity, particularly when targeting large and complex biomolecules such as proteins or cells. The reproducibility of MIP synthesis, especially on a large scale, is another significant hurdle, as small variations in the polymerization process can lead to inconsistencies in the performance of the MIPs. Additionally, while carbon-based materials offer excellent properties such as high surface area and conductivity, their biocompatibility and toxicity remain areas of concern, especially for in vivo applications. Ensuring that these materials do not elicit adverse immune responses or toxicity over extended periods is crucial for their safe use in medical devices and therapies [16–19].

To address these challenges, ongoing research is focusing on optimizing the polymerization process to enhance the reproducibility and scalability of MIP production. This includes exploring novel polymerization techniques and functional monomers that can improve the selectivity and stability of MIPs. In terms of biocompatibility, surface modification of carbon-based MIPs, such as coating with biocompatible materials or the incorporation of biodegradable components, is being investigated to reduce toxicity and enhance the safety profile of these materials. Future research directions also include the development of hybrid MIPs that combine the strengths of different materials, such as incorporating metal nanoparticles or polymers with carbon-based MIPs to create multifunctional systems with enhanced properties. As these challenges are addressed, the potential for carbon-based MIPs in biomedical applications is expected to expand significantly, paving the way for more innovative and effective healthcare solutions.

7.7 Conclusion

This chapter explored the role of carbon-based MIPs in biomedical applications, emphasizing their advantages in drug delivery, biosensing, and tissue engineering. Carbon-based MIPs offer enhanced performance due to their high surface area, conductivity, and biocompatibility, enabling precise and efficient therapeutic and diagnostic solutions. We also discussed the challenges related to toxicity and biocompatibility, highlighting strategies such as surface functionalization and the development of biodegradable materials to mitigate these issues. Lastly, we identified emerging trends and ongoing challenges in advancing the use of carbon-based MIPs in healthcare. Carbon-based MIPs hold great promise for transforming biomedical applications by offering targeted, efficient, and safer solutions in drug delivery, diagnostics, and tissue regeneration. While challenges remain, particularly in scalability and long-term safety, continued research and interdisciplinary collaboration are essential. As these materials evolve, they are poised to significantly impact the future of personalized and effective healthcare.

References

[1] Afzali M., Mostafavi A., & Shamspur T. (2020). A novel electrochemical sensor based on magnetic core@ shell molecularly imprinted nanocomposite (Fe3O4@ graphene oxide@ MIP) for sensitive and selective determination of anticancer drug capecitabine, *Arabian Journal of Chemistry*. 13(8), 6626–6638.

[2] Han S., et al. (2020). Drug-loaded dual targeting graphene oxide-based molecularly imprinted composite and recognition of carcino-embryonic antigen, *RSC Advances*. 10(19), 10980–10988.

[3] Chen Q., et al. (2021). Preparation and evaluation of magnetic graphene oxide molecularly imprinted polymers (MIPs-GO-Fe3O4@ SiO2) for the analysis and separation of tripterine, *Reactive and Functional Polymers*. 169, 105055.

[4] Itoo A. M., et al. (2022). Multifunctional graphene oxide nanoparticles for drug delivery in cancer, *Journal of Controlled Release*. 350, 26–59.

[5] Zhao L., et al. (2020). Molecularly imprinted polymers doped with carbon nanotube with aid of metal-organic gel for drug delivery systems, *Pharmaceutical Research*. 37, 1–14.

[6] Zhang X., et al. (2022). Preparation of carbon nanotubes and polyhedral oligomeric-reinforced molecularly imprinted polymer composites for drug delivery of gallic acid, *International Journal of Pharmaceutics*. 615, 121476.

[7] Rahamathulla M., et al. (2021). Carbon nanotubes: Current perspectives on diverse applications in targeted drug delivery and therapies, *Materials*. 14(21), 6707.

[8] Er S., et al. (2021). Amino acids, peptides, and proteins: Implications for nanotechnological applications in biosensing and drug/gene delivery, *Nanomaterials*. 11(11), 3002.

[9] Bose S., et al. (2024). Overview of carbon nanotubes as drug delivery system, *Progress and Prospect of Nanocarriers*. Kashaw, S. K., Sau, S., Iyer, A., (eds.). Elsevier, Molecular Therapeutics Program, Barbara Ann Karmanos Cancer Institute, Detroit, MI, United States, Wayne State University, Detroit, MI, United States, 241–270. ISBN 978-0-12-819979-4, https://doi.org/10.1016/C2019-0-01213-7.

[10] Nasiriani T., et al. (2024). Gelatin-based carbon quantum dot-molecularly imprinted polymer: Safe photoluminescent core-shell nano-carrier for the pH-responsive anticancer drug delivery, *International Journal of Biological Macromolecules*. 278, 134669.

[11] Sobiech M., et al. (2021). Quantum and carbon dots conjugated molecularly imprinted polymers as advanced nanomaterials for selective recognition of analytes in environmental, food and biomedical applications, *TrAC Trends in Analytical Chemistry*. 142, 116306.

[12] Keçili R., Hussain C. G., & Hussain C. M. (2023). Fluorescent nanosensors based on green carbon dots (CDs) and molecularly imprinted polymers (MIPs) for environmental pollutants: Emerging trends and future prospects, *Trends in Environmental Analytical Chemistry*. 40, e00213.

[13] Shi Y., et al. (2020). Molecularly imprinting polymers (MIP) based on nitrogen doped carbon dots and MIL-101 (Fe) for doxorubicin hydrochloride delivery, *Nanomaterials*. 10(9), 1655.

[14] Kalambate P. K., et al. (2021). Nanomaterials-based electrochemical sensors and biosensors for the detection of non-steroidal anti-inflammatory drugs, *TrAC Trends in Analytical Chemistry*. 143, 116403.

[15] Torrinha Á., et al. (2022). (Bio) sensing strategies based on ionic liquid-functionalized carbon nanocomposites for pharmaceuticals: Towards greener electrochemical tools, *Nanomaterials*. 12(14), 2368.

[16] Torrinha Á., et al. (2020). Application of nanostructured carbon-based electrochemical (bio) sensors for screening of emerging pharmaceutical pollutants in waters and aquatic species: A review, *Nanomaterials*. 10(7), 1268.

[17] Kaya S. I., Cetinkaya A., & Ozkan S. A. (2022). Carbon nanomaterial-based drug sensing platforms using state-of-the-art electroanalytical techniques, *Current Analytical Chemistry*. 18(1), 79–101.

[18] Wu H., et al. (2021). Construction of a selective non-enzymatic electrochemical sensor based on hollow nickel nanospheres/carbon dots–chitosan and molecularly imprinted polymer film for the detection of glucose, *New Journal of Chemistry*. 45(46), 21676–21683.

[19] Azad S., et al. (2022). A novel contemporary molecular imprinting technique for non-enzymatic selective glucose detection, *Optics & Laser Technology*. 148, 107786.

[20] Zidarič T., et al. (2023). The development of an electropolymerized, molecularly imprinted polymer (MIP) sensor for insulin determination using single-drop analysis, *Analyst*. 148(5), 1102–1115.

[21] Jahangiri-Manesh A., et al. (2022). Molecularly imprinted polymer-based chemiresistive sensor for detection of nonanal as a cancer related biomarker, *Microchemical Journal*. 173, 106988.

[22] Carneiro L. P. T., et al. (2022). An all-in-one approach for self-powered sensing: A methanol fuel cell modified with a molecularly imprinted polymer for cancer biomarker detection, *Journal of Electroanalytical Chemistry*. 906, 116009.

[23] Suzaei F. M., et al. (2022). Molecularly imprinted polymers for selective extraction/microextraction of cancer biomarkers: A review, *Microchimica Acta*. 189(7), 255.

[24] Pandey H., et al. (2020). Carbon nanomaterials integrated molecularly imprinted polymers for biological sample analysis: A critical review, *Materials Chemistry and Physics*. 239, 121966.

[25] Wasilewski T., et al. (2024). Molecularly imprinted polymers for the detection of volatile biomarkers, *TrAC Trends in Analytical Chemistry*. 177, 117783.

[26] Kang M. S., et al. (2023). Molecularly imprinted polymers (MIPs): Emerging biomaterials for cancer theragnostic applications, *Biomaterials Research*. 27(1), 45.

[27] Preda D., et al. (2022). Recent trends in the development of carbon-based electrodes modified with molecularly imprinted polymers for antibiotic electroanalysis, *Chemosensors*. 10(7), 243.

[28] Singhal A., et al. (2022). Multifunctional carbon nanomaterials decorated molecularly imprinted hybrid polymers for efficient electrochemical antibiotics sensing, *Journal of Environmental Chemical Engineering*. 10(3), 107703.

[29] Zheng X., et al. (2021). Towards clean and safe water: A review on the emerging role of imprinted polymer-based electrochemical sensors, *Sensors*. 21(13), 4300.

[30] Pintavirooj C., et al. (2022). Biosensors for Klebsiella pneumoniae with molecularly imprinted polymer (MIP) technique, *Sensors*. 22(12), 4638.

[31] Tigari G., et al. (2022). Research developments in carbon materials based sensors for determination of hormones, *Journal of Electrochemical Science and Engineering*. 12(1), 3–23.

[32] Pareek S., et al. (2022). Ultra-sensitive nano-molecular imprinting polymer-based electrochemical sensor for Follicle-Stimulating Hormone (FSH) detection, *Biochemical Engineering Journal*. 180, 108329.
[33] Mayet A. M., et al. (2024). Molecularly imprinted polymers for biosensing of hormones in food safety and biomedical analysis: Progress and perspectives, *Materials Today Chemistry*. 35, 101899.
[34] Chi H., & Liu G. (2023). Carbon nanomaterial-based molecularly imprinted polymer sensors for detection of hazardous substances in food: Recent progress and future trends, *Food Chemistry*. 420, 136100.
[35] Singh M., et al. (2020). Recent advancement of carbon nanomaterials engrained molecular imprinted polymer for environmental matrix, *Trends in Environmental Analytical Chemistry*. 27, e00092.
[36] Ma X., et al. (2021). Carbon-based nanocomposite smart sensors for the rapid detection of mycotoxins, *Nanomaterials*. 11(11), 2851.
[37] Suriyaprakash J., et al. (2021). Molecular engineered carbon-based sensor for ultrafast and specific detection of neurotransmitters, *ACS Applied Materials & Interfaces*. 13(51), 60878–60893.
[38] Song H., et al. (2023). Carbon-based electrochemical sensors for in vivo and in vitro neurotransmitter detection, *Critical Reviews in Analytical Chemistry*. 53(5), 955–974.
[39] Liu R., et al. (2022). Recent trends in carbon-based microelectrodes as electrochemical sensors for neurotransmitter detection: A review, *TrAC Trends in Analytical Chemistry*. 148, 116541.
[40] Elugoke S. E., et al. (2021). Molecularly imprinted polymers (MIPs) based electrochemical sensors for the determination of catecholamine neurotransmitters–Review, *Electrochemical Science Advances*. 1(2), e2000026.
[41] Ye G., et al. (2020). Nanomaterial-based scaffolds for bone tissue engineering and regeneration, *Nanomedicine*. 15(20), 1995–2017.
[42] Vedhanayagam M., et al. (2021). Carbon dots-mediated fluorescent scaffolds: Recent trends in image-guided tissue engineering applications, *International Journal of Molecular Sciences*. 22(10), 5378.
[43] Sharker S. M. (2024). Emerging carbon-based nanomaterials for biomedical and bioengineering applications, *Nanomaterials for Biomedical and Bioengineering Applications*. Javed, R., Chen, J. T., Khalil, A.T. (eds.). Springer, Singapore, 203–219. ISBN 978-981-97-0220-6, https://doi.org/10.1007/978-981-97-0221-3.
[44] Govindarajan D., et al. (2023). Graphene: A multifaceted carbon-based material for bone tissue engineering applications, *ACS Omega*. 9(1), 67–80

Hemlata Vashisht, Sudershan Kumar, and Ajai Kumar

Chapter 8
Analytical and industrial applications of carbon-based molecularly imprinted polymers

Abstract: Carbon-based molecularly imprinted polymers (C-MIPs) have emerged as advanced materials that integrate the high selectivity of molecular imprinting with the unique properties of carbon-based nanomaterials such as graphene, carbon nanotubes (CNTs), and activated carbon. This chapter presents a comprehensive overview of the evolution of C-MIPs, their synthesis strategies, functional monomers, and structural advantages in comparison to conventional MIPs. Special attention is given to their enhanced performance in various analytical and industrial applications, including environmental pollutant detection, pharmaceutical residue analysis, and biomedical diagnostics. The incorporation of carbon nanostructures improves surface area, mechanical stability, and signal transduction, making C-MIPs highly suitable for electrochemical sensing and solid-phase extraction. Recent trends such as surface imprinting, nano-template approaches, and integration with artificial intelligence (AI) tools for imprint design are also discussed. The chapter concludes with a critical assessment of current challenges and future prospects in scalable synthesis and real-world deployment of C-MIPs in advanced analytical technologies.

Keywords: Carbon-based MIPs, molecular imprinting, graphene, carbon nanotubes, electrochemical sensors, environmental monitoring, pharmaceutical analysis, biomedical diagnostics

8.1 Introduction

This chapter focuses on **carbon-based molecularly imprinted polymers (C-MIPs)**, a cornerstone of evolving scientific research. Molecular imprinting is a process that creates template-shaped voids within polymer matrices, offering controlled selectivity and affinity. Inspired by the "lock-and-key" mechanism used by enzymes to recognize substrates, this technique relies on cavities in polymers that are complementary in

Hemlata Vashisht, Department of Chemistry, Kirori Mal College, University of Delhi, Delhi 110007, India
Sudershan Kumar, Ajai Kumar, Department of Chemistry, Hindu College, University of Delhi, Delhi 110007, India

https://doi.org/10.1515/9783111503325-008

shape and functionality to a specific template molecule. Substrates with matching features bind selectively to these cavities, while others are excluded.

The chapter explores the synthesis, characteristics, and applications of C-MIPs, emphasizing their roles in analytical and industrial domains. Additionally, it highlights recent advancements in this field and outlines potential future developments.

8.1.1 Key terms

- **Molecularly imprinted polymers (MIPs):** Synthetic polymers designed with molecular recognition sites.
- **Carbon-based molecularly imprinted polymers (C-MIPs):** MIPs incorporating carbon materials for enhanced functionality.
- **Carbon nanotubes (CNTs):** Cylindrical carbon structures with high strength, electrical conductivity, and surface area.

8.1.2 Core concept: molecular imprinting

Molecular imprinting is a process in which functional monomers self-assemble around a template molecule, followed by polymerization and cross-linking. Once the template molecule is partially or completely removed, the resulting polymer matrix contains cavities tailored to selectively recognize the template molecule.

This technique has evolved significantly over recent decades, finding applications in **drug delivery, separations, sensing, and catalysis**. For example, the form-selective cavities in MIPs have proven effective in catalytic processes.

8.1.3 Historical development of molecular imprinting

- Polyakov [1]: Demonstrated molecular imprinting by polymerizing sodium silicate with ammonium carbonate in the presence of benzene, resulting in silica with enhanced benzene absorption properties.
- Dickey [2]: Applied molecular imprinting to precipitate silica gels in the presence of organic dyes, showcasing the selectivity of imprinted silica for template dyes.
- **Patrikeev's contributions:** Enhanced bacterial growth using imprinted silica and demonstrated its utility in chromatography, including thin-layer chromatography (TLC) and high-performance liquid chromatography (HPLC).
- Wulff and Klotz [3]: Advanced the field by incorporating covalent functional groups into imprinted polymer cavities, enabling specific molecular recognition.

- **Mosbach group:** Pioneered non-covalent imprinting, where functional groups were introduced into imprinted cavities through non-covalent interactions, thereby broadening the range of molecular imprinting techniques.

8.1.4 Relevance of the topic to broader themes or goals

MIPs are versatile materials with recognition sites tailored to specific target molecules. Their high specificity makes them invaluable in applications such as **sensing, separation, and catalysis.**

C-MIPs integrate carbon materials – such as **graphene**, **CNTs**, and **activated carbon** – into MIP structures. This integration enhances their properties by leveraging the **large surface area, electrical conductivity, and mechanical strength** of carbon materials. These features make C-MIPs especially suited for high-sensitivity and durable applications, such as:
- Advanced sensing technologies
- Robust separation techniques
- Catalytic processes requiring high stability

By bridging molecular recognition and the remarkable properties of carbon materials, C-MIPs contribute to innovation in both analytical science and industrial solutions. This chapter establishes their role in advancing technologies and highlights their potential to address emerging challenges.

8.2 Origin and evolution of molecular imprinting

8.2.1 Definition and explanation of key concepts

Molecular imprinting is a process that creates template-shaped cavities in a polymer matrix, allowing for the selective recognition of specific molecules. This method mimics biological molecular recognition, such as the interaction between enzymes and substrates or antibodies and antigens.

The foundation of molecular imprinting was first laid conceptually in the 1930s when **Linus Pauling** hypothesized that antibodies could form complementary structures to antigens, highlighting the potential for synthetic molecular recognition. However, it was not until the 1970s that practical applications of synthetic MIPs were developed.

- **Early milestones in molecular imprinting**
  a. [3]: Wulff and colleagues created the first polymer networks capable of recognizing and binding specific molecules. Their pioneering work demon-

strated the feasibility of incorporating functional monomers into polymers, thereby creating a solid foundation for MIP technology.

b. [4]: Klaus Mosbach and his team introduced the use of non-covalent interactions in molecular imprinting, significantly advancing the field. This approach allowed greater flexibility in synthesizing MIPs by forming recognition sites through non-covalent bonds, such as hydrogen bonding, ionic interactions, and hydrophobic forces, between the template and monomers.

8.2.2 Development of traditional MIPs

Throughout the 1980s and 1990s, researchers improved the synthesis and functionality of classical MIPs. New polymerization techniques were developed, including:

- **Bulk polymerization:** Producing large polymer blocks that can be ground into fine particles.
- **Suspension polymerization:** Creating spherical MIP beads for improved applications in liquid-phase processes.
- **Surface imprinting:** Enhancing accessibility and selectivity by forming recognition sites on the surface of the polymer matrix.

- **Notable advancements:**
  a. **1985 – Mosbach and Andersson [30]:**
  They introduced the **solid-phase synthesis** of MIPs, a method that improved the uniformity of recognition sites and binding characteristics, setting a benchmark for future MIP development.
  b. **1993 – Takeuchi and Haginaka [29]:**

They demonstrated the use of MIPs as stationary phases in chromatography. This innovation proved MIPs' potential for selective separation and analysis of complex mixtures, establishing them as a powerful tool in analytical chemistry.

8.2.3 The emergence of carbon-based MIPs (C-MIPs)

The early 2000s marked the integration of **carbon materials** into MIPs, driven by the unique properties of carbon, including:

- High **surface area** for increased template binding.
- Superior **electrical conductivity** for applications in sensors and electronics.
- Exceptional **mechanical strength** for durability in harsh conditions.

8.2.4 Key developments

a. **2003 – Lanza and Sellergren [31]:**
A landmark study using activated carbon in MIPs was published to selectively extract environmental contaminants. This research highlighted the enhanced adsorption capabilities and selectivity of carbon-based MIPs (C-MIPs), paving the way for their application in environmental remediation.

– **Graphene and CNTs:**
Subsequent studies incorporated advanced carbon nanomaterials, such as graphene and CNTs, into MIP matrices. These materials enhanced the functionality of MIPs, particularly in **electrochemical sensors**, where sensitivity and rapid response are critical.

8.2.4.1 Examples and case studies

a. **Environmental remediation**
   Carbon-based MIPs have been used to detect and remove pollutants, such as pesticides, heavy metals, and pharmaceuticals, from water sources.
b. **Biomedical applications**
   Carbon-integrated MIPs have been applied in drug delivery systems for controlled and selective drug release, improving therapeutic outcomes.
c. **Analytical chemistry**
   MIPs with CNTs have enhanced the sensitivity and specificity of sensors used for detecting biomolecules, such as glucose and cholesterol, in clinical samples.

- **Emerging areas of research**
  - Integration of MIPs with nanotechnology and artificial intelligence for advanced applications.
  - Development of hybrid materials combining carbon-based MIPs with other functional materials.
  - Application in personalized medicine and biomarker discovery.

- **Challenges in analytical and industrial applications**
  - Limitations of current synthetic methods.
  - Scalability and reproducibility of carbon-based MIPs.
  - Template leaching and environmental safety concerns.
  - Optimizing selectivity and sensitivity for real-world samples.

Conclusion and Broader Relevance

The evolution of molecular imprinting from early theoretical concepts to advanced carbon-based MIPs underscores its importance in scientific and industrial applications. By combining the selectivity of MIPs with the exceptional properties of carbon materials, researchers have opened new avenues in ***environmental science, healthcare, and materials engineering****,* addressing critical societal challenges with innovative solutions.

8.3 Synthesis of carbon-based molecularly imprinted polymers (C-MIPs)

8.3.1 Exploration of related ideas

The synthesis of **C-MIPs** builds upon the foundational principles of molecular imprinting (Section 8.1) by incorporating carbon materials into the polymer matrix. This integration leverages the unique properties of carbon, such as its high surface area, electrical conductivity, and mechanical strength, to enhance the performance of MIPs. Various methods have been developed to synthesize C-MIPs, ensuring precise recognition, improved sensitivity, and robustness.

- **Synthesis methods: in situ polymerization**

In this approach, carbon materials (e.g., graphene, CNTs) are dispersed in a monomer solution containing the template molecule. Polymerization is initiated, embedding the carbon material uniformly within the polymer matrix.

- **Advantages:** Uniform carbon distribution enhances the structural integrity and functional performance of the C-MIP.
- **Applications:** Widely used for creating sensors that detects biomolecules or environmental contaminants.

- **Surface imprinting**

  This method involves functionalizing carbon materials with monomers capable of interacting with the template molecule. The template is then introduced, and polymerization occurs, forming a thin polymer layer on the carbon surface.
- **Advantages:** High surface area and accessibility to imprinted sites enhance sensitivity and selectivity.
- **Applications:** Used in catalysis and sensing technologies for rapid molecular recognition.

- **Bulk polymerization and subsequent carbon incorporation**

  Here, a bulk MIP is first synthesized. Carbon components are then incorporated through:

  - **Grinding:** The MIP is ground into fine particles and combined with a carbon material.
  - **Chemical modification:** Carbon materials are chemically attached to the MIP surface to enhance compatibility and performance.
  - **Applications:** Suitable for industrial-scale applications due to cost-efficiency and scalability.

- **Functionalization of carbon materials**
  Functionalization improves the interaction between carbon materials, monomers, and template molecules. Common methods include:
  - **Oxidation:** Adds oxygen-containing groups (e.g., hydroxyl, carboxyl) to enhance dispersion and interaction.
  - **Silane coupling:** Adds silane groups to improve compatibility with the polymer matrix.
  - **Polymer grafting:** Grows polymer chains directly on the carbon surface, enhancing integration and functionality.

8.3.2 Properties of carbon-based MIPs

The integration of carbon materials into MIPs offers several advantages:
- **High surface area:** Provides more binding sites for target molecules, thereby improving recognition efficiency.
- **Electrical conductivity:** Enables the development of electrochemical sensors with superior signal-transduction capabilities.
- **Mechanical strength and thermal stability:** Make C-MIPs suitable for harsh industrial environments.
- **Chemical versatility:** Functional groups on carbon materials allow customization for specific applications.
- **Enhanced selectivity and sensitivity:** Carbon-material interactions with target molecules amplify the polymer's molecular recognition capabilities.

8.3.3 Non-imprinted polymer

NIP stands for **non-imprinted polymer**. It is the control polymer synthesized without a template molecule during the polymerization process. NIPs are often used in studies involving MIPs to compare their performance and assess the specificity of molecular recognition Table 8.1.

Table 8.1: Differences between NIP (non-imprinted polymer), MIP (molecularly imprinted polymer), and C-MIP (carbon-based molecularly imprinted polymer).

S. no.	Aspect	NIP (non-imprinted polymer)	MIP (molecularly imprinted polymer)	C-MIP (carbon-based molecularly imprinted polymer)
1.	Definition	A polymer synthesized without a template molecule	A polymer is synthesized using a template molecule to create specific binding sites.	A MIP is enhanced with carbon-based materials (e.g., graphene, carbon nanotubes, fullerenes) for improved performance.
2.	Binding sites	Lacks specific binding sites; binding is non-specific	Contains specific binding sites tailored to the template molecule.	Contains enhanced binding sites due to the integration of carbon materials.
3.	Selectivity	Non-selective toward any molecule	High selectivity for the template molecule or structurally similar compounds.	Superior selectivity due to carbon's ability to enhance interactions (e.g., π-π stacking).
4.	Materials used	Basic polymer matrix without a template	Polymer matrix imprinted with a specific template molecule.	Polymer matrix combined with carbon materials for increased surface area and conductivity.
5.	Performance	Low binding capacity and no specific recognition	High binding capacity with specific molecular recognition.	Superior binding capacity due to higher surface area, conductivity, and interaction sites.
6.	Surface area	Limited surface area	Moderate surface area is based on the polymer structure.	High surface area achieved through carbon-based nanomaterials.
7.	Durability	Moderate durability	Durable for specific applications.	High thermal and chemical stability, attributed to carbon-based enhancements.
8.	Sensitivity	Minimal sensitivity; prone to non-specific adsorption	High sensitivity to the target molecule.	Enhanced sensitivity is due to carbon's excellent conductivity and interaction properties.
9.	Applications	Used as a control to measure non-specific binding	Used in sensors, drug delivery, separation, and detection systems.	Applied in high-performance sensors, environmental monitoring, and advanced analytical techniques.
10.	Integration with nanotechnology	Not applicable	Limited.	Integrated with nanomaterials for hybrid and multifunctional applications.

Table 8.1 (continued)

S. no.	Aspect	NIP (non-imprinted polymer)	MIP (molecularly imprinted polymer)	C-MIP (carbon-based molecularly imprinted polymer)
11.	Cost	Relatively low cost	Moderate cost, depending on the synthesis method.	Higher costs due to advanced materials and synthesis processes.

8.3.3.1 Key characteristics of NIPs

Non-specific binding:
Unlike MIPs, NIPs do not have specific recognition sites tailored to a template molecule. Any binding that occurs is due to non-specific interactions, such as van der Waals forces or hydrogen bonding Figure 8.2.

Control material:
NIPs are used as a baseline to evaluate the selective binding capacity of MIPs. The difference in binding between MIPs and NIPs highlights the molecular recognition efficiency of the MIPs Figure 8.3.

Simpler structure:
The polymer matrix of NIPs lacks the imprinted cavities or binding sites that MIPs possess, making them less selective.

8.3.4 Applications of NIPs

- **Control in analytical studies:** They are employed to differentiate between specific and non-specific adsorption in MIPs Figure 8.1.
- **Reference for optimization:** NIPs are used as a reference to optimize the synthesis and performance of MIPs.
- **Non-specific adsorption studies:** To understand and minimize non-specific binding in analytical and industrial processes.

8.3.5 Adsorption capacity comparison

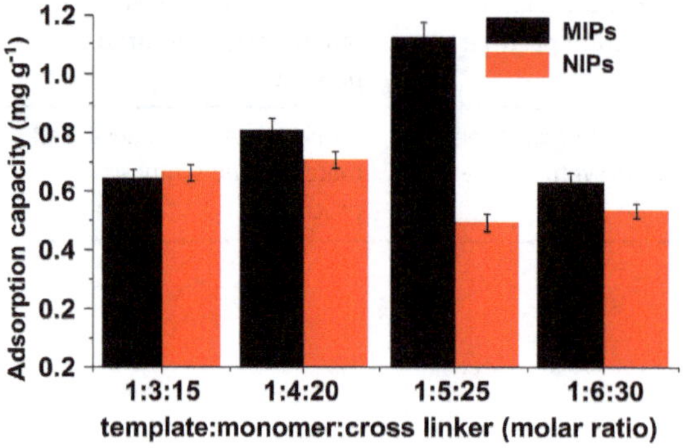

Figure 8.1: The graph shows the adsorption capacities of traditional NIPs versus MIPs, demonstrating the latter's superior performance due to increased surface area and binding sites.

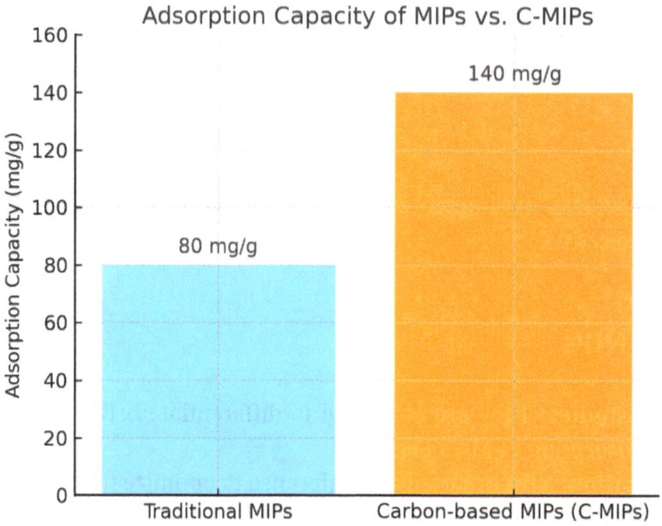

Figure 8.2: The graph demonstrates the adsorption capacities of traditional MIPs and carbon-based MIPs (C-MIPs). As shown, C-MIPs exhibit significantly higher adsorption capacity (140 mg/g vs. 80 mg/g for traditional MIPs), attributed to their increased surface area and additional binding sites, which enhance molecular recognition and adsorption efficiency.

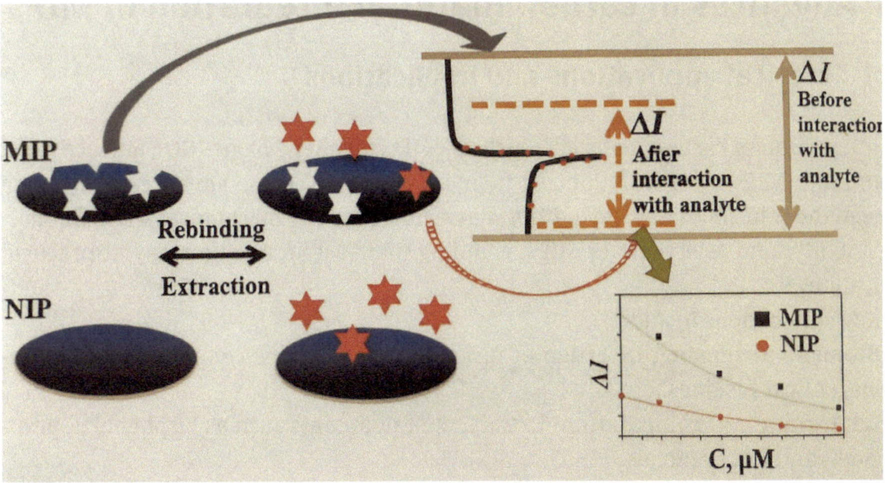

Figure 8.3: Signal response (current vs. analyte concentration) for sensors using traditional NIPs and MIPs, with MIPs showing higher signal intensity and sensitivity.

- **Conductivity enhancement in electrochemical sensors**

Summary

- NIP: Baseline polymer with no specific recognition; used as a control.
- MIP: A polymer with specific molecular recognition for the template molecule; widely used in separation and sensing.
- C-MIP: A more advanced version of MIP, enhanced with carbon materials, offering superior performance in terms of selectivity, sensitivity, and durability.

Building on the foundational principles of molecular imprinting discussed in Section 1, this section highlights how the incorporation of carbon materials has transformed traditional MIPs into highly functional materials with broader applications. Carbon-based MIPs represent a significant advancement, combining the selectivity of traditional MIPs with the structural and functional benefits of carbon materials.

8.4 Advances in carbon material integration in MIPs

8.4.1 Practical applications and implications

The integration of carbon materials, such as CNTs, graphene oxide (GO), and activated carbon, into MIPs has revolutionized their functionality. These advancements, made between the mid-2000s and early 2010s, have enhanced the mechanical strength, thermal stability, and selectivity of MIPs, enabling their use in a wide array of practical applications.

Key applications include:

- **Biomedical sensing:** Detection of proteins, small organic compounds, and pharmaceutical residues.
- **Environmental remediation:** Extraction of pollutants such as bisphenol A, pesticides, and heavy metals.
- **Food safety:** Monitoring contaminants, such as antibiotics, in food samples.
- **Catalysis:** Development of more robust catalysts for industrial applications.

- **Traditional MIPs vs. carbon-based MIPs (C-MIPs):**
- **Selectivity:** Traditional MIPs rely solely on the polymer matrix for molecular recognition, while C-MIPs integrate carbon materials to enhance binding capacity and specificity.
- **Sensitivity:** Carbon materials like GO and CNTs improve signal transduction in sensing applications, making C-MIPs more sensitive than traditional MIPs.
- **Durability:** C-MIPs exhibit superior mechanical strength and thermal stability due to the structural properties of carbon materials, which traditional MIPs lack.

- **Alternate methods for molecular recognition:**
- **Bioreceptors:** Techniques using antibodies or enzymes provide high selectivity but suffer from limited stability and reusability. C-MIPs offer comparable selectivity while being reusable and more robust in harsh environments.
- **Metal-organic frameworks (MOFs):** MOFs show promise for molecular recognition but are costlier and less scalable than C-MIPs for large-scale applications.
- **Early advances in carbon integration**
- **2007 – Song et al. [32]:**

Song and colleagues demonstrated the incorporation of CNTs into MIPs for protein recognition. This study highlighted CNTs' ability to enhance the mechanical strength and binding performance of MIPs, setting a precedent for future work.

Implications: Improved durability of MIPs in biomedical applications such as biosensors.

- **2010 – Liu et al.** [33]:

Liu and colleagues utilized GO in MIPs to detect small organic molecules. The high surface area and functional ability of GO significantly enhanced sensitivity and selectivity.

Applications: Analytical chemistry for detecting trace-level pollutants.

- **Recent developments and advanced applications**
- **[5]:** Guo and colleagues used **in situ polymerization** to embed GO in MIPs for the selective extraction of **bisphenol A**. This study demonstrated high extraction efficiency and specificity, proving the potential of GO-based C-MIPs for environmental remediation.
- **[6]: Liu et al.** developed CNT-based MIPs using **surface imprinting** to detect **tetracycline** in milk samples. This method improved the accessibility and efficiency of binding sites, facilitating rapid and accurate detection.

Impact: Enhanced food safety monitoring systems for trace contaminants.

- **Improved synthesis techniques**
- **In situ polymerization:**
  Allows uniform distribution of carbon materials within the polymer matrix, increasing structural integrity and binding performance.

- **Surface imprinting:**
  Creates accessible imprinted sites on the surface of carbon materials, enhancing sensitivity for applications in catalysis and sensing.

- **Post-synthesis modifications:**
  Functionalizing carbon materials after MIP formation adds versatility, allowing customization for specific applications, such as multi-analyte detection.

- **Emerging technologies in carbon integration**
- **Nanocomposites:** Combining carbon materials with metal nanoparticles to create hybrid MIPs with improved electrochemical and optical properties.
- **3D printing of MIPs:** Facilitates precise control over the shape and structure of C-MIPs, improving efficiency and scalability.
- **AI-driven optimization:** Leveraging machine learning to optimize synthesis parameters for tailored C-MIPs.

Conclusion: The integration of carbon materials into MIPs has addressed the limitations of traditional MIPs, offering superior sensitivity, selectivity, and durability. By combining innovative synthesis methods with advanced carbon nanomaterials, researchers have broadened the scope of MIP applications in critical areas such as environmental monitoring, healthcare, and food safety.

8.4.2 Expanded applications and future directions

a. **Environmental monitoring:**
   C-MIPs are used to detect pollutants, such as pesticides and heavy metals, with high sensitivity.
b. **Biomedical sensing:**
   Integration of C-MIPs into biosensors enables the detection of biomarkers for diseases like diabetes and Alzheimer's, enhancing early diagnosis capabilities.
c. **Industrial catalysis:**
   C-MIPs are increasingly employed in catalytic processes requiring specificity and thermal stability, such as polymer synthesis and fuel production.
d. **Future research:**

Emerging techniques like **3D printing** and **machine learning** are being combined with C-MIPs to optimize synthesis methods and enhance applications, paving the way for next-generation molecular recognition materials.

8.5 Industrial and analytical applications for carbon-based molecularly imprinted polymers (C-MIPs)

8.5.1 Analytical applications for carbon-based MIPs

CMIPs are widely employed in sensor technology due to their superior selectivity and sensitivity. The incorporation of carbon elements into MIPs improves their performance in numerous types of sensors, making them extremely effective at detecting a wide range of analytes.

- **Sensors**

Sensors are electrical devices that can detect minute differences in input energy and convert it into the same or different forms of energy as output signals [7]. They are essential in a wide range of disciplines, including environmental monitoring, health inspections, and industrial applications.

– **Electrochemical sensors**

Electrochemical sensors are frequently employed to detect and monitor a variety of common factors. These sensors use the electrical conductivity of carbon materials, with the MIP serving as an identification element and the carbon material enhancing signal transduction. Applications include the detection of contaminants, biomarkers, and medicinal substances. The sensing device reacts to chemical ions or physical movement, and the sensor's electrical response provides a signal, which is often associated with the electrochemical gradient, current, and resistance [7].

– **Optical sensors**

Carbon-based MIPs can be used in optical sensors to monitor changes in optical properties (e.g., fluorescence, absorbance) following the binding of the target molecule. These sensors are employed in environmental monitoring and biological diagnostics, providing highly sensitive detection of specific compounds via changes in optical signals.

– **Mass-sensitive sensors**

Carbon-based MIPs are used to functionalize mass-sensitive sensors, such as quartz crystal microbalances (QCMs) and surface acoustic wave (SAW) devices, allowing them to detect mass changes caused by target molecule binding. These sensors are particularly useful for sensing and monitoring volatile organic compounds (VOCs), as they provide precise measurements of mass fluctuations at the molecular level.

8.5.2 Industrial applications for carbon-based molecularly imprinted polymers (C-MIPs): environmental monitoring

C-MIPs have emerged as effective tools in a variety of industrial applications due to their great selectivity, sensitivity, and resilience. Their ability to detect and remove pollutants from air, water, and soil has established them as key components of environmental monitoring systems. This chapter investigates the use of C-MIPs in environmental monitoring, with an emphasis on air quality, water quality, and soil quality, as well as the opportunities for future developments and problems.

– **Air quality monitoring**

Carbon-based MIPs are critical for monitoring air quality because they detect gaseous pollutants such as VOCs, nitrogen oxides (NO_x), and sulfur dioxide (SO_2). These sensors enable real-time monitoring with great sensitivity and selectivity.

– **Detecting VOCs**

VOCs cause significant air pollution and pose health hazards. C-MIPs have been successfully employed to detect VOCs due to their high sensitivity and selectivity.

Key studies

a. Wang et al. [8] developed a graphene-based MIP sensor for detecting benzene. The sensor demonstrated good sensitivity and selectivity, highlighting the potential of C-MIPs for air quality monitoring.
b. Li et al. [9] reported the application of CNT-based MIPs to detect toluene in indoor air. The study emphasized the sensor's improved binding capacity and short response time, making it appropriate for real-time air quality monitoring.

– **Detecting hazardous gases**
C-MIPs are also used to detect dangerous chemicals, including nitrogen oxides (NO_x) and sulfur dioxide (SO_2), which contribute significantly to air pollution and have negative consequences for both health and the environment.

Key studies

a. Chen et al. [10] used GO-based MIPs to detect NO_2 in the air. The sensor demonstrated high sensitivity and selectivity, as well as a low detection limit and rapid response time.
b. Zhang et al. [11] developed CNT-based MIPs for SO_2 detection, which demonstrated great sensitivity and stability under a variety of environmental conditions.

– **Water quality monitoring**
Carbon-based MIPs are used in water treatment procedures to selectively adsorb and remove pollutants such as heavy metals, herbicides, and pharmaceutical residues. The use of carbon materials enhances adsorption capacity and reusability.

– **The detection of heavy metals**
Heavy metals (Pb, Hg, and Cd) are harmful pollutants that endanger aquatic ecosystems and human health. C-MIPs have been used to selectively detect and remove heavy metals from water.

Key studies

a. Yang et al. [12] developed GO-based MIPs for detecting lead(II) ions in water. The study found that the MIPs have a high adsorption capacity, selectivity, and reusability, making them appropriate for water purification applications.
b. Liu et al. [13] developed CNT-based MIPs for detecting Hg(II) ions, achieving great sensitivity and selectivity while maintaining a low detection limit. The MIPs were

also proven to be reusable, with their performance remaining consistent throughout numerous cycles.

- **Detection of organic pollutants**

C-MIPs have been used to identify organic contaminants in water, including pesticides, medications, and endocrine-disrupting compounds (EDCs).

Key studies

a. Huang et al. [14] used GO-based MIPs for the selective detection of EDCs in water. The study demonstrated the MIPs' strong binding capability and selectivity for several EDCs.
b. Zhao et al. [15] reported the application of CNT-based MIPs for pesticide detection in water samples, demonstrating great sensitivity and selectivity in detecting low pesticide concentrations.

- **Soil quality monitoring**

C-MIPs are also utilized to extract contaminants from soil, providing a selective and effective approach to soil remediation. Pesticide, herbicide, and heavy metal detection are some of the applications.

- **Detection of pesticides and herbicides**

Pesticides and herbicides are frequently used in agriculture, but they can contaminate the soil, compromising soil health, and potentially entering the food chain.

Key studies

a. Chen et al. [16] used graphene-based MIPs to detect atrazine, a prevalent herbicide, in soil samples. The MIPs demonstrated good binding capacity and selectivity.
b. Wang et al. [17] reported the use of CNT-based MIPs to detect chlorpyrifos, a commonly used pesticide. The study confirmed the MIPs' high sensitivity and selectivity.

- **The detection of heavy metals**

C-MIPs are also used to identify heavy metals in soil samples, thereby reducing the risks these contaminants pose to plants, animals, and humans.

Key studies

a. Liu et al. [6] developed GO-based MIPs for detecting Cd(II) ions in soil, demonstrating good selectivity and sensitivity.
b. Zhang et al. [18] developed CNT-based MIPs for detecting Pb(II) ions in soil, demonstrating good binding capacity and selectivity.

8.5.3 Use of carbon-based molecularly imprinted polymers in the pharmaceutical industry

The pharmaceutical industry is a crucial sector that focuses on the discovery, development, and production of drugs. This industry demands cutting-edge technologies for drug analysis, quality control, drug delivery, and chiral separation. C-MIPs have emerged as a viable tool in the pharmaceutical business, thanks to their exceptional selectivity, sensitivity, and durability. The incorporation of carbon materials such as graphene, CNTs, and activated carbon into MIPs significantly enhances their characteristics, making them suitable for a variety of medicinal applications. This chapter delves into the various applications of C-MIPs in the pharmaceutical sector, emphasizing their importance in drug analysis, quality control, drug delivery, and chiral separation.

– ***Benefits of C-MIPs***

– **High selectivity and sensitivity**

C-MIPs are engineered with specialized recognition sites targeted to medicinal molecules. The addition of carbon components, such as graphene and CNTs, improves the selectivity and sensitivity of these polymers. Carbon materials' high surface area and functional ability lead to more binding sites and better interaction with target molecules.

– **Increased surface area and binding capacity**

Carbon materials have a large surface area, which increases the number of available binding sites in C-MIPs. This enhancement results in increased binding capacity and more efficient detection and separation of medicinal substances. The large surface area also allows for quick adsorption and desorption processes, which are critical for real-time analysis and quality control.

– **High robustness and reusability**

Carbon materials' mechanical strength and chemical stability provide C-MIPs with durability and reusable properties. These qualities ensure that C-MIPs can withstand harsh environments and frequent use without significantly degrading performance, making them excellent for long-term pharmaceutical applications.

– **Applications in drug analysis**

The detection and quantification of active pharmaceutical ingredients (APIs).

C-MIPs have been successfully applied to the selective detection and quantification of APIs in complicated mixtures. Their great selectivity and sensitivity make them ideal for measuring API levels in varied formulations and guaranteeing regulatory compliance.

Example study

- Chen et al. [19] developed graphene-based MIPs to detect acetaminophen in pharmaceutical formulations. The sensor demonstrated good sensitivity and selectivity, with a low detection limit, highlighting C-MIPs' potential for accurate API analysis.

– **Monitoring of drug metabolites**

C-MIPs are also used to monitor drug metabolites in biological samples, which helps with pharmacokinetic investigations and therapeutic medication monitoring. Their ability to selectively bind and detect metabolites in complex matrices is critical for studying medication metabolism and efficacy.

Example study

- Zhao et al. [20] reported using CNT-based MIPs to identify theophylline metabolites in urine samples. The sensor demonstrated increased binding capacity and selectivity, making it ideal for real-time monitoring of drug metabolites.

– **Applications in quality control**

Purity testing for pharmaceutical products: Ensuring the purity of pharmaceutical goods is critical for patient safety and regulatory compliance. C-MIPs are used to detect and quantify contaminants and degradation products in pharmaceutical formulations, ensuring that they meet quality specifications.

Example study

- Wang et al. [8] developed GO-based MIPs to detect ibuprofen contaminants in pharmaceutical products. The MIPs demonstrated remarkable sensitivity and selectivity, detecting trace quantities of contaminants, and indicating their potential for quality control applications.

– **Stability testing**

Stability testing is an important part of pharmaceutical research, as it assesses how the quality of a medicinal substance or product changes over time under different environmental circumstances. C-MIPs can be used to evaluate the stability of pharmaceutical substances, providing useful information for shelf-life estimation and formulation improvement.

Example study

– Liu et al. [21] used CNT-based MIPs for the stability assessment of aspirin in pharmaceutical formulations. The study confirmed the MIPs' great sensitivity and selectivity for detecting degradation products, emphasizing their relevance in stability testing.

– **Applications in drug delivery**

– **Controlled release systems**

C-MIPs have been used in the creation of controlled release systems, which are intended to release medications at a regulated pace over a predetermined time. Carbon materials' high surface area and functionalizability enable precise control of drug loading and release kinetics.

Example study

Zhao et al. [22] created CNT-based MIPs for regulated anticancer medication release. The study found that adding CNTs increased drug-loading capacity and release kinetics, making the system appropriate for targeted drug administration.

– **Targeted drug delivery**

Targeted drug delivery aims to deliver medications to specific tissues or cells, reducing side effects and enhancing therapeutic efficacy. C-MIPs can be customized to recognize and bind to specific target molecules, enabling medications to be delivered to the desired locations.

Example study

Liu et al. [23] used GO-based MIPs for targeted medication delivery to cancer cells. The study highlighted the MIPs' remarkable selectivity and binding capacity, indicating their potential for targeted therapy.

- **Applications in chiral separation**

- **Enantiomeric separation**

Chiral separation is critical in the pharmaceutical industry because different enantiomers of a chiral medication may have distinct therapeutic and adverse effects. C-MIPs are utilized to selectively separate enantiomers, resulting in enantiomerically pure medicines.

Example study

Wang et al. [24] used graphene-based MIPs to separate ibuprofen enantiomers. The study showed that C-MIPs can separate enantiomers with great selectivity and efficiency, underlining their potential for chiral separation applications.

- **Purification of chiral compounds**

C-MIPs are also used to purify chiral substances throughout the production process, guaranteeing that the finished product is devoid of undesired enantiomers. Their high selectivity and binding capability make them ideal for large-scale purification applications.

Example study

Chen et al. [25] used CNT-based MIPs to purify chiral amino acids. The study indicated that MIPs are highly efficient and selective in eliminating undesired enantiomers, making them ideal for large-scale purification.

- **Uses of C-MIPs in biomedical sciences**

Biomedical applications necessitate high precision, sensitivity, and selectivity in detecting and interacting with biological molecules. C-MIPs offer considerable benefits in these areas because of their large surface area, outstanding electrical conductivity, and strong mechanical qualities. Researchers have developed improved materials for biomedical applications by incorporating carbon components such as graphene, CNTs, and activated carbon into MIPs. This chapter explores the applications of C-MIPs in biosensing, drug administration, and diagnostic imaging, emphasizing their advantages and potential future advancements.

- ***Benefits of C-MIPs***

- **High selectivity and sensitivity**

C-MIPs are designed to have specialized recognition sites for target biomolecules. The addition of carbon compounds increases surface area and binding capacity, resulting in higher selectivity and sensitivity. This is especially important in biomedical applications, where identifying low amounts of biomarkers or medicinal substances is critical.

- **Enhanced surface area and conductivity**

Carbon compounds have a high surface area and excellent electrical conductivity, which make them ideal for building sensitive biosensors. The large surface area increases the number of binding sites, while the conductivity improves signal transduction, resulting in extremely sensitive detection devices.

- **Robustness and biocompatibility**

Carbon materials' mechanical strength and chemical stability provide C-MIPs with durability and reusable properties. Furthermore, carbon materials, like graphene and CNTs, have demonstrated high biocompatibility, making them appropriate for a variety of biomedical applications.

- **Applications in biosensing**

- **Detecting biomarkers**

C-MIPs have been successfully utilized to selectively detect biomarkers, which are essential for illness diagnosis and monitoring. Their high selectivity and sensitivity allow them to detect biomarkers at low concentrations in complex biological samples.

Example study

Chen et al. [19] used graphene-based MIPs to selectively identify cancer biomarkers in serum samples. The sensor demonstrated good sensitivity and selectivity, with a low detection limit, highlighting C-MIPs' potential for early cancer diagnosis.

- **Detection of pathogens**

Rapid and precise pathogen identification is critical for preventing and controlling infectious illnesses. C-MIPs have been used to selectively identify bacterial and viral pathogens, providing benefits in terms of sensitivity and specificity.

Example study

Zhao et al. [26] reported using CNT-based MIPs to identify *E. coli* bacterium. The sensor demonstrated increased binding capacity and selectivity, making it ideal for real-time pathogen detection in clinical situations.

- **Applications in drug delivery**

- **Controlled release systems**

C-MIPs have been used in the creation of controlled release systems, which are intended to release medications at a regulated pace over a predetermined time. Carbon materials' high surface area and functional ability enable precise control of drug loading and release kinetics.

Example study

Zhao et al. [27] created CNT-based MIPs for regulated anticancer medication release. The study found that adding CNTs increased drug-loading capacity and release kinetics, making the system appropriate for targeted drug administration.

- **Targeted drug delivery**

Targeted drug delivery aims to deliver medications to specific tissues or cells, reducing side effects and enhancing therapeutic efficacy. C-MIPs can be customized to recognize and bind specific target molecules, enabling medications to be delivered to the intended locations.

Example study

Liu et al. [23] used GO-based MIPs for targeted medication delivery to cancer cells. The study highlighted the MIPs' remarkable selectivity and binding capacity, indicating their potential for targeted therapy.

- **Applications for diagnostic imaging: MRI (magnetic resonance imaging)**

C-MIPs have been investigated as contrast agents for MRI, which can enhance the imaging of specific tissues or cells. Carbon compounds can improve MRI contrast and specificity, enabling earlier identification and diagnosis of illnesses.

Example study

Wang et al. [24] developed CNT-based MIPs with gadolinium as MRI contrast agents. The study found that C-MIPs have improved contrast and specificity, making them appropriate for imaging malignant tissues.

- **Fluorescence imaging**

Fluorescence imaging is commonly employed in biomedical research to visualize cellular and molecular processes. C-MIPs can be created with fluorescent tags to enable the selective imaging of target molecules within biological systems.

Example study

Chen et al. [25] used GO-based MIPs with fluorescent tags to image glucose levels in living cells. The study demonstrated exceptional selectivity and sensitivity, allowing real-time monitoring of glucose levels within cells.

- **Applications of C-MIPs in the detection of biomarkers**

Biomarker detection is critical in the diagnosis, monitoring, and management of a wide range of disorders. Proteins, nucleic acids, metabolites, and other small molecules can all be used as biomarkers. Traditional biomarker detection techniques, such as enzyme-linked immunosorbent assays (ELISAs) and mass spectrometry, can necessitate sophisticated, time-consuming procedures and costly equipment. C-MIPs have emerged as a promising alternative, with good selectivity, sensitivity, and robustness for biomarker detection. This article investigates the use of C-MIPs to detect biomarkers, highlighting their benefits, synthesis methods, and major research confirming their efficacy.

- **Benefits of C-MIPs**

- **High selectivity and sensitivity**

C-MIPs are built with biomolecule-specific recognition sites. The addition of carbon compounds, such as graphene, CNTs, and activated carbon, improves the selectivity and sensitivity of these polymers. Carbon materials' high surface area and functionalizability lead to more binding sites and better interaction with target molecules.

- **Enhanced surface area and conductivity**

Carbon compounds have a high surface area and excellent electrical conductivity, which make them ideal for building sensitive biosensors. The large surface area increases the number of binding sites, while the conductivity improves signal transduction, resulting in extremely sensitive detection devices.

– **Robustness and biocompatibility**

Carbon materials' mechanical strength and chemical stability provide C-MIPs with durability and reusable properties. Furthermore, carbon materials like graphene and CNTs have demonstrated high biocompatibility, making them appropriate for a variety of biomedical applications.

– **Synthesis methods for C-MIP**

In situ polymerization

In situ polymerization entails scattering carbon materials in a monomer solution that contains the template molecule. The polymerization process begins, incorporating the carbon element into the polymer matrix. This approach ensures that the carbon material and polymer interact uniformly and strongly.

Example study

Guo et al. [5] developed GO-based MIPs for cancer biomarker detection by in situ polymerization. The homogeneous dispersion of GO within the polymer matrix resulted in increased adsorption capacity and selectivity in the C-MIPs.

– **Surface imprinting**

Surface imprinting entails modifying carbon materials with monomers that interact with the target molecule. The template molecule is introduced, followed by polymerization, which results in a thin polymer coating on the carbon surface. This approach increases the surface area and accessibility of imprinted spots.

Example study

Chen et al. [19] used surface imprinting to create graphene-based MIPs for the selective detection of cardiac biomarkers. The thin polymer coating on the graphene surface resulted in excellent binding capacity and rapid adsorption kinetics, proving the utility of surface imprinting for biomarker detection.

– **Sol-gel process**

The sol-gel technique converts a solution (sol) into a solid (gel) phase to produce porous materials. Incorporating carbon elements into the sol-gel process can result in C-MIPs with high porosity and surface area.

Example study

Gao et al. [28] employed the sol-gel technique to incorporate silane-functionalized CNTs into MIPs to detect protein biomarkers. Due to the sol-gel process's high porosity and surface area, the resulting C-MIPs demonstrated good mechanical strength, stability, and enhanced adsorption capacity.

- **Applications for biomarker detection**

- **Cancer biomarkers**

Cancer biomarkers are chemicals that can detect the presence or progression of cancer. Detecting these biomarkers is critical for early detection and monitoring therapy effectiveness. C-MIPs have been successfully used for the selective identification of cancer biomarkers in a variety of biological samples.

Example study

Chen et al. [19] created graphene-based MIPs for the selective detection of alpha-fetoprotein (AFP), a liver cancer biomarker, in serum samples. The sensor demonstrated good sensitivity and selectivity, with a low detection limit, highlighting C-MIPs' potential for early cancer diagnosis.

- **Cardiovascular biomarkers**

Cardiac biomarkers are proteins that are released into the bloodstream when the heart is damaged or stressed. Detecting these indicators is critical for detecting and treating cardiovascular disease. C-MIPs have been used to selectively identify cardiac biomarkers, providing benefits in terms of sensitivity and specificity.

Example study

Liu et al. [6] described the application of CNT-based MIPs to detect cardiac troponin I (cTnI) in blood samples. The sensor demonstrated improved binding capacity and selectivity, making it ideal for real-time monitoring of cardiac events.

- **Neurological biomarkers**

Neurological biomarkers are molecules that represent the degenerative processes of neurological illnesses, including Alzheimer's and Parkinson's. C-MIPs have been utilized to selectively identify these biomarkers, allowing for early detection and tracking of illness progression.

Example study

Wang et al. [8] used GO-based MIPs to selectively detect amyloid-beta (Aβ) peptides, biomarkers for Alzheimer's disease. The MIPs demonstrated remarkable sensitivity and selectivity, detecting even low amounts of Aβ peptides in cerebral fluid.

– **Biomarkers for infectious diseases**

Biomarkers for infectious diseases can detect the presence of pathogens or the host's reaction to infection. C-MIPs have been used to selectively identify these biomarkers, resulting in speedy and accurate detection of infectious illnesses.

Example study

Zhao et al. [18] used CNT-based MIPs to detect CRP, a biomarker of inflammation and infection, in serum samples. The sensor demonstrated good sensitivity and selectivity, making it ideal for monitoring inflammatory responses and infections.

– **Applications of C-MIPs in pathogen detection**

Pathogen detection is critical for diagnosing and managing infectious diseases, ensuring food safety, and monitoring environmental contamination levels. Traditional pathogen detection methods, such as culture-based procedures and polymerase chain reaction (PCR), are often time-consuming, labor-intensive, and require specialized equipment. C-MIPs represent a promising alternative due to their high selectivity, sensitivity, and robustness. Researchers have developed enhanced materials for pathogen detection by incorporating carbon materials, including graphene, CNTs, and activated carbon, into MIPs. This article explores the use of C-MIPs in pathogen detection, highlighting their advantages, synthesis methods, and key research demonstrating their efficacy.

– **Benefits of C-MIPs**

– **High selectivity and sensitivity**

C-MIPs are designed to have specialized recognition sites for target pathogens. The addition of carbon compounds increases surface area and binding capacity, resulting in higher selectivity and sensitivity. This is critical for detecting low levels of pathogens in complex biological and environmental samples.

– **Enhanced surface area and conductivity**

Carbon compounds have a high surface area and excellent electrical conductivity, which make them ideal for building sensitive biosensors. The large surface area increases the number of binding sites, while the conductivity improves signal transduction, resulting in extremely sensitive detection devices.

- **High robustness and reusability**

Carbon materials' mechanical strength and chemical stability provide C-MIPs with durability and reusable properties. These characteristics ensure that C-MIPs can resist harsh conditions and frequent usage without significantly deteriorating in performance, making them excellent for long-term pathogen detection applications.

- **Synthesis methods for C-MIP**

In situ polymerization.

In situ polymerization entails dispersing carbon materials in a monomer solution that contains the template molecule. The polymerization process begins, incorporating the carbon element into the polymer matrix. This approach ensures that the carbon material and polymer interact uniformly and strongly.

Example study

Guo et al. [5] used in situ polymerization to create GO-based MIPs for selectively detecting *Escherichia* bacteria. The homogeneous dispersion of GO within the polymer matrix resulted in increased adsorption capacity and selectivity in the C-MIPs.

- **Surface imprinting**

Surface imprinting is the process of functionalizing carbon materials using monomers that interact with the target pathogen. The template molecule is injected, followed by polymerization, which results in a thin polymer coating on the carbon surface. This approach increases the surface area and accessibility of imprinted spots.

Example study

Chen et al. [19] used surface imprinting to create graphene-based MIPs for the selective detection of Staphylococcus aureus. The thin polymer coating on the graphene surface resulted in excellent binding capacity and rapid adsorption kinetics, proving the efficacy of surface imprinting for pathogen detection.

- **Sol-gel process**

The sol-gel technique converts a solution (sol) into a solid (gel) phase to produce porous materials. Incorporating carbon elements into the sol-gel process can result in C-MIPs with high porosity and surface area.

Example study

Gao et al. [28] employed the sol-gel method to incorporate silane-functionalized CNTs into MIPs for Salmonella detection. Due to the sol-gel process's high porosity and surface area, the resulting C-MIPs demonstrated good mechanical strength, stability, and enhanced adsorption capacity.

- **Applications for pathogen detection**

- **Bacterial pathogens**

Bacterial pathogens, including *E. coli*, *S. aureus*, and *Salmonella*, are frequent causes of foodborne illnesses and infections. Detecting these infections quickly and reliably is critical to food safety and public health.

Example studies

Guo et al. [5] designed GO-based MIPs for the specific detection of *E. coli*. The sensor demonstrated good sensitivity and selectivity with a low detection limit, highlighting C-MIPs' potential for real-time monitoring of bacterial contamination in food and water samples.

Chen et al. [19] described the application of graphene-based MIPs to detect *S. aureus* in clinical specimens. The MIPs demonstrated increased binding capability and selectivity, making them useful for identifying bacterial infections in hospital settings.

Gao et al. [28] used CNT-based MIPs to detect *Salmonella* in food samples. The study showed great sensitivity and specificity, demonstrating the usefulness of C-MIPs in ensuring food safety.

- **Viral pathogens**

Viral diseases, such as influenza and human papillomavirus (HPV), are serious health hazards that necessitate quick and precise detection procedures. C-MIPs have been used to selectively identify viral pathogens, offering advantages in terms of sensitivity and specificity.

Example studies

a. Zhao et al. [20] created CNT-based MIPs for the selective detection of influenza viruses. The sensor demonstrated increased binding capacity and selectivity, making it ideal for real-time monitoring of viral infections in clinical settings.

b. Liu et al. [23] described the application of GO-based MIPs for HPV detection in clinical samples. The study emphasized the MIPs' excellent selectivity and binding capacity, suggesting their potential for the early detection of viral infections.

– **Fungal pathogens**

Fungal pathogens, such as *Candida albicans*, are prevalent sources of infection, especially in immunocompromised individuals. Detecting these pathogens quickly and accurately is critical for successful treatment and management.

Example study

Wang et al. [8] created GO-based MIPs for the selective detection of *Candida albicans* in clinical specimens. The MIPs were highly sensitive and selective, detecting low amounts of fungal pathogens, making them appropriate for diagnosing fungal infections.

– **Parasitic pathogens**

Parasitic pathogens, such as *Plasmodium* species, which cause malaria, must be detected quickly and accurately to be treated and controlled effectively. C-MIPs have been used to selectively identify these pathogens, providing benefits in terms of sensitivity and specificity.

Example study

Chen et al. [10] used graphene-based MIPs to identify *Plasmodium falciparum* in blood samples. The sensor displayed good sensitivity and selectivity, making it ideal for diagnosing malaria in endemic areas.

8.6 Key tools and techniques in carbon-based molecularly imprinted polymers' (C-MIPs) research and applications

8.6.1 Tools and techniques

8.6.1.1 Synthesis methods

- In situ polymerization: Uniform dispersion of carbon materials (e.g., graphene, CNTs) in monomer solutions containing the target template molecule. Ensures strong interactions between carbon and the polymer.
- Surface imprinting: Functionalization of carbon materials with template molecules, followed by polymerization, to form a thin, accessible polymer layer.
- Sol-gel process: Produces porous materials through a solution-to-solid phase transformation. Incorporates carbon elements for enhanced porosity and surface area.

8.6.2 Analytical techniques

- **Electrochemical analysis:**
- Cyclic voltammetry (CV): Evaluates electrochemical behavior and sensor efficiency.
- Electrochemical impedance spectroscopy (EIS): Assesses binding interactions and conductivity enhancements.

- **Spectroscopic methods:**
- UV-Vis spectroscopy: Measures template-molecule binding.
- Fluorescence spectroscopy: Detects changes in optical signals for imaging and sensing.

- **Chromatography:**
- HPLC: Used with C-MIPs for high-selectivity separations.
- TLC: Demonstrates template-specific interactions.

- **Characterization tools:**
- Scanning electron microscopy (SEM): Examines surface morphology and imprinting efficiency.
- Transmission electron microscopy (TEM): Visualizes the internal structure of C-MIPs.
- Fourier-transform infrared spectroscopy (FTIR): Confirms functional group interactions in MIPs.

- X-ray diffraction (XRD): Studies crystallinity and structural changes.
- Thermogravimetric analysis (TGA): Assesses thermal stability of synthesized C-MIPs.

8.6.3 Application-specific methods

- **Biosensing:**
  - Use of QCMs and SAW devices for mass-sensitive sensing.
  - Integration with electrochemical transducers for real-time pathogen or biomarker detection.
- **Environmental monitoring:**
  - Adsorption studies to evaluate pollutant-binding efficiency.
  - Contaminant removal studies using water and soil samples.

8.6.4 Steps and guidelines for using C-MIPs

- **Synthesis of C-MIPs:**
- Select the target molecule (template) and a suitable carbon material (e.g., graphene and CNTs).
- Functionalize carbon materials if required (e.g., oxidation, silane coupling).
- Mix carbon materials with a monomer solution containing the template molecule.
- Initiate polymerization using UV, thermal, or chemical methods.
- Remove the template molecule (e.g., via washing or solvent extraction) to create specific binding cavities.

- **Characterization:**
- Analyze morphology and structural integrity using SEM and TEM.
- Confirm molecular recognition sites using FTIR and TGA.
- Assess the thermal and chemical stability of C-MIPs.

- **Integration into applications:**
- Functionalize C-MIPs for specific use (e.g., attach fluorescent tags for imaging).
- Incorporate into devices (e.g., electrochemical sensors and optical readers).
- Test performance under application-specific conditions (e.g., pH, temperature, and salinity for environmental monitoring).

- **Performance testing:**
- Conduct binding assays to determine selectivity and sensitivity.
- Perform real-sample analyses (e.g., in biological fluids, water, or air).
- Validate reproducibility and reusability under operational conditions.

8.6.5 Summary of key methods and frameworks

- **Key methods:**
- Synthesis: In situ polymerization, surface imprinting, and sol-gel methods.
- Analysis: CV, EIS, UV-Vis, FTIR, and fluorescence spectroscopy.
- Applications: Integration into biosensors, pollutant removal systems, and drug-delivery mechanisms.

8.6.6 Framework for implementation

- Design and synthesis: Choose target molecules and carbon materials based on application needs.
- Characterization and optimization: Confirm imprinting quality and optimize material properties.
- Integration: Adapt C-MIPs for specific analytical or industrial tools.
- Testing and validation: Ensure high performance in real-world conditions, and refine the design based on results.

8.7 Challenges and limitations of carbon-based molecularly imprinted polymers (C-MIPs)

8.7.1 Challenges

- **Template removal**
- Issue: Effective removal of the template molecule is critical for creating selective recognition sites. Incomplete removal can impair polymer performance, reducing selectivity and sensitivity.
- Real-world example: In water purification applications, residual templates may interfere with pollutant detection or removal, reducing the system's reliability.

- **Scalability**
- Issue: Developing scalable and cost-effective synthesis methods for C-MIPs is challenging, especially for industrial and environmental monitoring applications.
- Real-world example: The batch production of C-MIPs for large-scale environmental pollutant removal is constrained by high costs and the complexity of synthesis processes.

- **Reusability and stability**
- Issue: Ensuring the long-term reusability of C-MIPs without compromising their performance is essential for practical applications.
- Real-world example: C-MIPs used in biosensors for clinical diagnostics may degrade after multiple usage cycles, leading to inconsistent readings.

- **Biocompatibility and safety**
- Issue: For biomedical applications, ensuring that C-MIPs are biocompatible and non-toxic is essential. Long-term studies are required to assess potential immunological or toxicological effects.
- Real-world example: CNT-based C-MIPs used in drug delivery systems may exhibit cytotoxicity due to residual chemical groups or incomplete purification.

- **Environmental and economic considerations**
- Issue: Many current synthesis methods involve the use of non-eco-friendly solvents or chemicals, raising concerns about sustainability and environmental impact.
- Real-world example: The use of organic solvents in the synthesis of C-MIPs for environmental applications contradicts the goal of reducing environmental pollution.

- **Standardization and regulatory issues**
- Issue: The lack of standardized regulatory frameworks for the design, testing, and application of C-MIPs limits their industrial and clinical adoption.
- Real-world example: Pharmaceutical applications of C-MIPs for drug quality control face delays due to insufficient regulatory clarity and certification processes.

8.7.2 Limitations

- **Limited selectivity in complex mixtures**

Description: In highly complex biological or environmental samples, C-MIPs may exhibit cross-reactivity with non-target molecules, thereby reducing their efficiency.

- Example: Biosensors using C-MIPs for biomarker detection may generate false positives in serum samples with high protein content.

- **Thermal and mechanical limitations**
- Description: Although carbon materials enhance stability, some MIP matrices are prone to degradation under extreme temperatures or pressures.
- Example: C-MIPs used in catalytic processes may fail when exposed to industrial-scale conditions requiring high thermal stability.

- **Cost of advanced materials**
- Description: Incorporating high-quality carbon materials like CNTs or graphene can make C-MIPs prohibitively expensive for widespread use.
- Example: Water treatment plants using graphene-based C-MIPs face financial constraints due to the high cost of raw materials.

8.7.3 Integration with technology

- Description: Incorporating C-MIPs into microfluidics, lab-on-a-chip devices, and wireless sensors requires precise engineering, thereby increasing production complexity.
- Example: Lab-on-a-chip devices with embedded C-MIPs for real-time pathogen detection face delays in commercialization due to engineering challenges.

- **Performance variability**
- Description: Small inconsistencies in synthesis or functionalization can lead to performance variability between batches.
- Example: Environmental monitoring sensors using different batches of C-MIPs may exhibit fluctuating sensitivity, complicating data interpretation.

8.7.4 Examples of real-world challenges

- **Environmental monitoring:**

Scenario: Detecting trace-level pollutants, such as bisphenol A in river water.

Challenge: C-MIPs must operate efficiently in samples with high turbidity and a variety of interfering substances.

- **Pharmaceutical industry:**

Scenario: Ensuring the stability of drug delivery systems based on C-MIPs for cancer therapies.

Challenge: Prolonged-release in biological environments can alter polymer integrity and reduce efficacy.

- **Biomedical diagnostics:**

Scenario: Detecting cardiac biomarkers in emergency settings using C-MIP biosensors.

Challenge: Ensuring rapid detection while maintaining high specificity in a competitive binding environment.

- **Pathogen detection:**

Scenario: Deploying C-MIPs for detecting airborne pathogens in public spaces.

Challenge: High sensitivity is required under varying humidity and temperature conditions without compromising the device's portability.

8.7.5 Future perspectives to address challenges

- **Developing green synthesis methods:** Utilize eco-friendly solvents and minimize waste in MIP production to reduce environmental impact.
- **Exploring alternative carbon materials:** Investigate low-cost, sustainable carbon materials, such as biochar, as substitutes for graphene and CNTs.
- **Standardization of protocols:** Establish industry-wide standards for the synthesis, characterization, and application of C-MIPs.
- **Advanced engineering solutions:** Leverage 3D printing and AI optimization for scalable, precise, and efficient C-MIP production.
- **Collaborative research:** Foster partnerships between academia, industry, and regulatory bodies to accelerate innovation and adoption of C-MIPs in diverse sectors.

By addressing these challenges and leveraging future advancements, C-MIPs can achieve their full potential in analytical, environmental, pharmaceutical, and biomedical applications.

8.8 Conclusion

C-MIPs represent a transformative innovation in the fields of environmental monitoring, pharmaceutical quality control, biomedical diagnostics, and pathogen detection. By integrating the remarkable properties of carbon materials – such as high surface area, electrical conductivity, and mechanical stability – with the precision of molecular imprinting, C-MIPs offer unparalleled selectivity, sensitivity, and robustness for diverse applications.

This chapter explores the origins, synthesis, applications, and future directions of C-MIPs, highlighting their ability to address complex challenges in analytical and industrial sciences. While C-MIPs have shown tremendous promise, addressing key challenges such as template removal, scalability, and biocompatibility will be critical for their broader adoption. The ongoing development of advanced synthesis techniques, hybrid materials, and integration into emerging technologies will pave the way for future innovations.

As the field progresses, C-MIPs are poised to become indispensable tools for tackling global challenges in environmental sustainability, public health, and industrial efficiency. Their versatility and adaptability make them a cornerstone of future technological advancements.

8.9 Questions for reflection and discussion

a. **Scalability and cost:** What steps can be taken to develop cost-effective and scalable production methods for C-MIPs, especially for large-scale industrial and environmental applications?
b. **Hybrid materials:** How can the synergistic properties of hybrid carbon-based materials be optimized to improve the performance of C-MIPs in specific applications?
c. **Integration with advanced technologies:** What are the potential barriers to incorporating C-MIPs into emerging technologies such as wearable sensors, lab-on-a-chip devices, or wireless networks? How can these barriers be overcome?
d. **Regulatory and safety considerations:** How can researchers and industry professionals collaborate to establish standardized guidelines for the safe and effective use of C-MIPs in pharmaceuticals and biomedical sciences?
e. **Environmental impact:** Given the reliance on carbon materials and chemical processes in their synthesis, how can the environmental footprint of C-MIPs be minimized while maintaining their functionality?

As this chapter concludes, consider the broader implications of C-MIPs in solving societal challenges. Future chapters will explore the integration of C-MIPs into emerging technologies, advanced material design for enhanced functionality, and real-world case studies that showcase their impact. By building on this foundational understanding, we can unlock the full potential of C-MIPs to shape a sustainable and innovative future.

References

[1] Andersson, L. I., et al. (1995). Molecular imprinting of amino acid derivatives in macroporous polymers. *Analytica Chimica Acta*, 302, 15–27. https://doi.org/10.1016/0003-2670(94)00346-X.
[2] Sellergren, B. (2000). *Molecularly Imprinted Polymers: Man-Made Mimics of Antibodies and Their Applications in Analytical Chemistry*. Elsevier Science, Amsterdam. ISBN: 9780444504411.
[3] Pichon, V. (2002). Solid-phase extraction using molecularly imprinted polymers. *Analytical and Bioanalytical Chemistry*, 382(2), 428–438. https://doi.org/10.1007/s00216-005-3211-y.

[4] Wulff, G. (1995). Molecular imprinting in cross-linked materials with the aid of molecular templates —a way towards artificial antibodies. *Angewandte Chemie International Edition in English*, 34(17), 1812–1832. https://doi.org/10.1002/anie.199518121.

[5] Haupt, K., & Mosbach, K. (2000). Molecularly imprinted polymers and their use in biomimetic sensors. *Chemical Reviews*, 100(7), 2495–2504. https://doi.org/10.1021/cr990099w.

[6] Mayes, A. G., & Whitcombe, M. J. (2005). Synthetic strategies for the generation of molecularly imprinted organic polymers. *Advanced Drug Delivery Reviews*, 57(12), 1742–1778. https://doi.org/10.1016/j.addr.2005.09.009.

[7] Alexander, C., et al. (2006). Molecular imprinting science and technology: A survey of the literature for chemical sensors and separations. *Journal of Molecular Recognition*, 19(2), 106–180. https://doi.org/10.1002/jmr.760.

[8] Say, R., et al. (2003). Imprinted polymers for bioseparation. *TrAC Trends in Analytical Chemistry*, 22(10), 668–677. https://doi.org/10.1016/S0165-9936(03)01109-3.

[9] Moczko, E., et al. (2013). Hydrogel-based molecularly imprinted polymer sensors. *Biosensors and Bioelectronics*, 49, 536–546. https://doi.org/10.1016/j.bios.2013.05.057.

[10] Chen Y., et al. (2018). Graphene oxide-based molecularly imprinted polymers for NO_2 detection.

[11] Zhang W., et al. (2020). CNT-based MIPs for SO2 detection.

[12] Sellergren, B. (ed.). (2000). *Molecularly Imprinted Polymers: Man-Made Mimics of Antibodies and Their Applications in Analytical Chemistry*. Amsterdam, Elsevier Science.

[13] Liu, Y., et al. (2019). CNT-based MIPs for mercury detection in water. *Talanta*, 200, 272–279. https://doi.org/10.1016/j.talanta.2019.03.043.

[14] Huang, X., et al. (2017). Graphene oxide-based MIPs for endocrine-disrupting chemicals detection. *Sensors and Actuators B: Chemical*, 253, 929–937. https://doi.org/10.1016/j.snb.2017.06.063.

[15] Zhao, L., et al. (2021). CNT-based MIPs for controlled release of anticancer drugs. *Journal of Controlled Release*, 331, 456–464. https://doi.org/10.1016/j.jconrel.2021.01.022.

[16] Chen, Y., et al. (2020). Graphene-based MIPs for environmental monitoring of quercetin. *Journal of Hazardous Materials*, 382, 121085. https://doi.org/10.1016/j.jhazmat.2019.121085.

[17] Wang, P., et al. (2021). CNT-based MIPs for chlorpyrifos detection in soil. *Environmental Pollution*, 268, 115951. https://doi.org/10.1016/j.envpol.2020.115951.

[18] Zhang, T., et al. (2019). CNT-based MIPs for lead detection in soil. *Analytica Chimica Acta*, 1073, 65–73. https://doi.org/10.1016/j.aca.2019.05.032.

[19] Chen, X., et al. (2016). Graphene-based molecularly imprinted polymers for the selective detection of acetaminophen in pharmaceutical formulations. *Biosensors and Bioelectronics*, 77, 512–518. https://doi.org/10.1016/j.bios.2015.10.031.

[20] Zhao, Y., et al. (2019). CNT-based molecularly imprinted polymers for the detection of theophylline metabolites in urine samples. *Journal of Pharmaceutical and Biomedical Analysis*, 172, 69–75. https://doi.org/10.1016/j.jpba.2019.04.026.

[21] Liu, B., et al. (2018). Graphene oxide-based MIPs for cadmium detection in soil. *Journal of Environmental Chemical Engineering*, 6(2), 2026–2034. https://doi.org/10.1016/j.jece.2018.03.034.

[22] Zhao, X., et al. (2021). CNT-based MIPs for pesticides detection in water. *Chemosphere*, 277, 130241. https://doi.org/10.1016/j.chemosphere.2021.130241.

[23] Liu, Q., et al. (2020). Graphene oxide-based molecularly imprinted polymers for targeted drug delivery to cancer cells. *Advanced Functional Materials*, 30(5), 1905985. https://doi.org/10.1002/adfm.201905985.

[24] Wang, L., et al. (2018). CNT-based molecularly imprinted polymers functionalized with gadolinium for use as MRI contrast agents. *ACS Applied Materials & Interfaces*, 10(32), 27007–27015. https://doi.org/10.1021/acsami.8b07096.

[25] Chen, X., et al. (2019). Graphene oxide-based molecularly imprinted polymers with fluorescent tags for the imaging of glucose in live cells. *Biosensors and Bioelectronics*, 135, 161–167. https://doi.org/10.1016/j.bios.2019.03.059.
[26] Zhao, X., et al. (2019). CNT-based molecularly imprinted polymers for the detection of *E. coli* bacteria. *Analytical and Bioanalytical Chemistry*, 411, 7475–7483. https://doi.org/10.1007/s00216-019-02108-5.
[27] Zhao, Z., et al. (2021). CNT-based molecularly imprinted polymers for controlled release of anticancer drugs. *Materials Science and Engineering: C*, 121, 111817. https://doi.org/10.1016/j.msec.2020.111817.
[28] Tang, Y., Li, M., Gao, X., Liu, X., Ma, Y., Li, Y., Xu, Y., & Li, J. (2016). Preconcentration of the antibiotic enrofloxacin using a hollow molecularly imprinted polymer, and its quantitation by HPLC. *Microchimica Acta*, 183(2), 589–596.
[29] Takeuchi, T., & Haginaka, J. (1993). Selective separation and molecular recognition using molecularly imprinted polymers. *Journal of Chromatography A*. 641, 15–22. https://doi.org/10.1016/0021-9673(93)80558-R.
[30] Andersson, L. I., & Mosbach, K. (1985). Molecular imprinting of amino acid derivatives for peptide recognition, *Journal of Chromatography A*. 348, 19–26. https://doi.org/10.1016/S0021-9673(01)95335-2.
[31] Lanza, F., & Sellergren, B. (2003). The application of molecular imprinting technology to solid-phase extraction, *Chromatographia*. 57(S1), 377–384. https://doi.org/10.1007/BF024921430.
[32] Song, X., Xu, X., Wang, Y., & Zhang, W. (2007). Molecularly imprinted polymers embedded with carbon nanotubes for protein recognition. *Biosensors and Bioelectronics*, 22(13), 3146–3151. https://doi.org/10.1016/j.bios.2006.12.012.
[33] Liu, B., Liu, Y., Yang, Y., & Liu, Y. (2010). *Graphene oxide-based molecularly imprinted polymers for selective recognition and determination of chloramphenicol in milk*, Analytical Chemistry, 82(18), 8222–8230. https://doi.org/10.1021/ac101485j.

Milad Tavassoli*, Reza Abedi-Firoozjah, Behnam Bahramian, Ali Ehsani, and Ehsan Sadeghi

Chapter 9
Sensing applications of carbon-based molecularly imprinted polymers in agriculture

Abstract: The presence of various harmful substances in agricultural food and the environment (such as pesticides, antibiotics, heavy metals, microorganisms, toxins, etc.) poses a significant risk to human health, prompting a focused effort on the development of simple, rapid, and selective hazard analysis tools for agricultural products. Molecularly imprinted polymers have garnered increasing attention due to their facile preparation and high selectivity. Notably, carbon nanomaterials, owing to their uncomplicated preparation, cost-effectiveness, large specific surface area, and high conductivity, present themselves as promising sensing substrate carriers. Consequently, the amalgamation of carbon nanomaterials with molecularly imprinted polymers has attracted considerable interest. This chapter provides a comprehensive summary of the development, composition, and preparation methods of carbon-based molecularly imprinted polymers and outlines the latest research advancements in the utilization of carbon nanomaterials for detecting various agricultural food and environmental hazards using sensors. Additionally, it elucidates the practical applications of carbon nanomaterial-based molecularly imprinted polymer sensors, discusses existing challenges and future

Acknowledgments: This chapter was approved and supported by the Student Research Committee, Yasuj University of Medical Sciences, Yasuj, Iran.

***Corresponding author: Milad Tavassoli**, Yasuj University of Medical Sciences, Yasuj, Iran, e-mail: mtavassoli2006@gmail.com, ORCID: 0000-0003-2033-5695; Department of Nutrition, Faculty of Health and Nutrition Sciences, Yasuj University of Medical Sciences, Yasuj, Iran, e-mail: r.abedi1878@gmail.com, ORCID: 0000-0002-0705-7283

Reza Abedi-Firoozjah, Department of Food Science and Technology, School of Nutrition Sciences and Food Technology, Kermanshah University of Medical Sciences, Kermanshah, Iran, e-mail: bahramianbehnam1372@gmail.com, ORCID: 0000-0001-7611-9970

Behnam Bahramian, Ali Ehsani, Department of Food Science and Technology, Faculty of Nutrition and Food Sciences, Tabriz University of Medical Sciences, Tabriz, Iran, e-mail: ehsani@tbzmed.ac.ir, ORCID: 0000-0003-0391-9218

Ehsan Sadeghi, Department of Food Science and Technology, School of Nutrition Sciences and Food Technology, Kermanshah University of Medical Sciences, Kermanshah, Iran, e-mail: bahramianbehnam1372@gmail.com, ORCID: 0000-0001-7611-9970; Research Center for Environmental Determinants of Health (RCEDH), Health Institute, Kermanshah University of Medical Sciences, Kermanshah, Iran, e-mail: ehsan.sadeghi59@yahoo.com, ORCID: 0000-0002-1096-2380

https://doi.org/10.1515/9783111503325-009

trends, and delineates ongoing endeavors aimed at developing novel and efficient carbon nanomaterial-based molecularly imprinted polymer sensing platforms.

Keywords: Molecular imprinted polymers, carbon nanomaterial, sensor, detection, agriculture

9.1 Introduction

Ensuring food safety is crucial worldwide as it has a direct and significant impact on the health and well-being of people everywhere [1, 2]. The assessment of food quality is crucial to determine its suitability for consumption and distribution in the market [3–5]. Common contaminants found in agricultural food and the environment include pesticides, antibiotics, heavy metals, microorganisms, mycotoxins, and toxins, making it essential to employ detection methods to identify potential contamination and allergens. While traditional detection methods such as high-performance liquid chromatography-mass spectrometry offer accurate measurements, they have limitations such as complex sample pretreatment, lengthy detection time, and reliance on expensive equipment and expertise, restricting their use in certain settings [6–8]. Hence, there is a growing need for the development of simpler, rapid, cost-effective, and portable analytical methods for food safety detection [3, 9]. Nanotechnology, a rapidly advancing field focused on materials and systems at the nanoscale, presents a promising opportunity to address these challenges effectively. Its impact extends across the entire food chain, revolutionizing various aspects including production, processing, transportation, packaging, storage, delivery, and safety [10, 11].

Advances have been made in different sensor methods, including electrochemical, fluorescent, and mass-sensitive techniques. These methods provide cost-effective, highly selective, sensitive, reliable, and fast detection options that are suitable for onsite testing. They have been widely adopted and proven to be effective in monitoring food safety [3, 12, 13]. Moreover, sensor technology in food safety is increasingly showing promise in rapidly identifying harmful residues. Sensor technology relies on sensitivity elements and conversion elements to identify desired substances, with the choice of sensitive elements playing a critical role. Molecularly imprinted polymers (MIPs) are particularly valuable in sensing due to their distinct characteristics, making them highly practical in sensor analysis [14, 15]. MIPs offer several advantages as sensitive elements in sensors, including easy manufacturing, adaptable templates, and the ability to be customized for specific analytes at a low cost. MIPs are known for their structural predictability, recognition specificity, and application versatility, and they possess greater physical strength and thermal stability compared to other natural receptors [3, 16]. As a result, molecularly imprinted sensors based on MIPs have attracted significant attention as recognition elements and have become a focal point of international research. Advanced sensor technology relies heavily on MIPs, and the efficiency and responsiveness of MIP-based sensors are directly linked to the number of imprints on their surfaces [17, 18]. As shown in Scheme 9.1, Issues related to insufficient imprinting cavities on MIPs sensors can be overcome by utilizing nanomaterials, which create more binding sites, improve mass transfer, enhance binding capacity, and address challenges such as low binding of target molecules, inefficient elution, and low sensitivity. Carbon nanomaterials have garnered significant attention due to their ease of preparation, cost-effectiveness, large specific surface area, and high electrical conductivity [19, 20].

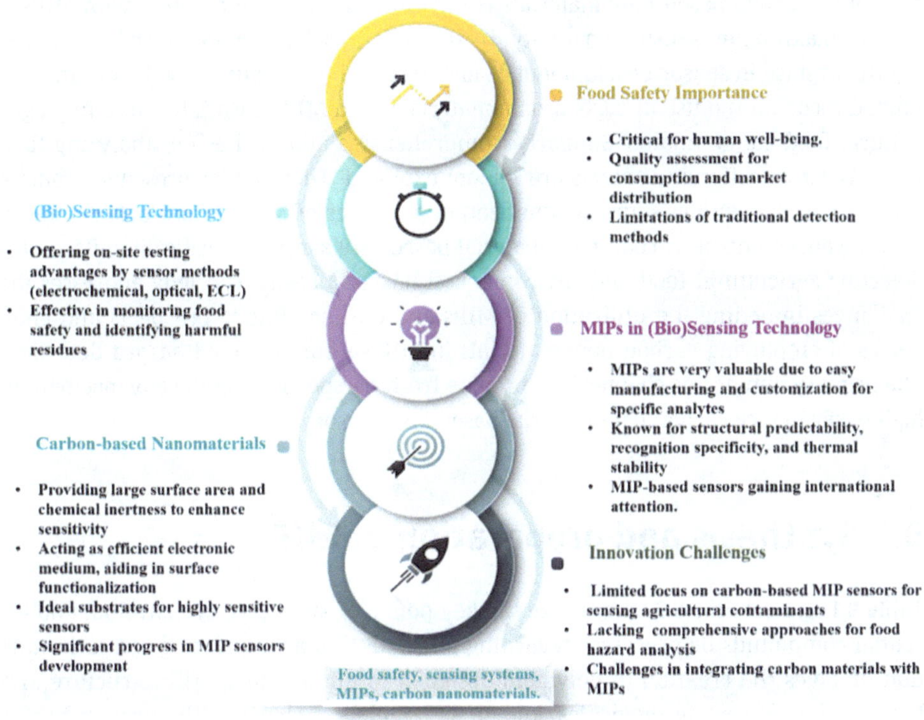

Scheme 9.1: Informative illustration on food safety, sensing technologies, MIPs, and carbon nanomaterials.

They can act as absorbers and carriers for electrode materials and play a crucial role in a wide range of applications including catalysis and sensing. Extensive research has been dedicated to exploring the production techniques of carbon nanomaterials and their interactions with other materials to further elevate the performance of MIP sensors. As a result, there has been significant progress in the development of MIP sensors based on carbon nanomaterials, and it is anticipated that sensors utilizing MIPs prepared with carbon nanomaterials will soon gain prominence in the sensor market [21–23].

Several recent articles delve into the applications of carbon nanomaterials in sensor technology. These publications emphasize how incorporating carbon nanomaterials into MIPs boosts sensor sensitivity by leveraging their chemical inertness and large surface area. Moreover, carbon nanomaterials act as an efficient electronic medium, significantly enhancing sensor sensitivity. They are also conducive to surface functionalization, allowing for versatile combinations with small organic molecules, polymers, DNA, proteins, enzymes, metal nanoparticles (NPs), and metal oxide NPs. This leads to improved chemical properties and reactivity, making carbon nanomaterials ideal substrates for MIPs in the creation of highly sensitive sensors. However, the production of functional carbon-based nanomaterials is intricate, and their integration with MIPs is vital to maximizing sensor sensitivity. In summary, carbon nanomaterials show immense promise in sensor development. While the existing literature is valuable, there is limited focus on the use of carbon nanomaterial-based MIP sensors for detecting agricultural food contaminants. Similarly, comprehensive approaches for analyzing food hazards across different contexts are seldom explored. This chapter presents a concise overview of the synthesis and polymerization processes of MIPs. It also delves into recent advancements in carbon nanomaterial-based MIP sensors specifically tailored for detecting agricultural food and environmental hazards. Lastly, it briefly addresses the challenges impeding the utilization of MIPs and outlines future prospects and challenges of employing carbon nanomaterials in MIP sensors for food hazard detection. The primary aim of this chapter is to provide fresh perspectives on the advancement of high-performance carbon nanomaterial-based MIP sensors.

9.2 Synthesis and preparation of MIP

Table 9.1 and Scheme 9.2 summarize the key points of synthesis, preparation, and essential compounds or processes regarding MIPs. Molecular imprinting is a technique that involves the creation of polymers designed to mimic the specific structure and binding sites of certain molecules, known as template molecules. The process begins with the interaction between the template molecule and a functional monomer, forming either covalent or non-covalent bonds [24–26]. This interaction creates multiple binding sites that are then utilized in the polymerization process, which involves the use of a cross-linking agent. After the template molecule is removed, imprinted cavities are left behind, matching the spatial configuration of the template molecule and creating multiple specific binding sites within the polymer [14, 27].

The production of imprinted particles in micrometer measurements often involves grinding the imprinted polymer monolith, leading to particles with irregular shapes and sizes. Consequently, these particles need to be sieved to achieve the desired size, resulting in low yields and making them unsuitable for standard industrial manufacturing processes [28, 29]. There is a growing need for simple, cost-effective techniques to create MIP particles with a narrow size range and high yield. Additionally, the increasing applications of MIPs require smaller diameters in the nanoscale range. Achieving this requires the use of higher cross-linker ratios, typically ranging between 60% and 90%, to facilitate early MIP NP formation during polymerization [24, 30, 31]. Studies have shown that photo-initiated polymerization produces MIPs with superior recognition properties compared to thermal initiation, as higher polymerization temperatures reduce selectivity and cross-linking. Free radical polymerization is widely used due to its extensive range of monomers and precise control over reaction conditions. However, managing particle size, shape, and distribution during free-radical polymerization poses a challenge. This can be addressed by using controllable "living" radical polymerization [32, 33].

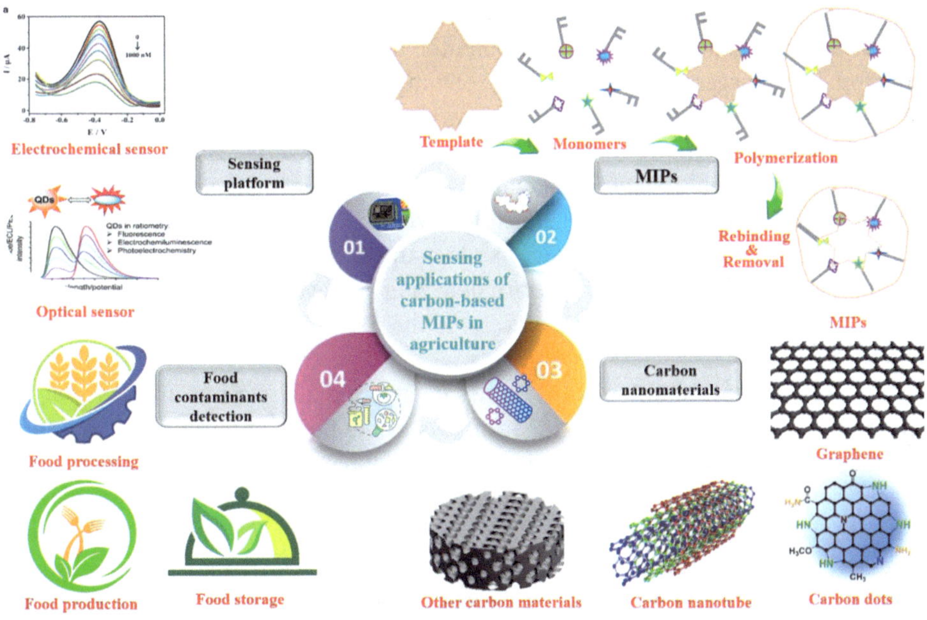

Scheme 9.2: General schematic information on the synthesis process, preparation, and important applications of MIPs.

9.2.1 Synthesis

MIPs are advanced materials created using a template-based method and polymerization. These polymers have 3D structures achieved through specific polymerization techniques [24]. To create MIPs, template species and a functionalized monomer are arranged and polymerized with a cross-linker in a porogenic solvent. The resulting molecular structure needs to be highly specific and directed toward the target. During polymerization, a monomer encapsulates the template and is fixed by copolymerization with a cross-linking monomer. After thorough washing, the template molecules are removed, leaving cavities with the same shape as the target molecule, enabling the polymer to bind to the target molecule [34, 18].

The synthesis of MIP is a complex process heavily dependent on the choice of template, functional monomer, cross-linker, initiators, and porogens. Creating high-performance MIP materials involves considering various factors that influence the polymerization process, such as the nature of the template molecules, types of monomers, and cross-linking agents. Selecting the right functional monomer is crucial as it must interact effectively with the template. Traditionally, methacrylic acid (MAA), 3-aminopropyltriethoxysilane (APTES), and 4-vinylpyridine (4-VP) have been commonly used as functional monomers, despite potential food hazards. Concurrently, appropriate cross-linking agents like ethylene glycol dimethacrylate (EGDMA) and trimethylolpropane trimethacrylate are vital for enhancing the imprinting sites and determining the affinity between imprinted molecules and templates [24, 35]. However, the limited number of traditional functional monomers limits the selectivity and further application of MIPs. Therefore, the introduction of new functional monomers such as pyrrole (Py), *o*-phenylenediamine (*o*-PD), acrylamide, and dopamine (DA) has expanded the capabilities of MIPs, especially in hazard detection applications. These new monomers do not require additional cross-linkers due to their strong electropolymerization ability [36, 29].

The components of MIP play vital roles in enabling the selective and sensitive detection of target molecules. The monomer acts as the building block, supplying functional groups that initiate the formation of the complex and creating specific empty spaces tailored for the target template molecule. Selection of the right monomer is crucial for the selectivity and sensitivity of MIP-based sensors [16, 15, 29]. Various types of interactions occurring during MIP preparation directly impact the sensitivity of the targeted analyte deposited on the electrode surface, enhancing the analytical signal. The stability of the complex formed relies on the strength and equilibrium of the molar ratio between the monomer and the template. A 1:4 ratio of the template and monomer has been found to be favorable in MIP synthesis. MAA stands out as a "universal functional monomer" due to its ability to act as a hydrogen-bond donor and acceptor, as well as its stable results in ionic interactions. The choice of cross-linker profoundly influences the morphology and mechanical stability of the polymer matrix [24, 25, 27]. As the cross-linker is removed, it creates voids that should complement the target molecule in terms of functional groups and structure. The quantity and type of cross-linker directly impact the selectivity and binding ability of MIP. Maintaining a high cross-linker ratio is crucial to allow the formation of a consistently porous material and retain the appropriate mechanical strength of the resulting complex material [37, 38]. The selection of initiators plays a key role in initiating the free polymerization reaction in the presence of the template, with their choice depending on the polymerization reaction involved. The porogenic solvent used in free radical polymerization affects the bonding between the functional monomer and the templates, especially in non-covalent interactions. The choice of porogenic solvent determines the properties and morphology of the polymer [39, 40]. Commonly used porogens include toluene, chloroform, acetonitrile, or dichloromethane. Following polymerization, complete removal of the template is necessary to create desired voids and spaces for the affinity of the analyte in the polymer complex. Washing with methanol or using dilute aqueous acid or appropriate solvents is carried out to ensure the breaking of non-covalent interactions between the targeted template and the adhering sites, thus ensuring the precision of the final analytical results [41].

The precipitation polymerization process is an efficient method for producing uniform spherical polymer particles with high yield and purity. It involves cross-linking polymerization in diluted monomer solutions containing template species to create MIP NPs [42]. Innovative approaches such as post-dilution polymerization and initiating polymerization in concentrated monomer solutions have been proposed to enhance selectivity and minimize template use. The method has been successfully used in the precipitation of TiO2 sol-gel NPs with water [43]. Various types of polymers, including hydrogen-bonding, charged, and hydrophobic functional monomers, are utilized in this process. One recent material design concept involves the creation of thermally responsive copolymer NP hydrogels for protein-imprinted NPs. These hydrogels exhibit reduced affinity toward the template as the temperature increases, protecting the protein from thermal stress [44, 45].

Emulsion polymerization is a potent method for synthesizing NPs with small sizes and results in MIP NPs with enhanced binding site uniformity. Two types of emulsion polymerization, mini-emulsion, and micro-emulsion, are used for creating stable emulsions with uniform droplet sizes. Mini-emulsion polymerization produces stable emulsions with uniform droplet size and employs stabilizers to create particles in the dispersed phase with sizes ranging from 50 to 1,000 nm [46, 47]. However, the process involves too many synthesis steps for commercial applications. Surfactant-free emulsion polymerization has been developed as a relatively straightforward and rapid process that does not significantly compromise the selectivity and uniformity of particles. Nevertheless, emulsion polymerization requires a larger number of chemicals, leading to more complex washing procedures and occasionally reduced material purity [48, 49].

In the process of in situ polymerization, porous support membranes are utilized to create surface-imprinted MIP NPs within their pores. By altering the high-viscosity polymerization solvent, it becomes feasible to generate particles as small as several hundred nanometers [50]. These MIP NPs, integrated into the support material, have applications in membrane separation. Moreover, nano- and micro-structured materials like polycarbonate or alumina membranes can function as sacrificial molds for producing diverse shaped MIPs, such as microrods and nanowires. The utilization of supporting materials facilitates the development of well-defined MIPs with high surface areas [51, 52]. Nevertheless, there is a need for further research to synthesize nanospheres, as only a few successful nanosized MIPs based on template synthesis have been validated thus far. Despite this, it is reasonable to expect a significant increase in the number of successful nanosized MIPs in the future [53, 51].

9.2.2 Preparation

MIPs can be created using three different methods: covalent, semi-covalent, and non-covalent approaches [54, 55]. The covalent approach involves creating reversible covalent bonds between the template and monomers before polymerization. After the polymerization, the template is removed by breaking the covalent bonds, which are then re-formed when the analyte re-binds. This method provides a mostly uniform population of binding sites, reducing non-specific binding [50, 56]. However, designing a suitable template-monomer complex with easily reversible covalent bond formation and cleavage under mild conditions is challenging. The semi-covalent approach strikes a middle ground, where the template is also covalently bound to a functional monomer, but template re-binding is based on non-covalent interactions [56, 57]. The non-covalent approach, pioneered by Arshady and Mosbach, relies on creating relatively weak non-covalent interactions (e.g., hydrogen bonding and ionic interactions) between the template molecule and selected monomers before polymerization [58]. This method is the most commonly used for MIP preparation due to its simplicity and the availability of a wide variety of commercially available monomers. However, an equilibrium process influences the template-monomer interactions, requiring a large amount of monomer to shift the equilibrium toward the formation of the template-monomer complex. As a result, excess free monomers are randomly integrated into the polymeric matrix, leading to the creation of non-selective binding sites [59, 60].

In recent years, the use of various templates has resulted in the development of different MIP preparation techniques. The diverse properties of MIPs obtained from these methods underscore the importance of selecting the appropriate preparation technique [13, 50]. The two most commonly used methods for detecting food hazards are precipitation polymerization and electropolymerization. Precipitation polymerization is typically employed to attach functional monomers, templates, cross-linking agents, and initiators to the surface of carriers [61]. The resulting MIPs are uniform and loose, eliminating the need for grinding, which could otherwise damage the binding site during bulk polymerization. Additionally, precipitation polymerization allows effective control of the distribution of MIPs and the aggregation of NPs on the surface [7, 62]. Electropolymerization involves the copolymerization of a template on the carrier's surface to create a MIP film. This method enables the regulation of MIP film thickness by controlling the number or length of polymerization cycles. It is highly adaptable and allows for fast and sensitive molecular reactions of the analyte, achieving uniform deposition on carriers of various shapes and sizes [7, 63]. The choice of carrier significantly impacts MIP preparation for both precipitation polymerization and electropolymerization. Carbon nanomaterials, with their high specific surface area and stability, facilitate increased contact between analytes and MIPs, resulting in rapid binding dynamics. They also serve as convenient, efficient, and cost-effective sensing substrates, and their stability in sensing functional interfaces can be effectively controlled through functionalization [64, 11].

The production of MIP involves several essential components. The monomer plays a crucial role by providing functional groups that encourage complex formation and by creating specific cavities tailored to the target molecule. Choosing the right monomer is vital for the effectiveness of MIP-based sensors, influencing their selectivity and sensitivity [62, 63]. Understanding the various interactions involved in MIP production is important as these interactions directly impact the sensitivity of the targeted analyte's deposition on the electrode, enhancing the signal. The stability of MIP synthesis is influenced by the equilibrium and strength of the molar ratio of template-to-monomer, with an optimal condition being a ratio of approximately 1:4. Due to its hydrogen-bond donor and acceptor properties as well as stable ionic interactions, methyl acrylic acid is considered an exceptional "universal functional monomer." The selection of the cross-linker is pivotal as it determines the polymer's morphology and mechanical stability [64, 29]. Removing the cross-linker should leave voids that are functionally and structurally complementary to the target molecule. The quantity and type of cross-linker used significantly impact MIP selectivity and binding properties. Thus, maintaining a high cross-linker ratio is essential for creating a permanently porous material while ensuring adequate mechanical strength [20, 11, 28]. All MIP initiators, in the presence of the template, can induce a free polymerase chain reaction. The selection of these initiators is contingent upon the polymerization reaction. The porogenic solvent is essential in free radical polymerization as it influences the bonding between functional monomers and templates. In MIP synthesis, toluene, chloroform, acetonitrile, and dichloromethane are commonly used porogens. Following polymerization, the attached template is completely removed, resulting in the formation of desired cavities for the analyte. However, the residual template can impact the accuracy of the final result. To completely eliminate the template, non-covalent interactions between the template and the adhering sites are disrupted through washing with methanol (as a continuous extracting solvent), diluted aqueous acid, or an appropriate solvent [7, 13].

9.3 Application of carbon-based MIPs in agriculture

Nanomaterials are being utilized extensively to develop highly sensitive sensors as a result of their exceptional properties, including quantum size and surface interface effect. In particular, different types of carbon nanomaterials have proven to be immensely valu-

Table 9.1: The key points of synthesis, preparation, and essential compounds or process regarding MIPs.

Mode	Description
Essential compounds or process	– Monomers provide functional groups for complex formation and tailored cavities for target molecules in MIP production. – Removing the template and achieving proper voids are crucial for MIP functioning and accuracy. – Initiators, porogenic solvents, and washing methods play key roles in the synthesis and purification of MIPs.
Synthesis of MIP	– Molecular imprinting creates polymers mimicking specific molecules using template molecules. – Imprinted cavities are created after removing the template, matching its spatial configuration – Techniques for creating MIP particles with narrow size range and high yields are in demand.
Preparation of MIP	– Various methods like covalent, semi-covalent, and non-covalent approaches are used for MIP creation. – Selection of monomers, cross-linking agents, and initiators significantly influences MIP performance. – Precipitation and electropolymerization are common methods for detecting food hazards using MIPs. – -Proper choice of carriers impacts MIP preparation efficacy in these techniques.

able in the detection and isolation of harmful substances in food due to their unique structures and high sensitivities. These attributes make them outstanding substrates for the development of MIP sensors, which rely on changes in response signal conversion elements resulting from the interaction between the target hazardous substance and the MIP. The distinctive structures and exceptional stability of carbon nanomaterials have enabled them to effectively detect and separate hazardous substances in food.

Moreover, these sensors can be classified into various types based on their conversion elements, which encompass electrochemical, optical, and mass-sensitive-based signals. Their application provides a convenient, cost-effective, and rapid sensing platform for the assessment of food hazards. Recent research further suggests that MIP sensors based on carbon nanomaterials are capable of meeting regulatory requirements for the detection of harmful substances in agriculture and diverse food products, thereby opening up significant industrial applications. This article delves into the diverse applications of MIP sensors based on carbon nanomaterials in identifying agricultural and food product hazards. Various carbon materials such as graphene (GR), carbon dots (CDs), carbon nanotubes (CNTs), and other carbon nanomaterials have been the subject of research for MIP sensors. These sensors have been utilized to detect diverse compounds present in agricultural and food products using both electrochemical and optical techniques, which will be detailed in the following discussion.

9.3.1 Carbon-based MIP sensors by electrochemical methods

Electrochemical sensors are being increasingly used in chemical analysis due to their high sensitivity, low cost, quick response, and easy modification and miniaturization. A particularly exciting development is the use of electrochemical sensors based on MIPs, which combine the advantages of MIPs with those of electrochemical sensors. These sensors offer high sensitivity and selectivity, as well as chemical and mechanical stability, reusability, and low limits of detection (LOD). They are easy to prepare, cost-effective, and can be miniaturized. The characteristics of these sensors are influenced by the type of electrode (transducer) and the modifier(s) acting as a receptor, which form the sensing core of the sensor. In this review, the focus is placed solely on carbon-based electrodes and their modifiers. The extensive range of monomers, templates, and substrates available for use with MIP-based electrochemical sensors makes them highly applicable for detecting various substances. Research in this area is rapidly increasing, with a focus on detecting agrochemicals, heavy metals, phenolic compounds, biomolecules, as well as emerging pollutants like hormones, drugs, and drug metabolites. Furthermore, recent research has demonstrated the potential of electrochemical sensors created through molecularly imprinted technology in ensuring food and drug safety, particularly in detecting low levels of antibiotics, pesticides, heavy metals, toxins, and pathogens in contaminated food [1]. The utilization of specific nanomaterials is utilized to enhance sensitivity in MIP-based electrochemical methodologies. Carbon nanomaterials like GR, CDs, CNTs, and others have gained widespread use due to their exceptional performance and compatibility for both qualitative and quantitative analysis of agricultural and food product hazards. Table 9.2 illustrates how the combination of electrochemical techniques such as differential cyclic voltammetry (CV), pulse voltammetry, square wave voltammetry, amperometry, linear sweep voltammetry, and electrochemical impedance spectroscopy (EIS) with carbon nanomaterials shows great potential for food detection.

9.3.1.1 Graphene

Graphene is a remarkable two-dimensional crystal structure that represents the most common form of carbon allotropes. In recent years, there has been growing interest in utilizing graphene and graphene-related materials for creating sensors in food-related applications. This is due to the stable benzene ring structure and lack of band gap energy in graphene, which make it particularly well-suited for such purposes [65, 66]. In the structure of graphene, each carbon atom's sp^2 hybrid contributes electrons to the remaining p orbital to form macromolecular π bonds, resulting in high freedom of movement for the π electrons. Graphene's large surface area of 2,630 $m^2\ g^{-1}$ and planar surface on both sides make it ideal for adsorption purposes, while its high specific surface area, superior thermal conductivity, and elevated mechanical strength add to its appeal. These properties make graphene an excellent choice for enhancing the sensing capabilities of sensors through surface modification or by combining with NPs. As a result, graphene holds tremendous potential in the field of food detection [20, 7, 67].

Various contaminants in agricultural food and environment, such as pesticides, antibiotics, heavy metals, PAHs/PCBs, microorganisms, mycotoxins, etc. have been identified by MIP-based graphene methods. Citrinin is a mycotoxin produced as a secondary metabolite by various types of fungi commonly found in agricultural products. It is primarily formed during post-harvest handling and storage, and can persist until the food reaches the market. This presents a significant threat to both consumer health and business interests. Although there are traditional methods for detecting citrinin in foods, there is a need for simpler extraction and clean-up procedures, as well as faster and more selective detection methods [68]. Elfadil and coworkers introduced a novel electrochemical technique for detecting citrinin with remarkable speed. This method involved using graphene nanoflakes, which were produced by rapidly exfoliating graphite in a water-based solution without the use of solvents, with the assistance of sodium cholate. The team thoroughly examined the nano-sized flakes dispersed in water and enhanced their suitability as a sensing layer in screen-printed electrodes. When compared to other carbon materials, the graphite-sodium cholate combination displayed exceptional sensing abilities for citrinin. This enabled the consistent detection of citrinin at levels below the maximum residual limit (LOD = 5 μg L^{-1}) permitted in food, with a low variation of 4.5% ($n = 8$). In addition, the team efficiently synthesized a MIP in just 5 min using sonochemistry. This MIP was successfully utilized for the selective extraction and purification of citrinin from a variety of samples, including red rice, blueberry, turmeric, corn, wheat germ, and rice starch. The method consistently yielded precise recoveries (85.8–111.4%) for all the samples, with relative standard deviations less than 5.7% [69].

Researchers have developed MIP sensors based on graphene to detect pesticide residues in food. Zhao and colleagues developed a highly effective dummy template MIP (Fe_3O_4-NH_2@GO@MIPs) using Fe_3O_4-NH_2@GO as a carrier through the dummy surface molecular imprinting technique (DSMIT) for the first time. The Fe_3O_4-NH_2@GO@MIPs showed superior selectivity and adsorption properties compared to non-imprinted polymers (Fe_3O_4-NH_2@GO@NIPs) for five pyrethroid pesticides (Figure 9.1A). The adsorption of pyrethroids on Fe_3O_4-NH_2@GO@MIPs followed the Langmuir equation adsorption isotherm and pseudo-second-order kinetic models. These MIPs were used in magnetic solid phase extraction (MSPE) for extracting pyrethroids from fruit juice samples, and the quantification was done using gas chromatography with electron capture detection (GC-ECD). The developed MSPE method coupled with GC-ECD achieved satisfactory recovery rates (72.1–106.8%) with relative standard deviations (RSD) below 10.8%, and the LOD for the pesticides ranged from 0.007 to 0.015 μg L^{-1} [70].

Xie and colleagues devised an electrochemical MIP sensor by modifying a graphene surface with 4-vinyl benzoic acid (VBA) functional groups containing alkenyl

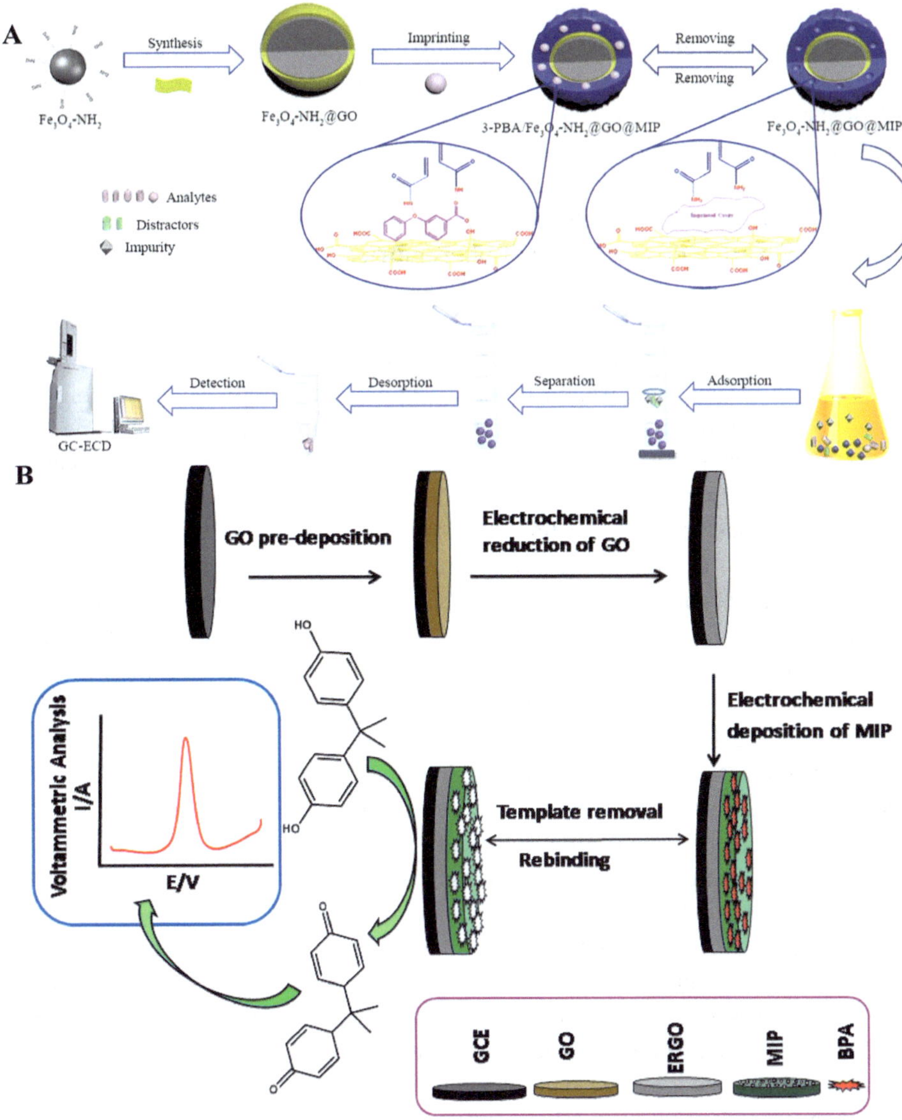

1

Figure 9.1: (A) Illustration in a schematic format of the creation of Fe_3O_4–NH_2@GO@MIP [70]. **(B)** The system of MIP|ERGO|GCE sensor for the detection of the BPA in milk [71].

and hydroxyl groups using bulk polymerization. This surface modification method utilized π-π interactions, and the resulting MIP/graphene was created by reacting a thiamethoxam (TMX) template with the cross-linking agent EGDMA and initiator AIBN at 60 °C for 24 h. The removal of the template resulted in the formation of cavities specific to TMX [72]. Furthermore, brown rice samples analyzed using this sensor displayed similar recoveries to those analyzed using HPLC. Additionally, the use of graphene hybrid nanomaterials has been shown to significantly improve sensor performance. Zhang and coworkers, for example, used AuNPs and ERGO to modify the SPCE layer-by-layer, and then created a cyhexatin (CYT) MIP through the self-assembly of *o*-aminothiophenol and CYT followed by electropolymerization. This resulting sensor exhibited enhanced analytical response compared to a sensor prepared using ERGO alone and successfully detected CYT in pear samples [73].

Flavonoids are a diverse group of plant nutrients known for their antioxidant and anti-inflammatory properties. They are named after their color-related chemistry but play various roles in the health of plants, animals, and humans. Quercetin, a flavonoid belonging to the flavanol subgroup, is particularly renowned for its health benefits. With a molecular weight of 302.24, quercetin exhibits slight solubility in water and ethanol but is more abundantly soluble in acetic acid and alkaline solutions. Its unique chemical structure, featuring adjacent hydroxyl groups and benzene rings, allows it to form complexes with metal ions and hinder their involvement in oxidation processes. These properties contribute to its substantial antioxidant capacity. Quercetin also demonstrates anti-inflammatory, anti-histamine, anti-swelling, anti-viral, and anti-cancer effects, as well as stabilizing cell membranes and impacting the functioning of the cardiovascular and hepatic systems. The recommended daily dosage of quercetin ranges from 4 to 68 mg, and its extraction and quantification through analytical methods are of paramount importance due to its valuable properties. In this regard, Mantashloo and Bahar developed a new inexpensive adsorbent by combining magnetic graphene quantum dots with a MIP coating. The adsorbent showed high selectivity in detecting quercetin, a medicinal compound. They optimized the adsorbent concentration, pH, and contact time and successfully detected quercetin in real samples of green tea, cumin, and thyme, with a LOD of 0.54 ng mL^{-1} and recoveries ranging from 97.61% to 102.11% [74].

Bisphenol A (BPA) is a common plastic additive found in food containers and packaging. It can migrate into food and disrupt the endocrine system, potentially leading to health issues. Due to numerous studies indicating negative health outcomes, there's been pressure to remove BPA from food and beverage applications worldwide. Palanisamy Karthika and coworkers introduced a new technique for detecting BPA utilizing an electrochemical sensor based on MIP (Figure 9.1B). They employed a glassy carbon electrode coated with electrochemically reduced graphene oxide. The team used density functional theory (DFT) at the B3LYP/6–31 + G (d,p) level to calculate the interaction between BPA and MIP at a molecular level. To create BPA-imprinted cavities, the electrode surface underwent electrochemical polymerization of pyrrole in the presence of the template molecule BPA, followed by the removal of entrapped BPA molecules from the polypyrrole film. The team employed a differential pulse voltammetry method with the MIP electrode to measure the concentration of BPA in both standard and real samples. Notably, the peak current showed a linear relationship with the logarithmic concentration of BPA within the range of 750 to 0.5 nmol L^{-1}, with a high correlation coefficient

($R^2 = 0.992$). Moreover, the LOD was established at 0.2 nmol L^{-1}. Furthermore, the team evaluated the sensor's reproducibility and repeatability [71].

A cutting-edge microfluidic chip has been developed that utilizes electrochemical techniques for the detection of carbofuran, a highly toxic insecticide. The chip employs a two-stage process, beginning with a MIP to capture and identify carbofuran. The captured carbofuran molecules are then transferred to a separate functional area containing graphene oxide-gold NPs (GO-AuNPs) and DNA aptamer (apt). By combining graphene oxide and gold NPs, the microfluidic chip is able to produce a strong electrochemical signal. The DNA aptamer specifically captures carbofuran, resulting in an electrical signal that can be measured to quantify the presence of carbofuran. The integration of the carbofuran-imprinted polymer with the second functional region provides enhanced selectivity, surpassing other methods. Furthermore, the presence of GO-AuNPs bolsters the electrochemical signal, improving the sensor's sensitivity. The response obtained from differential pulse voltammetry showed linearity within the concentration range of 0.2 to 50 nM for carbofuran, with a LOD of 67 pM at a potential of − 1.2 V [75].

9.3.1.2 Carbon nanotubes

CNTs have garnered significant attention in the field of nanoscience due to their unique structural properties and wide range of potential applications. These cylindrical structures are composed of graphene sheets, and can be categorized into single-walled CNTs (SWCNTs) and multi-walled CNTs (MWCNTs) based on the number of layers they contain. SWCNTs consist of a single-layer graphene sheet, while MWCNTs are made of double or multiple layers of graphene sheets held together by van der Waals forces. CNTs are valued for their distinctive properties, including high catalytic activity, exceptional mechanical strength, and high thermal stability. Additionally, their large specific surface areas and excellent biocompatibility make them easily modifiable using various methods. These remarkable characteristics make CNTs indispensable and fascinating materials for chemical analysis, particularly in the detection of harmful substances in agriculture and food products. Identification of various substances and contaminants in agricultural and food products has been done through CNTs-based MIP. Munawar and his team devised a cutting-edge sensor capable of detecting chloramphenicol (CAP) by utilizing 3D-CNTs@CuNPs@MIP, which were prepared through precipitation polymerization (Figure 9.2A). They harnessed aminated CNTs, known for their expansive specific surface area, as the focal point for depositing copper NPs. This was followed by a combination with MAA as the functional monomer and EGDMA as the cross-linking agent. The resultant 3D-CNTs@CuNPs@MIP exhibited the remarkable ability to manipulate mass flow and could be charged and animated akin to a nanomachine. The combined affinity and synergy led to the development of an exceptional selective sensing platform with a 100% response rate and an impressive LOD of 10 μM [76].

A flow-injection system was created with online amperometric detection at the MIP sensor to enable the selective and sensitive determination of profenofos (PFF), a moderately toxic organophosphate insecticide commonly found in Asian agriculture. PFF is classified as a moderately hazardous (toxicity class II) pesticide by the World Health Organization, making the detection of PFF crucial for farm-produced products. Amatatongchai and colleagues developed a simple flow-injection system using a molecularly-imprinted-polymer-coated carbon nanotube (3D-CNTs@MIP) amperometric sensor for PFF detection (Figure 9.2B). Silica was employed to functionalize CNTs, to improve their dissolution and electrocatalytic properties. The silica-functionalized CNT surfaces served to protect the CNT structure, and 3D-CNTs@MIP was obtained by grafting onto the CNTs@SiO_2 surface. Additionally, the inclusion of the CNTs enhanced the analysis signal of the MIP sensor, potentially attributed to synergistic catalysis with SiO_2, resulting in composites with improved conductivity. The 3D-imprinted sensor showed a linear response within the 0.01–200 μM range ($R^2 = 0.995$) with a LOD of 0.002 μM [77].

CNT-based MIP electrochemical sensors have been used to detect food additives, particularly for identifying sunset yellow (SY), an edible pigment. These sensors, specifically those utilizing acid-treated MWCNTs (COOH-MWCNTs) combined with acrylamide as a functional monomer, have shown a remarkable detection range from 50 to 1×10^5 nM for SY (Figure 9.3A). The inclusion of COOH-MWCNTs resulted in a significantly higher electrochemical signal compared to using a bare electrode without the functional layer, attributed to the presence of carboxyl and hydroxyl groups on the COOH-MWCNTs surfaces that facilitated the adhesion of the MIP layer [78]. Additionally, a novel MIP electrochemical sensor for SY detection involved the self-polymerization of dopamine (DA) on an MWCNTs matrix, enabling highly sensitive detection of SY due to the high electron transfer rate of the MWCNTs. In this case, SY partially inhibited the self-polymerization of DA on the MWCNTs surfaces, resulting in a lower detection limit. The combined approach using MWCNTs and MIP-Polydopamine (PDA) on the modified electrode significantly enhanced the SY redox peak current through a synergistic effect [79].

CNTs have proven to be versatile in their applications, including in the field of detecting mycotoxins such as aflatoxin B1 and deoxynivalenol (DON). DON, in particular, is considered a class 3 carcinogen with substantial toxic effects. To address this, researchers have developed MIP sensors using COOH-functionalized MWCNTs and the functional monomer *l*-arginine (Arg) through electropolymerization. These sensors were designed for the identification of DON in real wheat flour samples (Figure 9.3B). The research shows that, initially, a strong non-covalent bond was formed between Arg and DON through DFT. The non-steric hindering amino group of Arg and the hydroxyl group of DON interacted through hydrogen bonding to create a stable imprinted structure. The electrochemical performance of the resulting sensors was significantly improved after the deposition of COOH-MWCNTs on electrode substrates. This enhance-

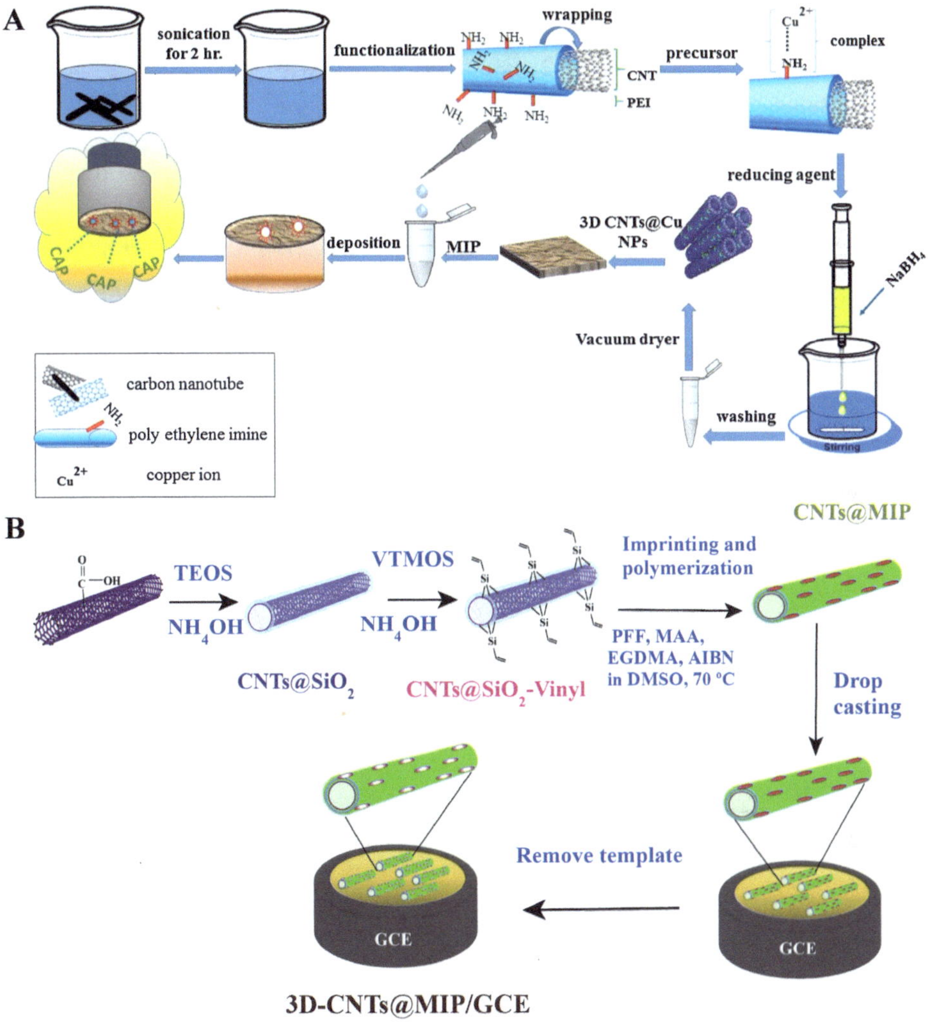

Figure 9.2: (A) Schematic description for creating 3D-imprinted nanohybrid for the detection of CAP [76]. **(B)** Creating 3D-CNTs@-MIP and making the MIP sensor for PFF [77].

ment is attributed to the increased active sites on the electrode surface and improved charge transfer efficiency. Subsequently, after electropolymerization in the presence of a template, the sensors were able to accurately and sensitively monitor DON levels. The system provided a positive outcome for detecting DON within a linear range of 0.1–70 μM, with a LOD of 0.07 μM in wheat flour samples [80]. Furthermore, CNT-based electrochemical sensors have been developed for the detection of natural toxins in food, such as histamine (HIS), an important biogenic amine commonly found in stored and fermented food products. While CNT nanomaterials have been widely utilized in

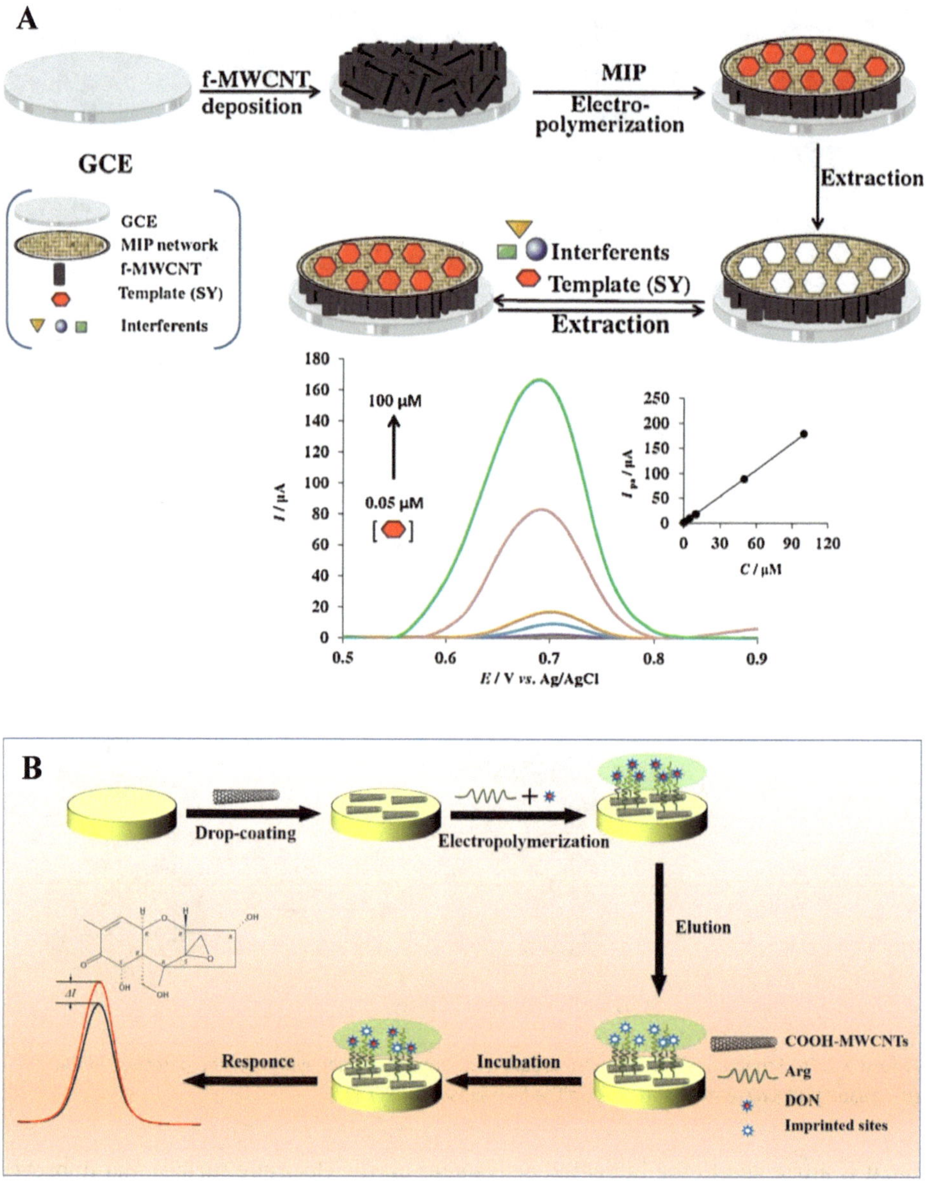

Figure 9.3: **(A)** Schematic of the MWCNTs/GCE system to detect SY [78]. **(B)** Schematic illustration for fabrication of antibody-like P-Arg-MIP/COOH-MWCNTs sensor for detection of DON [80].

MIP-based food hazard sensors, they have faced challenges in achieving a strong recognition effect. To address this, researchers combined MIP with aptamers (apt) to create an electrochemical sensor capable of dual recognition of HIS. In their approach, thioalcohol apt was loaded with HIS before polymerization. Gold NPs were used to functionalize carboxylated CNTs (AuNPs/cCNTs), which enhanced the electrochemical activity of glassy carbon electrodes (GCE). Subsequently, HIS was embedded in the MIP through *o*-phenylenediamine (*o*-PD) electropolymerization on the AuNPs/cCNTs/GCE surface to achieve a dual recognition effect. The collaboration between apt and MIP resulted in improved sensors with enhanced selectivity toward HIS compared to traditional imprinted sensors. The MIP-apt/AuNPs/cCNTs/GCE exhibited excellent performance, demonstrating linearity ranges of 0.46–35 nmol L^{-1} and 0.35–35 nmol L^{-1}. The LOD were determined to be 0.15 nmol L^{-1} using DPV and 0.11 nmol L^{-1} using EIS [81].

The use of CNTs as substrates provides several advantages. CNTs offer a large specific surface area and excellent biocompatibility, making them ideal for various applications. Furthermore, CNTs provide additional binding sites for MIPs, enhancing their functionality. By modifying the surface of CNTs, it becomes easier to combine them with functional monomers and template molecules, enabling the detection of various substances. MWCNTs are more commonly used compared to single-walled CNTs. Carboxyl functional groups with concentrated acids are frequently employed to modify CNTs. CNT-based MIPs consistently exhibit excellent characteristics for detecting hazardous substances in food matrices. They can be used directly as electrochemical sensor substrates or as deposition centers for nanomaterials.

9.3.1.3 Other carbon materials

A variety of carbon materials, including graphene quantum dots (GQDs), carbon dots (CDs), porous carbon, graphite carbon nitride (GCN), and carbon black (CB), have been employed as solid substrates for MIPs. These materials are known for their outstanding chemical inertness and extensive specific surface areas, making them particularly attractive for detecting harmful substances in agricultural and food products.

Metronidazole (MNZ) is an antibiotic used to treat infectious diseases caused by susceptible organisms. It can also be used as an additive in animal feed to promote growth. Exceeding a certain amount of MNZ in the human body can lead to seizures, peripheral neuropathy, and ataxia. Studies have indicated potential carcinogenic effects in animals, although there is insufficient evidence for its carcinogenic potential in humans. Accurate and sensitive determination of MNZ is crucial for human health and food safety [82, 83]. Ensafi et al. created an electrochemical sensor with high sensitivity and specificity by altering a glassy carbon electrode (GCE) to identify MNZ. Initially, they synthesized MIPs on GQDs using the sol-gel method. The proposed electrochemical sensor demonstrated two linear dynamic ranges of 0.005–0.75 μmol L^{-1} and 0.75–10.0 μmol L^{-1} under specific conditions, with a low detection limit of 0.52 nmol L^{-1} for MNZ detection [84]. Wei Guo and coworkers fabricated a highly effective method for the precise detection of patulin in fruit juice. They utilized a 2-oxindole dummy template and functional monomers, specifically ρ-aminothiophenol (ρ-ATP), in conjunction with highly sensitive electrochemical detection. To enhance the signal strength and improve the electron-transfer rate, CDs, and chitosan were used to modify materials on the glassy carbon electrode. The surface assembly of ρ-ATP and 2-oxindole was achieved through Au–S bond and hydrogen bond interactions. The researchers formed polymer membranes via electropolymerization in a solution containing ρ-ATP, $HAuCl_4$, tetrabutylammonium perchlorate (TBAP), and the template molecule 2-oxindole. This resulted in a specific cavity capable of adsorbing the target patulin after elution. The electropolymerization process was monitored and optimized using CV and DPV measurements. The MIP sensor demonstrated an impressive linear response range from 1×10^{-12} to 1×10 −9 mol L^{-1}, with a LOD of 7.57×10^{-13} mol L^{-1} [85].

A potentially harmful substance known as zearalenone (ZEA), which is a carcinogenic mycotoxin, is often found in various food products, particularly grains. Its adverse effects on human health include carcinogenicity, neurotoxicity, and estrogenic effects. To address this issue, Rehman and coworkers developed an innovative electrochemical sensor using a nanocomposite material consisting of GCN nanosheets adorned with bovine serum albumin@MnO_2 (g-C3N4NS/BSA@MnO_2) and MIPs to detect ZEA in rice samples (Figure 9.4A). The process involved synthesizing bulk g-C3N4 using a thermal polycondensation method, followed by ultrasonication to produce GCN nanosheets. The nanocomposite was then created through electrostatic forces between g-C3N4NS and BSA@MnO_2. Subsequently, a novel MIP-based electrochemical electrode containing g-C3N4NS/BSA@MnO_2 was fabricated with the presence of ZEA as the target analyte and pyrrole (Py) monomer using CV. Upon completion, the resulting electrochemical sensor demonstrated a linear range of 1–10 ng L^{-1} with a detection limit (LOD) of 0.25 ng L^{-1}. Furthermore, the MIP-based electrochemical sensor exhibited high selectivity, reproducibility, and stability, showcasing its potential for analyzing real samples such as rice [86]. Also, the unique properties of GCN make it highly stable in solvents due to the weak van der Waals interactions between its carbon nitride (CN) layers. To enhance its sensing capabilities, a novel GCN/hexagonal boron nitride nanocomposite was used in MIP-based electrochemical sensors to detect methyl parathion in pesticides. This sensor displayed exceptional selectivity and analytical performance and successfully detected methyl parathion in orange juice samples. The nanocomposite was characterized using CV and EIS. Additionally, an MP-imprinted voltammetric sensor on CN-HBN nanocomposite was developed in the presence of 100.0 mM pyrrole containing 25.0 mM MP through CV. The sensor exhibited a linearity range of 1.0×10^{-12}–2.0×10–9^{9} M and a LOD of 2.0×10–13^{13} M [87].

The unique properties of CB NPs have attracted considerable attention as a potent nanomaterial due to their abundant availability, large surface area-to-volume ratios, and numerous defect sites. Their cost-effectiveness in comparison to other carbon materials such as carbon nanotubes and graphene has broadened the possibilities for creating low-cost electrochemical sensors with exceptional electroanalytical performance. In addition to their direct use in modified electrodes, CB NPs can serve as a support for producing other nanostructures [94, 95]. For example, Arduini and coworkers documented the use of AuNPs on a CB NP layer, resulting in an electrocatalytic

Table 9.2: Some examples of different electrochemical sensing applications of carbon-based MIP in agriculture.

Sensing platform/ electrode	Target analyte	LOD	Linearity	Real sample	Reference
VBA/MIP/graphene@GCE	TMX	0.04 µM	0.5 to 20 µM	Grain	[72]
MIP/Rgo/AuNPs@SPCE	CYT	0.20 ng mL^{-1}	1 to 800 ng mL^{-1}	Pear	[73]
GQD/MIP@SPCE	Quercetin	0.54 ng mL^{-1}	5 to 220 ng mL^{-1}	Green tea, cumin, and thyme	[74]
MIP/polypyrrole film@DDE	BPA	0.2 nM	750 to 0.5 nM	Milk	[71]
AuNPs/GO/carbofuran-MIP@SPCE	Carbofuran	67 pM	0.2 to 50 nM	Chinese cabbage, chili, lettuce, tomato, apple, banana, tangerine, and watermelon	[75]
CNTs/CuNPs/MIP@GCE	CAP	10 µM	NS	Milk	[76]
3D-CNTs/MIP/SiO2@GCE	PFF	2 nM	0.01 to 200 µM	Spring onion, tomato, cabbage, Chinese cabbage, green pepper, and chili pepper	[77]
COOH-MWCNTs/MIP-acrylamide/PDA@GCE	SY	5.0 nM	50 to 1 × 10^5 nM	Candy, orange flavored jelly powder, peach juice powder, candy-coated chocolate, and beverage	[78]
COOH-MWCNTs/MIP/monomer l-arginine@GCE	DON	0.07 µM	0.1 to 70 µM	Wheat flour	[80]

(continued)

Table 9.2 (continued)

Sensing platform/ electrode	Target analyte	LOD	Linearity	Real sample	Reference
AuNPs/cCNTs/Apt/ MIP@GCE	HIS	0.15 nM	0.46 to 35 nM	Human plasma and canned tuna	[81]
GQDs/MIP@GCE	MNZ	0.52 nM	0.005 to 0.75 μM	Human blood plasma	[84]
ρ-ATP/Apt/MIP@GCE	Patulin	7.57×10^{-4} nM	1×10^{-3} nM to 1 nM	Apple juice	[85]
C3N4NS/BSA/MnO2/ MIP@ GCE	ZEA	0.25 ng L^{-1}	1 to 10 ng/L	Rice	[86]
CN-HBN/MIP@GCE	MP	2.0×10^{-4} nM	1.0×10^{-3} nM to 2.0 nM	Orange juice	[87]
FONPs/MIP/CB@ GCE	BPA	0.031 nM	0.1 nM and 50 μM	Water	[88]
AuNPs/MIP/CB@ GCE	BPA	8.8 nM	0.07 μM to 10 μM	Tap and mineral water	[89]
Poly(MAA-Co-EGDMA) MIP/SPCE	THC	0.18 ng mL^{-1}	NS	NS	[90]

MIPs@SPCE	Amoxicillin, cephalexin, cefazolin, penicillin G, and oxacillin	0.24 μg L^{-1}	0.24 to 0.56 μgL	Milk	[91]
CS-MWNTs/AuNPs/MIP@GCE	CAP	3.3×10^{-9} g L^{-1}	10^{-8} g L^{-1} to 10^{-2} g L^{-1}	Milk and water	[92]
GO/MIP@GCE	Amoxicillin	2.94×10^{-10}	5.0×10^{-10} to 9.1×10^{7}	NS	[93]

Abbreviations: VBA, 4-vinyl benzoic acid; CYT, cyhexatin; TMX, thiamethoxam; RGO, reduced graphene oxide; GQD, graphene quantum dots; BPA, bisphenol A; AuNPs, gold nanoparticles; GO, graphene oxide; CNTs, carbon nanotubes; CAP, chloramphenicol; PFF, profenofos; MWCNTs, multiwalled carbon nanotubes; SY, sunset yellow; PDA, polydopamine; DON, deoxynivalenol; HIS, histamine; Apt, aptamer; MNZ, metronidazole; GCE, glassy carbon electrode; ρ-ATP, ρ-aminothiophenol; ZEA, zearalenone; BSA, bovine serum albumin; CN, carbon nitride; MP, methyl parathion; FONPs, ferroferric oxide nanoparticles; CB, carbon black; DDE, doped diamond electrodes; NS, not stated; HBN, hexagonal boron nitride; THC, tetrahydrocannabinol; poly(MAA-Co-EGDMA), poly(methyl acrylic acid-co-ethylene glycol dimethacrylate); CS, chitosan; CPE, carbon paste electrodes.

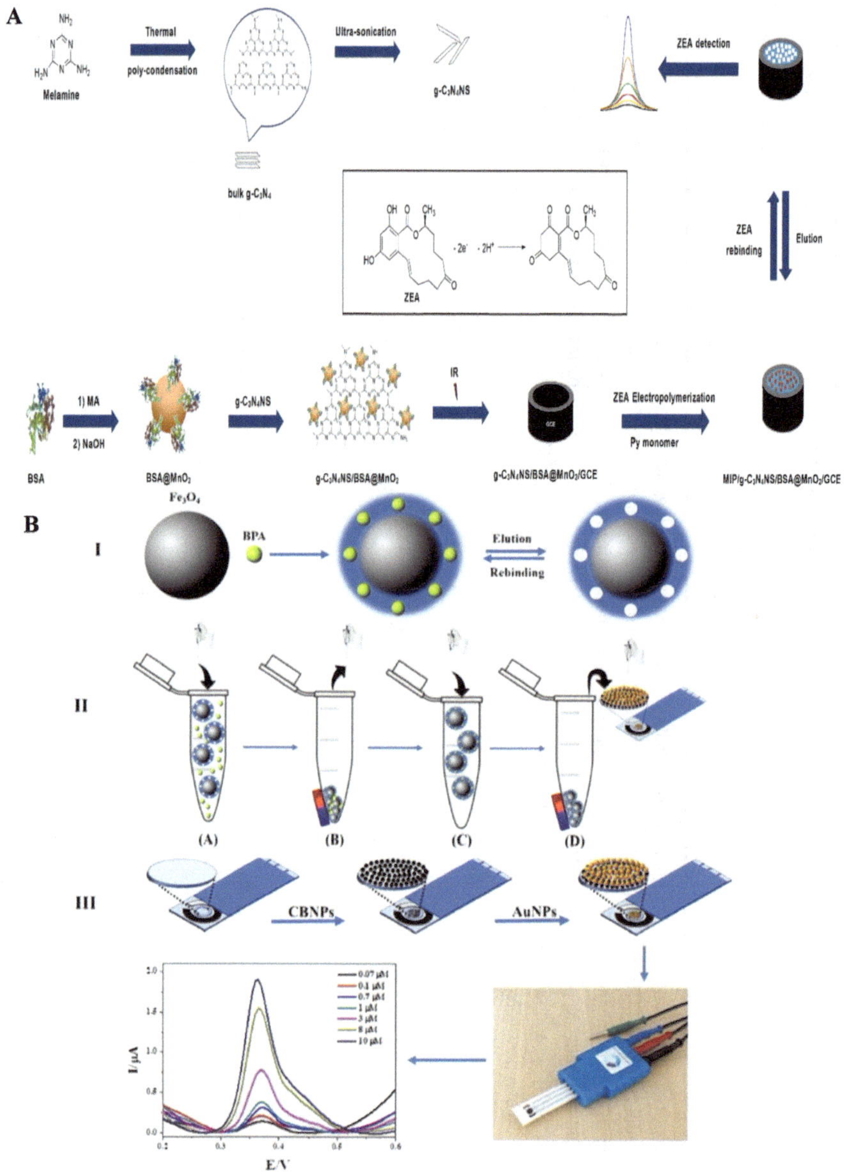

Figure 9.4: (A) Preparation plan of g-C3N4NS/BSA@MnO$_2$ nanocomposite and ZEA printed electrode for ZEA detection [86]. **(B)** Schematic explanation of the developed analytical approach of AuNPs/CBNPs/SPCE for detection of BPA [89].

effect on the redox reactions of glucose, hydrogen peroxide, and ascorbic acid [96]. Similarly, Hou et al. employed a comparable method by depositing ferroferric oxide NPs on the surface of CB NPs, which exhibited high electrocatalytic activity toward the oxidation of BPA [88]. In another study, Messaoud and coworkers devised a cutting-edge approach that combines ultrasound and magnets to produce a MIP coupled with an electrochemical sensor. This sensor is enhanced with a nanocomposite of CB NPs and AuNPs to achieve remarkably precise and sensitive detection of BPA (Figure 9.4B). The electrochemical sensor exhibits exceptional sensitivity for BPA, boasting a low LOD of 8.8 nM and an extensive linear range from 0.07 μM to 10 μM [89].

Numerous studies have been conducted to explore the potential of alternative carbon materials as substrates for electrochemical sensing. The impressive electronic conductivity and user-friendly nature of these materials suggest that they could be highly advantageous for the development of sensing platforms.

9.3.2 Carbon-based MIP sensors by optical methods

The first MIP-based sensor using optical surface ellipsometry method for vitamin K1 sensing was developed by Andersson et al. in the 1980s [97]. Since then, the combination of MIPs with various optical transduction mechanisms has advanced the field of analytical sensing. MIP-based optical sensors are designed to detect changes in the optical properties of a substance and convert this information into measurable signals. These sensors are currently showing great potential and are being widely used to identify potential risks in agriculture and food products [98, 99]. For example, carbon-based materials such as GQDs, GR, and CDs are being increasingly utilized due to their excellent biocompatibility and eco-friendliness. There is a growing interest in using carbon nanomaterial-based MIP optical sensors, with a focus on employing fluorescence as the primary detection method. As a result, MIP-based fluorescence sensors with highly sensitive properties have found widespread use in detection and/or determination processes. When the concentration of the target analyte is very low, the sensitivity of fluorescence sensors can be improved by enhancing the amplification and photomultiplier system or increasing the incident intensity. However, alternative approaches such as colorimetric, electrochemiluminescence (ECL), and surface-enhanced Raman scattering are also being explored, as shown in Table 9.3.

9.3.2.1 Graphene

GQDs are carbon fragments obtained from graphene, and they have diameters smaller than 100 nm. These are considered as quasi-zero-dimensional materials with unique physical and chemical properties. GQDs are known for their low toxicity, superior biocompatibility, large specific surface area, flexible chemical modification, and resistance to photobleaching, which makes them highly promising for use in establishing ECL sensors compared to traditional ECL luminescent materials such as ruthenium compounds and luminol, and semiconductor quantum dots. It is interesting to note that hybridizing GQDs with other nanomaterials results in fascinating properties and unexpected characteristics. Quantum dots have the ability to be blended with nanomaterials to create impressive electronic and visual properties. Additionally, the herbicide 2-methyl-4-chlorophenoxyacetic acid has been found to have various adverse effects on human health. A new type of glowing MoS_2-GQDs hybrid nanocomposite for producing MIP ECL sensors was described by Yang and coworkers, showcasing a blend of electrical and optical properties to enhance the sensor's capability. By incorporating MoS_2 nanosheets, the researchers were able to maintain intrinsic features, increase photon yield, improve fluorescence stability, and boost the ECL performance of the GQDs by 46 times. These sensors demonstrated a LOD of 5.5×10^3 nM, making them a fast, portable, and reliable option for testing pesticides and veterinary drugs in food samples [100].

The widespread presence of herbicides in the environment and our food supply presents a significant risk to human health. These chemicals can have both short-term and long-term negative effects on human health, including acute and chronic toxicity, impacts on reproduction and development, as well as the potential to cause genetic damage and cancer. Among the commonly used symmetrical triazine herbicides is prometryn, which can be particularly harmful when present in high concentrations and due to its persistent presence in the environment. As a result, it is critically important to regularly monitor the levels of prometryn in biological samples and food products to ensure the safety of consumers. Chumkaeo and coworkers fabricated a nanocomposite fluorescence sensor to detect property with high sensitivity and specificity. The sensor utilized melanin NPs and nitrogen-doped graphene quantum dots (N-GQDs) integrated into MIP to form the nanoprobe. This innovative sensor demonstrated a linear detection range from 1.0 to 100.0 μg kg^{-1} with a LOD of 0.63 μg kg^{-1} for prometryn. In real sample testing, the nanocomposite sensor achieved excellent recoveries ranging from 92.4% to 102.8% with very low RSDs of less than 5.4% [101].

Sulfadimethoxine, an antimicrobial agent belonging to the sulfonamide class, is commonly employed in veterinary medicine to combat bacterial infections in the urinary and respiratory systems, as well as in soft tissues [102]. Overuse of sulfadimethoxine can lead to its accumulation in meat, milk, and other food products, presenting a significant health risk to consumers. In response to concerns about sulfadimethoxine, the European Union has set a maximum residue limit of 100 μg kg^{-1} for sulfadimethoxine in meat. Consequently, there is a pressing need for the development of a precise and highly sensitive testing method for detecting sulfadimethoxine [103, 104]. Pongprom and coworkers developed a new composite nanoprobe called GQDs@PC@MIP. This nanoprobe has a unique structure that allows for the sensitive and precise detection of sulfadimethoxine using fluorescence. The nanoprobe incorporates graphene quantum dots and porous carbon into a MIP. When tested, the nanoprobe showed a decrease in fluorescence intensity that was directly proportional to the concentration of sulfadimethoxine in the range of 0.10–25.0 μg L^{-1}, with a very low LOD of 0.03 μg L^{-1}. The binding sites within the nanoprobe displayed a high degree of specificity for sulfadimethoxine, resulting in an impressive imprinting factor of 8.75. Furthermore, this innovative nanoprobe was successfully used to quantify sulfadimethoxine in milk, with recoveries ranging from 86.8% to 98.9% and very low RSDs below 5% (Figure 9.5) [105].

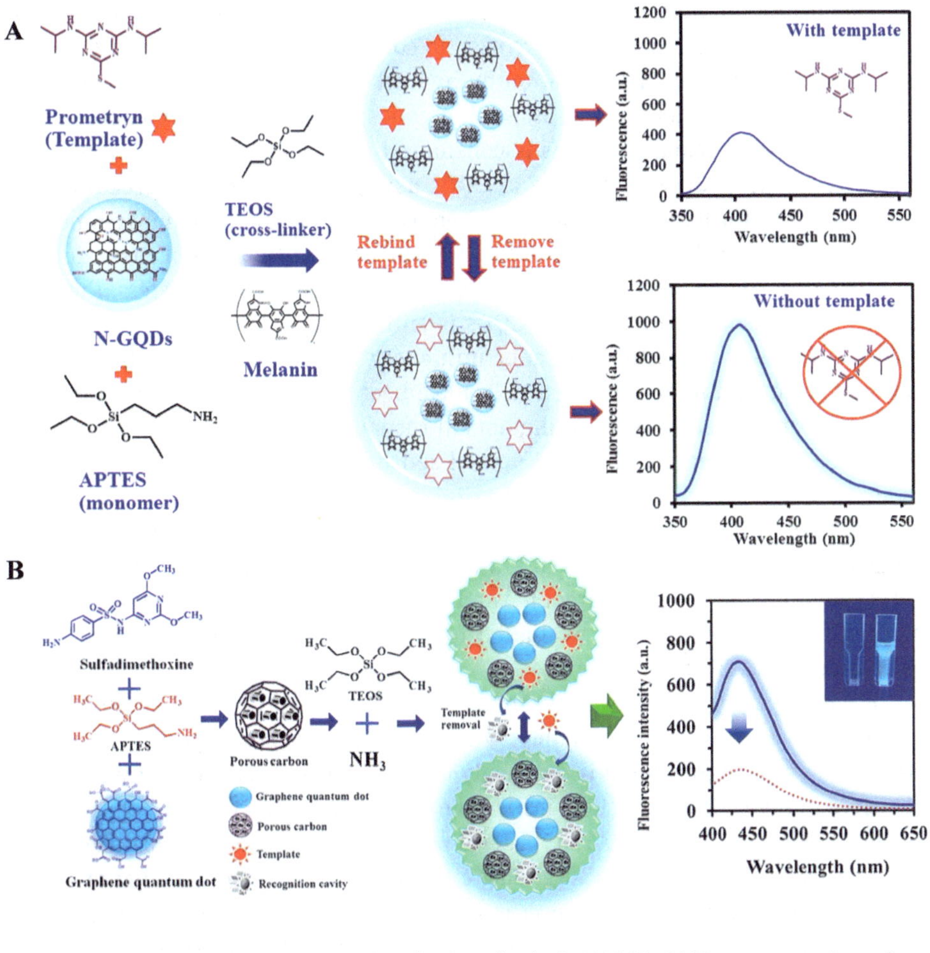

Figure 9.5: (A) Scheme of preparation and application of melanin@N-GQDs@MIP nanocomposite probe system for the detection of prometryn [101]. **(B)** The preparation strategy of the GQDs@PC@MIP nanoprobe system for the detection of sulfadimethoxine [105].

9.3.2.2 Carbon dots

Carbon quantum dots (CQDs) are a new type of carbon NPs with sizes ranging from 2 to 10 nm. They have gained attention for being less toxic than semiconductor quantum dots and for their stability, simple synthesis, eco-friendliness, low toxicity, and biocompatibility [106, 107]. CQDs have applications in biomedicine, catalysis, and sensing. Various synthesis techniques exist, including microwave, electrochemical, chemical oxidation, hydrothermal treatment, and ultrasonic methods. Surface passivation and functionalization are used to control CQD' characteristics. CQDs can be produced through self-passivation by hydrothermal carbonization at high temperatures or through microwave-assisted methods using various carbon precursors, and their hydrothermal synthesis is considered effortless, effective, and environmentally friendly. Green synthesis of CQDs using natural biomass precursors reduces the need for toxic and expensive chemicals and eliminates cumbersome post-synthesis processes [108, 109]. Modifying the surface and function of these dots can be utilized to regulate the physicochemical characteristics. Nonetheless, the effectiveness of sensors using CDs and GQDs is still an issue. Since food matrices are highly intricate, these sensors might be influenced by substances other than the target they are meant to detect. Thus, MIPs have been employed to enhance the precision and responsiveness of quantum dots in recent years. In this regard, Ensafi and colleagues developed a CQD-based optical sensor with a coating of MIPs to selectively and sensitively detect promethazine hydrochloride (PrHy). They applied a MIPs matrix to the CQDs' surface through sol-gel polymerization. The fluorescence intensity of the CQDs-MIPs showed a linear relationship with the concentration of PrHy ranging from 2.0 to 250 $\mu mol\ L^{-1}$ and had a detection limit of 0.5 $\mu mol\ L^{-1}$. The method's reproducibility was evaluated for 20 $\mu mol\ L^{-1}$ of PrHy, yielding a result with an RSD of 5.1% [110].

Regarding agriculture and food products, the CD-MIP sensors were effectively used to identify various pesticides. Acetamiprid (ACT) is a systemic broad-spectrum insecticide classified as a neonicotinoid. It is commonly utilized as a replacement for traditional insecticides in the control of sucking insects across a variety of crops [111]. While it is not deemed a human carcinogen, ACT is deemed potentially hazardous to human health. Recently, there have been reports suggesting a link between ACT and the onset of erectile dysfunction in men. Given these concerns, it is essential to develop a dependable and efficient analytical method for detecting ACT residues in both the environment and food. Shirani and colleagues developed a new optical sensor by creating a MIP on silane-doped carbon dots (Si-CDs) specifically for detecting ACT. They formed the MIP on the Si-CDs surface (MIP@Si-CDs) using a sol-gel process. The fluorescence signal of MIP@Si-CDs demonstrated a linear response to ACT concentration within the range of 7–107 nM, with a detection limit of 2 nM. Under optimal conditions, the precision for five replicate detections of 17 and 33 nM ACT was 2.7% and 3.3%, respectively [112]. Indoxacarb (IXC) belongs to the oxadiazine class of pesticides and is commonly utilized on a variety of crops including sweet corn (maize), cotton, lettuce, vegetables, as well as fruits like pears, apples, and cherries to control Lepidoptera, cockroaches, and ants [113]. The mode of action of IXC involves binding to sites on sodium channels, thereby inhibiting the entry of sodium ions into nerve cells. While it is considered safe for mammals due to its hydrophobic nature and low capacity to penetrate natural water systems, it has been noted to have adverse effects on non-target organisms such as bees and aquatic organisms like fish [114]. Shirani and coworkers developed an innovative optical sensor for IXC by using an environmentally friendly MIP coated on silica-CQDs (SiCQDs). They found that as the concentration of IXC increased within the range of 4–102 nM, there was a corresponding decrease in fluorescence. After optimizing the conditions, they were able to achieve a remarkable detection limit of 1 nM. The precision of the sensor was found to be 4.5% and 2.3% for five repeated measurements at 21 and 60 nM IXC, respectively. To test the sensor's effectiveness in quantifying IXC in environmental and biological samples, the team obtained recoveries in the range of 95–106% with a relative standard deviation of less than 6.0% [115].

A specialized sensor, known as a MIP sensor with double emission, was used to visually detect CAP with high sensitivity. In this innovative method, Jalili and Khataee created silica core-embedded blue-emitting CDs to serve as the reference signal, while yellow-emitting CDs confined within mesoporous silica acted as the response signal marker. When the probe was excited at 400 nm, it selectively decreased in intensity in the presence of CAP, causing a noticeable change in fluorescence color from yellow to blue. After refining the conditions, this technique was successfully used to detect CAP in milk samples, with a detection limit as low as 0.035 $\mu g\ L^{-1}$ [116]. In a similar study, Pan and colleagues developed a fluorescent probe for detecting CAP in food, using CDs-encapsulated luminescent metal-organic frameworks@surface MIP (CDs@MOF@SMIP) (Figure 9.6A). The fluorescent CDs acted as signal tags enclosed within metal-organic backbones (ZIF-8), resulting in luminescent MOF materials (CDs@ZIF-8). The fluorescence of the material decreased proportionally with CAP concentration in the range of 0.323 $\mu g\ L^{-1}$ (0.001 μM) to 8075.0 $\mu g\ L^{-1}$ (25.0 μM), featuring a LOD of 0.08 $\mu g\ L^{-1}$ [117]. Ceftazidime (CF) is an important antibiotic used to treat severe bacterial infections but it can have harmful side effects. Amiri and colleagues have developed a highly responsive optical nanosensor using CDs produced with citric acid and embedded in SMIP (Figure 9.6B). This sensor, called CD-SMIP, emits strong fluorescence at 420 nm when excited at 340 nm. The fluorescence intensity of CD-SMIP decreases when CF molecules are present. The probe can detect CF in a range of 0.18–1.27 $\mu g\ L^{-1}$, with an LOD of 0.06 $\mu g\ L^{-1}$ and RSD of 3.2% [118].

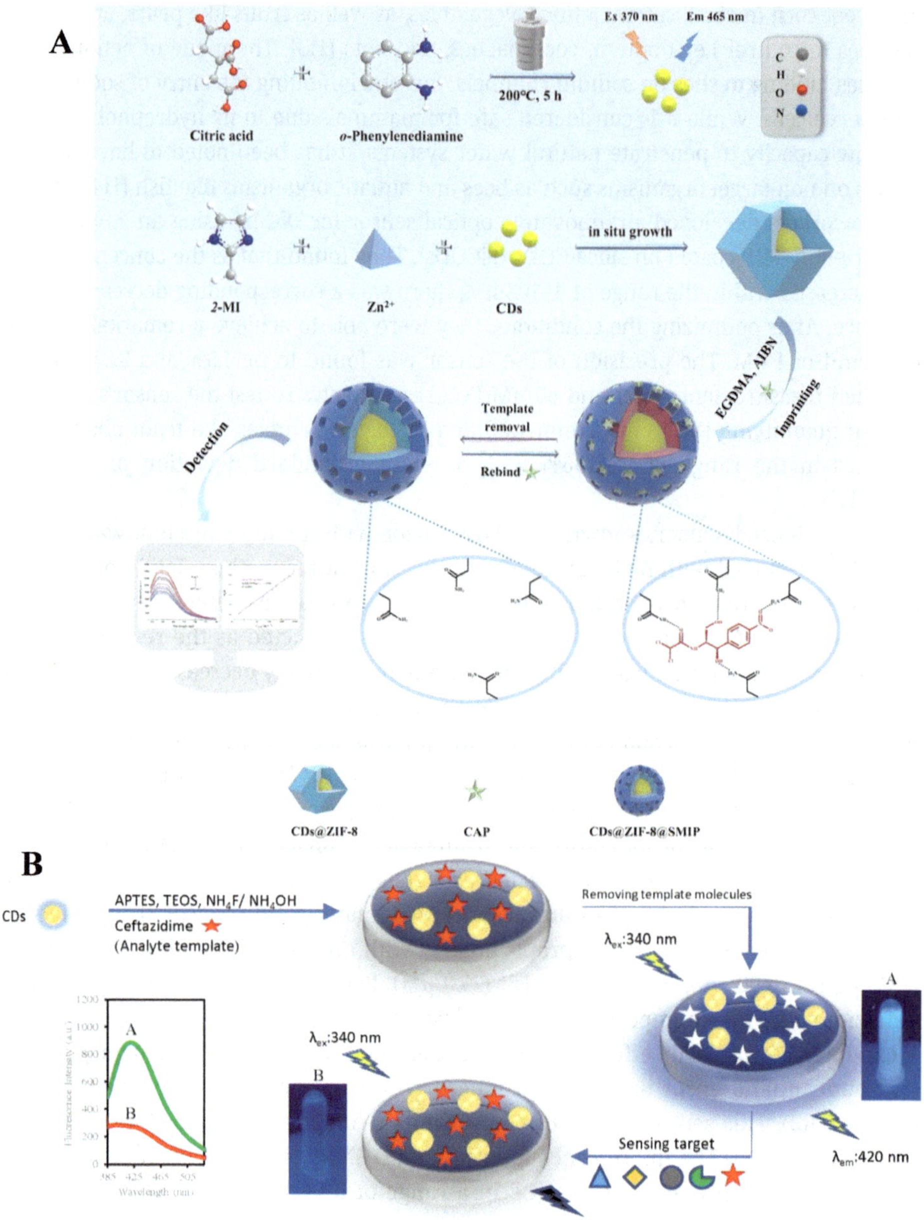

Figure 9.6: **(A)** Schematic of the CDs@ZIF-8@SMIP system for the detection of CAP [117]. (**B**) Scheme of preparation and application of CDs-SMIP for the detection of CF [118].

9.3.2.3 Other carbon materials

Several other studies have been conducted with carbon-based MIP sensors, such as CNTs, porous carbon, and GCN, to detect compounds and contaminants in agricultural and food products. Typically, these materials act as signal amplifiers to increase the effectiveness of luminophores and improve the detectability of sensors.

Yuphintharakun and coworkers developed a highly sensitive nanocomposite optosensor for detecting low levels of ciprofloxacin (Figure 9.7). The sensor consisted of MWCNT functionalized with carboxylic acid and CdTe quantum dots enclosed in a MIP (COOH@MWCNT-MIP-QDs). Under optimal conditions, the fluorescence intensity of the optosensor decreased proportionally with the concentration of ciprofloxacin in two linear dynamic ranges: 0.10–1.0 μg L^{-1} and 1.0–100.0 μg L^{-1}, achieving an impressively low detection limit of 0.066 μg L^{-1}. The imprinting factors for the two linear ranges were 17.67 and 4.28, respectively. This novel nanocomposite fluorescence probe was successfully used to analyze ciprofloxacin levels in chicken muscle and milk samples, with recoveries ranging from 82.6% to 98.4% [119].

To address the issue of high toxicity associated with traditional metal QDs, researchers have uncovered new fluorescent nanomaterials, including GCN (g-C3N4), GQD, and CDs. The production process for CDs and GQD is relatively intricate, leading to impurity in the final product [120, 121]. In comparison, g-C3N4 appears to be a promising substitute for other QDs due to its strong fluorescence, uncomplicated preparation process, low toxicity, and excellent biocompatibility. Similar to graphite, g-C3N4 is a visible-light-responsive material with a two-dimensional planar conjugation structure and a unique electronic band. Each layer of g-C3N4 is composed of a repeating tri-s-triazine ring held together by van der Waals interactions. Its distinct optical properties have made it a subject of interest [122, 123]. Xu and coworkers developed a highly sensitive fluorescent probe for the precise detection of quercetin, a crucial indicator of beverage freshness. The probe consists of silica ball encapsulated GCN (g-C3N4) modified with a MIP and was synthesized using reverse microemulsion. To measure quercetin levels in tea juice, grape juice, red wine, and black tea, they employed fluorometry with an excitation wavelength of 350 nm and an emission wavelength of 460 nm. The probe exhibited a linear response within the 10–1,000 ng mL^{-1} quercetin concentration range, with a LOD of 2.5 ng mL^{-1}. The recoveries ranged between 90.7% and 94.1%, and the relative standard deviations ranged from 2.1% to 5.5% [124].

In another study, Xu and colleagues developed an innovative fluorescent probe by combining GCN (g-C3N4) with molecularly imprinted silica. This probe is designed to specifically detect chlortetracycline (CTC) by creating numerous unique recognition sites for the target molecule. The researchers successfully applied this probe for milk analysis, achieving recoveries between 90.1% and 95.7% with an RSD of 1.8–2.8%. The LOD for CTC was determined to be 8 ng mL^{-1}. These results highlight the potential of this approach, which combines the sensitivity of fluorescence detection with the selectivity of an MIP [123].

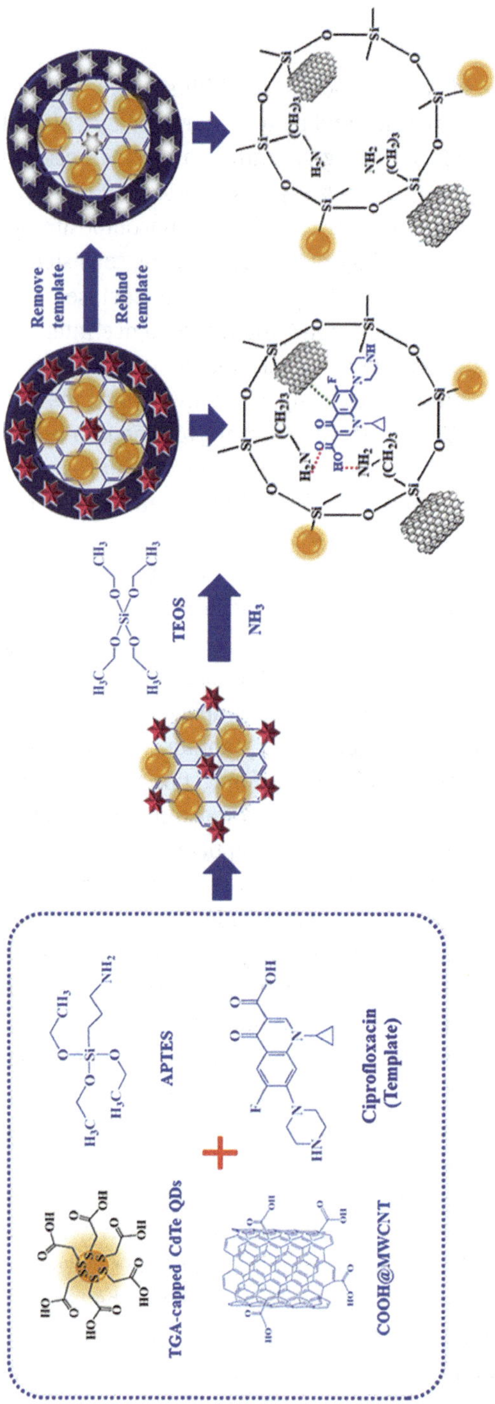

Figure 9.7: Scheme of preparation and application of nanocomposite COOH@MWCNT-MIP-QD optosensors for the detection of ciprofloxacin [119].

Table 9.3: Some examples of different optical sensing applications of carbon-based MIP in agriculture.

Sensing platform	Target analyte	LOD	Linearity	Real sample	Reference
MoS2-GQDs/MIP	2-Methyl-4-chlorophenoxyacetic acid.	5.5×10^3 nM	10 pM to 0.1 μM	Tap water, lake water, and oat samples	[100]
N-GQDs/MIP	Prometryn	0.63 μg kg^{-1}	1.0 to 100.0 μg kg^{-1}	Vegetable samples	[101]
CQDs@MIPs	PrHy	20 μM	2.0 to 250 μM	Orange juice	[110]
Si-CDs@MIP	ACT	2 nM	7 to 107 nM	Blood filtered serum	[112]
GQDs/PC@MIP	Sulfadimethoxine	0.03 μg L^{-1}	0.10 to 25.0 μg L^{-1}	Milk	[105]
Si-CDs@MIP	IXC	1 nM	4 to 102 nM	Water, apple, and tomato	[115]
Silica core-embedded blue-emitting CDs/MIP	CAP	0.035 μg L^{-1}	0.1 to 3 μg L^{-1}	Milk	[116]
CDs@MOF@SMIP	CAP	0.08 μg L^{-1}	0.323 μg L^{-1} to 8075.0 μg L^{-1}	Urine	[118]
COOH@MWCNT-MIP-QDs	Ciprofloxacin	0.066 μg L^{-1}	1.0 to 100.0 μg L^{-1}	Milk	[119]
GCN@MIP	Quercetin	2.5 ng mL^{-1}	10 to 1,000 ng mL^{-1}	Tea juice, grape juice, red wine, and black tea	[124]
GCN@MIP	CTC	8 ng mL^{-1}		Milk	[123]
GQDs@MIPs	TC	1 μg L^{-1}	1.0 to 10^4 μg L^{-1}	Milk	[125]
CDs@MIPs	5-Nitroimidazole	5 nM	20–5,000 nM	River water	[126]

(continued)

Table 9.3 (continued)

Sensing platform	Target analyte	LOD	Linearity	Real sample	Reference
N-CDs@MIPs	TBC	99.2 nM	1–12 μM	Milk	[127]
Mg/N co-doped CDs@MIPs	TC	0.79 ng mL^{-1}	5–100 ng mL^{-1}	Animal-derived foods	[128]
CDs@MIPs	Fluoroquinolones	0.062 ng mL^{-1}	0.03–2.60 ng mL^{-1}	Meat	[129]
B-CDs@MIPs	DA	5 nM	0.05 to 5 mM	Shellfish and lake water	[130]
Sulfur-doped-CQDs@MIPs	IXC	5 μg kg^{-1}	1–15 μg kg^{-1}	Vegetables and tea	[131]

Abbreviations: ECL, electrochemiluminescence; GQDs, graphene quantum dots; N-GQDs, nitrogen-doped graphene quantum dots; PrHy, promethazine hydrochloride; CQDs, carbon quantum dots; Si-CDs, silane-doped carbon dots; ACT, acetamiprid; IXC, indoxacarb; CAP, chloramphenicol; MOFs, metal-organic frameworks; MWCNT, multiwalled carbon nanotubes; GCN, g-C3N4; CTC, chlortetracycline; TC, tetracycline; TBC, tobramycin; N-CDs, nitrogen-doped carbon dots; LOD, limit of detection; DA, domoic acid; PC, porous carbo.

9.4 Conclusions and future perspectives

The development of MIPs poses several challenges due to the need for specific, selective properties tailored to particular applications. One major hurdle is the selection of water-soluble monomers with non-hydrogen bonding interactions in aqueous media. Another concern is the potential for template leaking, which can be addressed by imprinting dummy approach molecules for trace sample analysis. Selecting the most suitable monomer for effective adhesion to modified carbon material surfaces remains a challenging task [35, 16]. Furthermore, there is still much to discover about new carbon materials. It is crucial to determine the most appropriate polymerization method for MIP synthesis. The chapter aims to provide researchers with insights into the latest applications of carbon nanomaterial-based MIP sensors in food hazard detection. Furthermore, the text suggests that while evaluating different carbon nanomaterials for MIP sensor analysis, attention should also be given to sustainability challenges in hazardous substance detection, such as the development of imprinted polymers based on macromolecules. Additionally, identifying new strategic technologies for rapid detection is crucial for the industrialization and commercial application of MIP sensors. This chapter provides the utilization of MIP optical and electrochemical sensors integrated with carbon nanomaterials for the precise and sensitive detection of various threats to agriculture and food products. It also discusses the advantages of combining carbon nanomaterials with MIPs to enhance sensitivity and identify pesticides, antibiotics, heavy metals, microorganisms, toxins, and other agricultural contaminants. Recent advancements in carbon nanomaterial-based MIP sensors have focused on addressing the requirements for reliability, portability, accuracy, and ease of application in manufacturing MIP sensors. The text highlights new applications, including the development of highly sensitive MIP sensing platforms incorporating carbon nanomaterials, which are approaching commercialization. However, significant challenges remain. These include the incomplete understanding of the binding mechanism of template molecules, the limitations in selectivity, stability, and affinity of MIPs, inadequate research on comparing different carbon materials, and the complexity of food sample matrices which necessitate pretreatment before utilizing MIP sensors for detection.

To overcome these challenges and integrate the benefits of existing carbon nanomaterial-based MIP sensors, future research directions are outlined as follows:

- Emphasizing theoretical analysis to predict potential binding energy and interaction mechanisms to enhance MIP sensitivity.
- Developing innovative functional carbon nanomaterials and imprinting technologies to create MIP sensors with enhanced detection capabilities.
- Exploring multi-template MIP sensors for high-throughput food hazard analysis.
- Creating a MIP sensing platform with dual identification elements to improve identification ability.

- Integrating MIP sensors with advanced technologies such as 3D printing and smartphones for on-site detection capabilities.
- Addressing matrix interferences in food materials to achieve direct quantitative detection.

Carbon nanomaterials, including graphene, carbon nanotubes, and fullerene derivatives, are recognized for their efficient electronic properties. They function as excellent electronic media, which facilitates the transfer of charge and significantly boosts the sensitivity of sensors. Additionally, the large surface area of these nanomaterials permits extensive functionalization, making them highly adaptable for various applications. Researchers have been able to combine carbon nanomaterials with a diverse range of small organic molecules, polymers, biomolecules such as DNA, proteins, and enzymes, as well as with metal NPs and metal oxide NPs. This versatility enables the creation of highly reactive and tailored sensing platforms. The effective surface functionalization of carbon nanomaterials improves their chemical properties, enhancing their reactivity and making them ideal candidates for use as substrates in MIPs. This integration not only maximizes sensor sensitivity but also enables specific binding sites tailored for target molecules, which is critical for the development of highly sensitive sensors [75, 112]. However, it' is important to note that the production of functional carbon-based nanomaterials is a complex process. The intricate manufacturing techniques required to synthesize these materials, along with their integration into MIPs, are crucial for advancing sensor technology. Despite the promising advancements reflected in the current literature, there remains a notable gap regarding the application of carbon nanomaterial-based MIP sensors in the detection of agricultural food contaminants. Furthermore, there is a lack of comprehensive strategies that address the analysis of food hazards across various contexts, which is an area that warrants further exploration to enhance food safety and quality assurance measures.

References

[1] Moradi M., Azizi-Lalabadi M., Motamedi P., & Sadeghi E. (2021). Electrochemical determination of T2 toxin by graphite/polyacrylonitrile nanofiber electrode, *Food Science & Nutrition*. 91171–91179. https://doi.org/10.1002/fsn3.2097

[2] Sadeghi E., Rahimi F., Azizi Z., Kaki S., & Babakhanian A. (2023b). Fabrication of a sensitive electrochemical sensor based on hybrid polyamide/chromotropic acid nanofibers electrospun on glassy carbon electrode for Hg^{2+} sensing in drinking water and canned fish samples, *Food Chemistry*. 414135467. https://doi.org/10.1016/j.foodchem.2023.135467

[3] Abedi-Firoozjah R., Ebdali H., Soltani M., Abdolahi-Fard P., Heydari M., Assadpour E., Azizi-Lalabadi M., Zhang F., & Jafari S. M. (2024b). Nanomaterial-based sensors for the detection of pathogens and microbial toxins in the food industry; a review on recent progress, *Coordination Chemistry Reviews*. 500215545. https://doi.org/10.1016/j.ccr.2023.215545

[4] Roshanak S., Maleki M., Sani M. A., Tavassoli M., Pirkhezranian Z., & Shahidi F. (2023). The impact of cold plasma innovative technology on quality and safety of refrigerated hamburger: Analysis of microbial safety and physicochemical properties, *International Journal of Food Microbiology*. 388110066, https://doi.org/https://doi.org/10.1016/j.ijfoodmicro.2022.110066.

[5] Tiscali M., Khezerlou A., Khalilzadeh B., Ehsani A., & Kazemian H. (2023b). Aptamer-modified metal organic frameworks for measurement of food contaminants: A review, *Microchimica Acta*. 190371, https://doi.org/10.1007/s00604-023-05937-2.

[6] Ahangari H., Tavassoli M., Khezerlou A., Kiani Salmi N., Ehsani A., & Afshar Mogaddam M. R. (2023). A review on the determination of biogenic amines in fresh and processed fish products using HPLC, LC-MS/MS and other chromatographic methods, *Combinatorial Chemistry & High Throughput Screening*. 262598–262606, https://doi.org/10.2174/1386207326666230316141040.

[7] Li F., Yue S., Zhao Z., Liu K., Wang P., & Zhan S. (2024a). Application of molecularly imprinted polymers in the water environmental field: A review on the detection and efficient removal of emerging contaminants, *Materials Today Sustainability*. 27100904, https://doi.org/https://doi.org/10.1016/j.mtsust.2024.100904.

[8] Tavassoli M., Khezerlou A., Hamishehkar H., Ehsani A., & Khalilzadeh B. (2023a). An ultrasensitive aptamer-based fluorescent on/off system for trace amount evaluation of Yersinia enterocolitica in food samples, *Microchimica Acta*. 190253, https://doi.org/10.1007/s00604-023-05820-0.

[9] Moradi S., Azizi-Lalabadi M., Bagheri V., & Sadeghi E. (2020). Fabrication of electrospun sensor based on a synthesized component doped into PAN (polyacrylonitrile) nanofibers for electrochemical detection of zearalenone mycotoxin in foods simulant, *Sensing and Bio-Sensing Research*. 28100321.

[10] Khezerlou A., Tavassoli M., Alizadeh Sani M., Ghasempour Z., Ehsani A., & Khalilzadeh B. (2023a). Rapid and sensitive detection of tetracycline residue in food samples using Cr(III)-MOF fluorescent sensor, *Food Chemistry: X*. 20100883, https://doi.org/https://doi.org/10.1016/j.fochx.2023.100883.

[11] Li P., & Liu Z. (2024). Glycan-specific molecularly imprinted polymers towards cancer diagnostics: Merits, applications, and future perspectives, *Chemical Society Reviews*. 531870–531891, https://doi.org/10.1039/D3CS00842H.

[12] Khezerlou A., Tavassoli M., Khalilzadeh B., Ehsani A., & Kazemian H. (2023b). Metal-organic framework-based advanced sensing platforms for the detection of tetracycline in food and water samples, *Food Control*. 153109965, https://doi.org/https://doi.org/10.1016/j.foodcont.2023.109965.

[13] Liu Y., Wang L., Li H., Zhao L., Ma Y., Zhang Y., Liu J., & Wei Y. (2024). Rigorous recognition mode analysis of molecularly imprinted polymers – Rational design, challenges, and opportunities, *Progress in Polymer Science*. 150101790, https://doi.org/https://doi.org/10.1016/j.progpolymsci.2024.101790.

[14] Pandey H., Khare P., Singh S., & Singh S. P. (2020). Carbon nanomaterials integrated molecularly imprinted polymers for biological sample analysis: A critical review, *Materials Chemistry and Physics*. 239121966, https://doi.org/https://doi.org/10.1016/j.matchemphys.2019.121966.

[15] Preda D., David I. G., Popa D.-E., Buleandra M., & Radu G. L. (2022). Recent trends in the development of carbon-based electrodes modified with molecularly imprinted polymers for antibiotic electroanalysis, *Chemosensors*. 10, https://doi.org/10.3390/chemosensors10070243.

[16] Kamyab H., Chelliapan S., Tavakkoli O., Mesbah M., Bhutto J. K., Khademi T., Kirpichnikova I., Ahmad A., & Aljohani A. A. (2022). A review on carbon-based molecularly-imprinted polymers (CBMIP) for detection of hazardous pollutants in aqueous solutions, *Chemosphere*. 308136471, https://doi.org/https://doi.org/10.1016/j.chemosphere.2022.136471.

[17] Arabi M., Ostovan A., Bagheri A. R., Guo X., Li J., Ma J., & Chen L. (2020). Hydrophilic molecularly imprinted nanospheres for the extraction of rhodamine B followed by HPLC analysis: A green approach and hazardous waste elimination, *Talanta*. 215120933, https://doi.org/https://doi.org/10.1016/j.talanta.2020.120933.

[18] Wackerlig J., & Lieberzeit P. A. (2015). Molecularly imprinted polymer nanoparticles in chemical sensing – Synthesis, characterisation and application, *Sensors and Actuators B: Chemical*. 207144–207157, https://doi.org/https://doi.org/10.1016/j.snb.2014.09.094.

[19] Bagheri A. R., Arabi M., Ghaedi M., Ostovan A., Wang X., Li J., & Chen L. (2019). Dummy molecularly imprinted polymers based on a green synthesis strategy for magnetic solid-phase extraction of acrylamide in food samples, *Talanta*. 195390–195400, https://doi.org/https://doi.org/10.1016/j.talanta.2018.11.065.

[20] Chi H., & Liu G. (2023). Carbon nanomaterial-based molecularly imprinted polymer sensors for detection of hazardous substances in food: Recent progress and future trends, *Food Chemistry*. 420136100, https://doi.org/https://doi.org/10.1016/j.foodchem.2023.136100.

[21] Hassan Amini M., Alijani H., & Hossein Beyki M. (2022). Toxic cadmium selective sequestration from food samples using melamine anchored magnetic cellulose by surface imprinting route, *Food Chemistry*. 396133688, https://doi.org/https://doi.org/10.1016/j.foodchem.2022.133688.

[22] Jalalvand A. R., Karami M. M., Arkan E., & Sadeghi E. (2023). A novel triple templates molecularly imprinted biosensor assisted by first derivatives of second-order hydrodynamic differential normal pulse voltametric data and multi-way calibration methods for simultaneous biosensing of penicillin, tetracycline and amoxicillin in dairy products: A novel multi-disciplinary study, *Chemometrics and Intelligent Laboratory Systems*. 235104765. https://doi.org/10.1016/j.chemolab.2023.104765

[23] Udomsap D., Brisset H., Culioli G., Dollet P., Laatikainen K., Siren H., & Branger C. (2018). Electrochemical molecularly imprinted polymers as material for pollutant detection, *Materials Today Communications*. 17458–17465, https://doi.org/https://doi.org/10.1016/j.mtcomm.2018.10.019.

[24] Abedi-Firoozjah R., Alizadeh-Sani M., Zare L., Rostami O., Salim S. A., Assadpour E., Azizi-Lalabadi M., Zhang F., Lin X., & Jafari S. M. (2024a). State-of-the-art nanosensors and kits for the detection of antibiotic residues in milk and dairy products, *Advances in Colloid and Interface Science*. 328103164. https://doi.org/10.1016/j.cis.2024.103164

[25] Qin L., Liu W.-F., Liu X.-G., Yang Y.-Z., & Zhang L.-A. (2020). A review of nano-carbon based molecularly imprinted polymer adsorbents and their adsorption mechanism, *New Carbon Materials*. 35459–35485, https://doi.org/https://doi.org/10.1016/S1872-5805(20)60503-0.

[26] Sadeghi E., Kappers C., Chiumento A., Derks M., & Havinga P. (2023a). Improving piglets health and well-being: A review of piglets health indicators and related sensing technologies, *Smart Agricultural Technology*. 5100246, https://doi.org/https://doi.org/10.1016/j.atech.2023.100246.

[27] Yin R., Gao L., Qin D., Chen L., & Niu N. (2021). Preparation of porous carbon-based molecularly imprinted polymers for separation of triazine herbicides in corn, *Microchimica Acta*. 18923, https://doi.org/10.1007/s00604-021-05100-9.

[28] Poonia K., Raizada P., Singh A., Verma N., Ahamad T., Alshehri S. M., Khan A. A. P., Singh P., & Hussain C. M. (2022). Magnetic molecularly imprinted polymer photocatalysts: Synthesis, applications and future perspective, *Journal of Industrial and Engineering Chemistry*. 1131–14, https://doi.org/https://doi.org/10.1016/j.jiec.2022.05.029.

[29] Tse Sum Bui B., Mier A., & Haupt K. (2023). Molecularly imprinted polymers as synthetic antibodies for protein recognition: The next generation, *Small*. 192206453, https://doi.org/https://doi.org/10.1002/smll.202206453.

[30] Lamaoui A., García-Guzmán J. J., Amine A., Palacios-Santander J. M., & Cubillana-Aguilera L. (2021). Chapter 4 – Synthesis techniques of molecularly imprinted polymer composites, Sooraj M. P., Nair A. S., Mathew B., & Thomas S. (eds.) *Molecularly Imprinted Polymer Composites*. Woodhead Publishing, 49–91.

[31] Madikizela L. M., Nomngongo P. N., & Pakade V. E. (2022). Synthesis of molecularly imprinted polymers for extraction of fluoroquinolones in environmental, food and biological samples, *Journal of Pharmaceutical and Biomedical Analysis*. 208114447, https://doi.org/https://doi.org/10.1016/j.jpba.2021.114447.

[32] Liu Y., Huang Y., Liu J., Wang W., Liu G., & Zhao R. (2012). Superparamagnetic surface molecularly imprinted nanoparticles for water-soluble pefloxacin mesylate prepared via surface-initiated atom transfer radical polymerization and its application in egg sample analysis, *Journal of Chromatography A*. 124615–124621, https://doi.org/https://doi.org/10.1016/j.chroma.2012.01.045.
[33] Piletska E. V., Guerreiro A. R., Whitcombe M. J., & Piletsky S. A. (2009). Influence of the Polymerization conditions on the performance of molecularly imprinted polymers, *Macromolecules*. 424921–424928, https://doi.org/10.1021/ma900432z.
[34] Guerreiro A. R., Chianella I., Piletska E., Whitcombe M. J., & Piletsky S. A. (2009). Selection of imprinted nanoparticles by affinity chromatography, *Biosensors and Bioelectronics*. 242740–242743, https://doi.org/https://doi.org/10.1016/j.bios.2009.01.013.
[35] Arshad R., Rhouati A., Hayat A., Nawaz M. H., Yameen M. A., Mujahid A., & Latif U. (2020). MIP-based impedimetric sensor for detecting dengue fever biomarker, *Applied Biochemistry and Biotechnology*. 1911384–1911394, https://doi.org/10.1007/s12010-020-03285-y.
[36] Gao L., Han W., Yan Y., Li X., Li C., & Hu B. (2016). Expeditious quantitative analysis of λ-cyhalothrin depending on fluorescence quenching of fluorescent surface molecularly imprinted sensors, *Analytical Methods*. 82434–82440, https://doi.org/10.1039/C5AY02721G.
[37] Cui Y., He Z., Xu Y., Su Y., Ding L., & Li Y. (2021). Fabrication of molecularly imprinted polymers with tunable adsorption capability based on solvent-responsive cross-linker, *Chemical Engineering Journal*. 405126608, https://doi.org/https://doi.org/10.1016/j.cej.2020.126608.
[38] Muhammad T., Nur Z., Piletska E. V., Yimit O., & Piletsky S. A. (2012). Rational design of molecularly imprinted polymer: The choice of cross-linker, *Analyst*. 1372623–1372628, https://doi.org/10.1039/C2AN35228A.
[39] Zeng H., Wang Y., Liu X., Kong J., & Nie C. (2012). Preparation of molecular imprinted polymers using bi-functional monomer and bi-crosslinker for solid-phase extraction of rutin, *Talanta*. 93172–93181, https://doi.org/https://doi.org/10.1016/j.talanta.2012.02.008.
[40] Zhang X., Shen F., Zhang Z., Xing Y., & Ren X. (2014). Synthesis of a novel cross-linker doubles as a functional monomer for preparing a water compatible molecularly imprinted polymer, *Analytical Methods*. 69483–69489, https://doi.org/10.1039/C4AY01651C.
[41] Sibrian-Vazquez M., & Spivak D. A. (2003). Improving the strategy and performance of molecularly imprinted polymers using cross-linking functional monomers, *The Journal of Organic Chemistry*. 689604–689611, https://doi.org/10.1021/jo0352225.
[42] Schirhagl R., Qian J., & Dickert F. L. (2012). Immunosensing with artificial antibodies in organic solvents or complex matrices, *Sensors and Actuators B: Chemical*. 173585–173590, https://doi.org/https://doi.org/10.1016/j.snb.2012.07.036.
[43] Lieberzeit P. A., Afzal A., Rehman A., & Dickert F. L. (2007). Nanoparticles for detecting pollutants and degradation processes with mass-sensitive sensors, *Sensors and Actuators B: Chemical*. 127132–127136, https://doi.org/https://doi.org/10.1016/j.snb.2007.07.020.
[44] Beierle J. M., Yoshimatsu K., Chou B., Mathews M. A. A., Lesel B. K., & Shea K. J. (2014). Polymer nanoparticle hydrogels with autonomous affinity switching for the protection of proteins from thermal stress, *Angewandte Chemie International Edition*. 539275–539279, https://doi.org/https://doi.org/10.1002/anie.201404881.
[45] Ding X., & Heiden P. A. (2014). recent developments in molecularly imprinted nanoparticles by surface imprinting techniques, *Macromolecular Materials and Engineering*. 299268–299282, https://doi.org/https://doi.org/10.1002/mame.201300160.
[46] Priego-Capote F., Ye L., Shakil S., Shamsi S. A., & Nilsson S. (2008). Monoclonal behavior of molecularly imprinted polymer nanoparticles in capillary electrochromatography, *Analytical Chemistry*. 802881–802887, https://doi.org/10.1021/ac070038v.
[47] Slomkowski S., Alemán J. V., Gilbert R. G., Hess M., Horie K., Jones R. G., Kubisa P., Meisel I., Mormann W., Penczek S., et al. (2011). *Terminology of Polymers and Polymerization Processes in*

Dispersed Systems (IUPAC Recommendations 2011). Pure and Applied Chemistry 832229–832259, https://doi.org/doi:10.1351/PAC-REC-10-06-03.
[48] Congur G., Senay H., Turkcan C., Canavar E., Erdem A., & Akgol S. (2013). Estrone specific molecularly imprinted polymeric nanospheres: Synthesis, characterization and applications for electrochemical sensor development, *Combinatorial Chemistry & High Throughput Screening*. 16, 503–510.
[49] Vaihinger D., Landfester K., Kräuter I., Brunner H., & Tovar G. E. M. (2002). Molecularly imprinted polymer nanospheres as synthetic affinity receptors obtained by miniemulsion polymerisation, *Macromolecular Chemistry and Physics*. 2031965–2031973, https://doi.org/https://doi.org/10.1002/1521-3935(200209)203:13<1965::AID-MACP1965>3.0.CO;2-C.
[50] Martín-Esteban A. (2013). Molecularly-imprinted polymers as a versatile, highly selective tool in sample preparation, *TrAC Trends in Analytical Chemistry*. 45169–45181, https://doi.org/https://doi.org/10.1016/j.trac.2012.09.023.
[51] Li Y., Yang -H.-H., You Q.-H., Zhuang Z.-X., & Wang X.-R. (2006). Protein recognition via surface molecularly imprinted polymer nanowires, *Analytical Chemistry*. 78317–78320, https://doi.org/10.1021/ac050802i.
[52] Vandevelde F., Pantigny J., Pezron I., Gheber L. A., & Haupt K. (2008). Nanostructured molecularly imprinted polymer films as synthetic recognition layers, *International Journal of Nanotechnology*. 5757–5768, https://doi.org/10.1504/IJNT.2008.018695.
[53] Ceolin G., Orbán Á., Kocsis V., Gyurcsányi R. E., Kézsmárki I., & Horváth V. (2013). Electrochemical template synthesis of protein-imprinted magnetic polymer microrods, *Journal of Materials Science*. 485209–485218, https://doi.org/10.1007/s10853-013-7309-6.
[54] Hu Y., Pan J., Zhang K., Lian H., & Li G. (2013). Novel applications of molecularly-imprinted polymers in sample preparation, *TrAC Trends in Analytical Chemistry*. 4337–4352, https://doi.org/https://doi.org/10.1016/j.trac.2012.08.014.
[55] Turiel E., & Martín-Esteban A. (2010). Molecularly imprinted polymers for sample preparation: A review, *Analytica Chimica Acta*. 66887–66899, https://doi.org/https://doi.org/10.1016/j.aca.2010.04.019.
[56] Pichon V., Delaunay N., & Combès A. (2020). Sample preparation using molecularly imprinted polymers, *Analytical Chemistry*. 9216–9233, https://doi.org/10.1021/acs.analchem.9b04816.
[57] Spivak D. A. (2005). Optimization, evaluation, and characterization of molecularly imprinted polymers, *Advanced Drug Delivery Reviews*. 571779–571794, https://doi.org/https://doi.org/10.1016/j.addr.2005.07.012.
[58] Arshady R., & Mosbach K. (1981). Synthesis of substrate-selective polymers by host-guest polymerization, *Die Makromolekulare Chemie*. 182687–182692, https://doi.org/https://doi.org/10.1002/macp.1981.021820240.
[59] Gama M. R., & Bottoli C. B. G. (2017). Molecularly imprinted polymers for bioanalytical sample preparation, *Journal of Chromatography B*. 1043107–1043121, https://doi.org/https://doi.org/10.1016/j.jchromb.2016.09.045.
[60] Huang S., Xu J., Zheng J., Zhu F., Xie L., & Ouyang G. (2018). Synthesis and application of magnetic molecularly imprinted polymers in sample preparation, *Analytical and Bioanalytical Chemistry*. 4103991–4104014, https://doi.org/10.1007/s00216-018-1013-y.
[61] Gamal M., Imam M. S., Albugami A. S., Hunjur S. A., Aldhalmi A. K., AbdElrahman M., Ghoneim M. M., Ali H. M., Abdelwahab N. S., & Eissa M. S.(2024). Current advances in the implementation of magnetic molecularly imprinted polymers tailored for enrichment of target analytes in different environmental samples: An overview from a comprehensive perspective, *Trends in Environmental Analytical Chemistry*. 43(e00236), https://doi.org/https://doi.org/10.1016/j.teac.2024.e00236.

[62] Li Y., Luo L., Kong Y., Li Y., Wang Q., Wang M., Li Y., Davenport A., & Li B. (2024b). Recent advances in molecularly imprinted polymer-based electrochemical sensors, *Biosensors and Bioelectronics*. 249116018, https://doi.org/https://doi.org/10.1016/j.bios.2024.116018.

[63] Rapacz D., Smolinska-Kempisty K., & Wolska J. (2024). Molecularly imprinted polymers toward herbicides – Progress in development and the current state of the art depending on the polymerization environment – Short review, *Journal of Environmental Chemical Engineering*. 12112159, https://doi.org/https://doi.org/10.1016/j.jece.2024.112159.

[64] Jahanban-Esfahlan A., & Amarowicz R. (2024). Molecularly imprinted polymers for sensing/depleting human serum albumin (HSA): A critical review of recent advances and current challenges, *International Journal of Biological Macromolecules*. 266131132, https://doi.org/https://doi.org/10.1016/j.ijbiomac.2024.131132.

[65] Anirudhan T. S., & Anju S. M. (2019). Synthesis and evaluation of TiO_2 nanotubes/silylated graphene oxide-based molecularly imprinted polymer for the selective adsorption and subsequent photocatalytic degradation of 2,4-Dichlorophenoxyacetic acid, *Journal of Environmental Chemical Engineering*. 7103355, https://doi.org/https://doi.org/10.1016/j.jece.2019.103355.

[66] Pophali A., Singh S., & Verma N. (2020). Simultaneous hydrogen generation and COD reduction in a photoanode-based microbial electrolysis cell, *International Journal of Hydrogen Energy*. 4525985–4525995, https://doi.org/https://doi.org/10.1016/j.ijhydene.2020.01.053.

[67] Stoller M. D., Park S., Zhu Y., An J., & Ruoff R. S. (2008). Graphene-based ultracapacitors, *Nano Letters*. 83498–83502, https://doi.org/10.1021/nl802558y.

[68] Zhang H., Ahima J., Yang Q., Zhao L., Zhang X., & Zheng X. (2021). A review on citrinin: Its occurrence, risk implications, analytical techniques, biosynthesis, physiochemical properties and control, *Food Research International*. 141110075, https://doi.org/https://doi.org/10.1016/j.foodres.2020.110075.

[69] Elfadil D., Silveri F., Palmieri S., Della Pelle F., Sergi M., Del Carlo M., Amine A., & Compagnone D. (2023). Liquid-phase exfoliated 2D graphene nanoflakes electrochemical sensor coupled to molecularly imprinted polymers for the determination of citrinin in food, *Talanta*. 253124010, https://doi.org/https://doi.org/10.1016/j.talanta.2022.124010.

[70] Zhao Y., Du D., Li Q., Chen W., Li Q., Zhang Q., & Liang N. (2020). Dummy-surface molecularly imprinted polymers based on magnetic graphene oxide for selective extraction and quantification of pyrethroids pesticides in fruit juices, *Microchemical Journal*. 159105411, https://doi.org/https://doi.org/10.1016/j.microc.2020.105411.

[71] Karthika P., Shanmuganathan S., Viswanathan S., & Delerue-Matos C. (2021). Molecularly imprinted polymer-based electrochemical sensor for the determination of endocrine disruptor bisphenol-A in bovine milk, *Food Chemistry*. 363130287, https://doi.org/https://doi.org/10.1016/j.foodchem.2021.130287.

[72] Xie T., Zhang M., Chen P., Zhao H., Yang X., Yao L., Zhang H., Dong A., Wang J., & Wang Z. (2017). A facile molecularly imprinted electrochemical sensor based on graphene: Application to the selective determination of thiamethoxam in grain, *RSC Advances*. 738884–738894, https://doi.org/10.1039/C7RA05167K.

[73] Zhang C., Zhao F., She Y., Hong S., Cao X., Zheng L., Wang S., Li T., Wang M., Jin M., et al. (2019a). A disposable molecularly imprinted sensor based on Graphe@AuNPs modified screen-printed electrode for highly selective and sensitive detection of cyhexatin in pear samples, *Sensors and Actuators B: Chemical*. 28413–28422, https://doi.org/https://doi.org/10.1016/j.snb.2018.12.075.

[74] Mantashloo R., & Bahar S. (2023). Synthesis of magnetic graphene quantum dots based molecularly imprinted polymers for fluorescent determination of quercetin, *Microchemical Journal*. 185108233, https://doi.org/https://doi.org/10.1016/j.microc.2022.108233.

[75] Li S., Li J., Luo J., Xu Z., & Ma X. (2018). A microfluidic chip containing a molecularly imprinted polymer and a DNA aptamer for voltammetric determination of carbofuran, *Microchimica Acta*. 185295, https://doi.org/10.1007/s00604-018-2835-1.

[76] Munawar A., Tahir M. A., Shaheen A., Lieberzeit P. A., Khan W. S., & Bajwa S. Z. (2018). Investigating nanohybrid material based on 3D CNTs@Cu nanoparticle composite and imprinted polymer for highly selective detection of chloramphenicol, *Journal of Hazardous Materials*. 34296–106, https://doi.org/https://doi.org/10.1016/j.jhazmat.2017.08.014.

[77] Amatatongchai M., Sroysee W., Sodkrathok P., Kesangam N., Chairam S., & Jarujamrus P. (2019). Novel three-dimensional molecularly imprinted polymer-coated carbon nanotubes (3D-CNTs@MIP) for selective detection of profenofos in food, *Analytica Chimica Acta*. 107664–107672, https://doi.org/https://doi.org/10.1016/j.aca.2019.04.075.

[78] Arvand M., Zamani M., & Sayyar Ardaki M. (2017). Rapid electrochemical synthesis of molecularly imprinted polymers on functionalized multi-walled carbon nanotubes for selective recognition of sunset yellow in food samples, *Sensors and Actuators B: Chemical*. 243927–243939, https://doi.org/https://doi.org/10.1016/j.snb.2016.12.077.

[79] Yin -Z.-Z., Cheng S.-W., Xu L.-B., Liu H.-Y., Huang K., Li L., Zhai -Y.-Y., Zeng Y.-B., Liu H.-Q., Shao Y., et al. (2018). Highly sensitive and selective sensor for sunset yellow based on molecularly imprinted polydopamine-coated multi-walled carbon nanotubes, *Biosensors and Bioelectronics*. 100565–100570, https://doi.org/https://doi.org/10.1016/j.bios.2017.10.010.

[80] Li W., Diao K., Qiu D., Zeng Y., Tang K., Zhu Y., Sheng Y., Wen Y., & Li M. (2021). A highly-sensitive and selective antibody-like sensor based on molecularly imprinted poly(L-arginine) on COOH-MWCNTs for electrochemical recognition and detection of deoxynivalenol, *Food Chemistry*. 350129229, https://doi.org/https://doi.org/10.1016/j.foodchem.2021.129229.

[81] Mahmoud A. M., Alkahtani S. A., Alyami B. A., & El-Wekil M. M. (2020). Dual-recognition molecularly imprinted aptasensor based on gold nanoparticles decorated carboxylated carbon nanotubes for highly selective and sensitive determination of histamine in different matrices, *Analytica Chimica Acta*. 113358–113365, https://doi.org/https://doi.org/10.1016/j.aca.2020.08.001.

[82] Khezerlou A., Zolfaghari H., Forghani S., Abedi-Firoozjah R., Alizadeh Sani M., Negahdari B., Jalalvand M., Ehsani A., & McClements D. J. (2023c). Combining non-thermal processing techniques with edible coating materials: An innovative approach to food preservation, *Coatings*. 13830. https://doi.org/10.3390/coatings13050830

[83] Neghi N., Krishnan N. R., & Kumar M. (2018). Analysis of metronidazole removal and micro-toxicity in photolytic systems: Effects of persulfate dosage, anions and reactor operation-mode, *Journal of Environmental Chemical Engineering*. 6754–6761, https://doi.org/https://doi.org/10.1016/j.jece.2017.12.072.

[84] Ensafi A. A., Nasr-Esfahani P., & Rezaei B. (2018a). Metronidazole determination with an extremely sensitive and selective electrochemical sensor based on graphene nanoplatelets and molecularly imprinted polymers on graphene quantum dots, *Sensors and Actuators B: Chemical*. 270192–270199, https://doi.org/https://doi.org/10.1016/j.snb.2018.05.024.

[85] Guo W., Pi F., Zhang H., Sun J., Zhang Y., & Sun X. (2017). A novel molecularly imprinted electrochemical sensor modified with carbon dots, chitosan, gold nanoparticles for the determination of patulin, *Biosensors and Bioelectronics*. 98299–98304, https://doi.org/https://doi.org/10.1016/j.bios.2017.06.036.

[86] Rehman N. Ç., Özdemir N., Boyacıoğlu H., & Yola M. L. (2024). A novel molecularly imprinted electrochemical sensor based on graphitic carbon nitride nanosheets decorated bovine serum albumin@MnO_2 nanocomposite for zearalenone detection, *Journal of Food Composition and Analysis*. 125105857, https://doi.org/https://doi.org/10.1016/j.jfca.2023.105857.

[87] Kıran T. R., Atar N., & Yola M. L. (2019). A methyl parathion recognition method based on carbon nitride incorporated hexagonal boron nitride nanosheets composite including molecularly

imprinted polymer, *Journal of the Electrochemical Society*. 166H495, https://doi.org/10.1149/2.0331912jes.
[88] Hou C., Tang W., Zhang C., Wang Y., & Zhu N. (2014). A novel and sensitive electrochemical sensor for bisphenol A determination based on carbon black supporting ferroferric oxide nanoparticles, *Electrochimica Acta*. 144324–144331, https://doi.org/https://doi.org/10.1016/j.electacta.2014.08.053.
[89] Ben Messaoud N., Ait Lahcen A., Dridi C., & Amine A. (2018). Ultrasound assisted magnetic imprinted polymer combined sensor based on carbon black and gold nanoparticles for selective and sensitive electrochemical detection of Bisphenol A, *Sensors and Actuators B: Chemical*. 276304–276312, https://doi.org/https://doi.org/10.1016/j.snb.2018.08.092.
[90] Zhang Q., Berg D., & Mugo S. M. (2019c). Molecularly imprinted carbon-based electrodes for tetrahydrocannabinol sensing, *Inorganic Chemistry Communications*. 107107459, https://doi.org/https://doi.org/10.1016/j.inoche.2019.107459.
[91] Wang Y., Ma X., Peng Y., Liu Y., & Zhang H. (2021). Selective and fast removal and determination of β-lactam antibiotics in aqueous solution using multiple templates imprinted polymers based on magnetic hybrid carbon material, *Journal of Hazardous Materials*. 416126098, https://doi.org/https://doi.org/10.1016/j.jhazmat.2021.126098.
[92] Geng L., Sun J., Liu M., Huang J., Dong J., Guo Z., Guo Y., & Sun X. (2024). Molecularly imprinted polymers-aptamer electrochemical sensor based on dual recognition strategy for high sensitivity detection of chloramphenicol, *Food Chemistry*. 437137933, https://doi.org/https://doi.org/10.1016/j.foodchem.2023.137933.
[93] Güney S., Arslan T., Yanık S., & Güney O. (2021). An electrochemical sensing platform based on graphene oxide and molecularly imprinted polymer modified electrode for selective detection of amoxicillin, *Electroanalysis*. 3346–3356. https://doi.org/10.1002/elan.202060129
[94] Arduini F., Di Nardo F., Amine A., Micheli L., Palleschi G., & Moscone D. (2012). Carbon black-modified screen-printed electrodes as electroanalytical tools, *Electroanalysis*. 24743–24751, https://doi.org/https://doi.org/10.1002/elan.201100561.
[95] Vicentini F. C., Ravanini A. E., Figueiredo-Filho L. C. S., Iniesta J., Banks C. E., & Fatibello-Filho O. (2015). Imparting improvements in electrochemical sensors: Evaluation of different carbon blacks that give rise to significant improvement in the performance of electroanalytical sensing platforms, *Electrochimica Acta*. 157125–157133, https://doi.org/https://doi.org/10.1016/j.electacta.2014.11.204.
[96] Arduini F., Zanardi C., Cinti S., Terzi F., Moscone D., Palleschi G., & Seeber R. (2015). Effective electrochemical sensor based on screen-printed electrodes modified with a carbon black-Au nanoparticles composite, *Sensors and Actuators B: Chemical*. 212536–212543, https://doi.org/https://doi.org/10.1016/j.snb.2015.02.051.
[97] Andersson L. I., Mandenius C. F., & Mosbach K. (1988). Studies on guest selective molecular recognition on an octadecyl silylated silicon surface using ellipsometry, *Tetrahedron Letters*. 295437–295440, https://doi.org/https://doi.org/10.1016/S0040-4039(00)82889-0.
[98] Ansari S., & Masoum S. (2021). Recent advances and future trends on molecularly imprinted polymer-based fluorescence sensors with luminescent carbon dots, *Talanta*. 223121411, https://doi.org/https://doi.org/10.1016/j.talanta.2020.121411.
[99] Singh M., Singh S., Singh S. P., & Patel S. S. (2020). Recent advancement of carbon nanomaterials engrained molecular imprinted polymer for environmental matrix, *Trends in Environmental Analytical Chemistry*. 27e00092, https://doi.org/https://doi.org/10.1016/j.teac.2020.e00092.
[100] Yang Y., Fang G., Wang X., Zhang F., Liu J., Zheng W., & Wang S. (2017). Electrochemiluminescent graphene quantum dots enhanced by MoS2 as sensing platform: A novel molecularly imprinted electrochemiluminescence sensor for 2-methyl-4-chlorophenoxyacetic acid assay, *Electrochimica Acta*. 228107–228113, https://doi.org/https://doi.org/10.1016/j.electacta.2017.01.043.
[101] Chumkaeo P., Davis F., & Bunkoed O. (2024). A melanin-integrated graphene quantum dots embedded molecularly imprinted polymer: A nano-fluorescence sensing probe for trace

determination of prometryn, *Journal of Photochemistry and Photobiology A: Chemistry*. 447115247, https://doi.org/https://doi.org/10.1016/j.jphotochem.2023.115247.

[102] You H., Bai L., Yuan Y., Zhou J., Bai Y., & Mu Z. (2018). An amperometric aptasensor for ultrasensitive detection of sulfadimethoxine based on exonuclease-assisted target recycling and new signal tracer for amplification, *Biosensors and Bioelectronics*. 117706–117712, https://doi.org/https://doi.org/10.1016/j.bios.2018.06.011.

[103] Chen A., Jiang X., Zhang W., Chen G., Zhao Y., Tunio T. M., Liu J., Lv Z., Li C., & Yang S. (2013). High sensitive rapid visual detection of sulfadimethoxine by label-freeaptasensor, *Biosensors and Bioelectronics*. 42419–42425, https://doi.org/https://doi.org/10.1016/j.bios.2012.10.059.

[104] Chiesa L. M., Nobile M., Panseri S., & Arioli F. (2017). Antibiotic use in heavy pigs: Comparison between urine and muscle samples from food chain animals analysed by HPLC-MS/MS, *Food Chemistry*. 235111–235118, https://doi.org/https://doi.org/10.1016/j.foodchem.2017.04.184.

[105] Pongprom A., Chansud N., & Bunkoed O. (2022). A fluorescence sensor probe based on porous carbon, molecularly imprinted polymer and graphene quantum dots for the detection of trace sulfadimethoxine, *Journal of Photochemistry and Photobiology A: Chemistry*. 427113812, https://doi.org/https://doi.org/10.1016/j.jphotochem.2022.113812.

[106] Ensafi A. A., Hghighat Sefat S., Kazemifard N., Rezaei B., & Moradi F. (2017). A novel one-step and green synthesis of highly fluorescent carbon dots from saffron for cell imaging and sensing of prilocaine, *Sensors and Actuators B: Chemical*. 253451–253460, https://doi.org/https://doi.org/10.1016/j.snb.2017.06.163.

[107] Rezaei B., Irannejad N., Ensafi A. A., & Dinari M. (2016). Application of modified carbon quantum dots/multiwall carbon nanotubes/pencil graphite electrode for electrochemical determination of dextromethorphan, *IEEE Sensors Journal*. 162219–162227, https://doi.org/10.1109/JSEN.2016.2517005.

[108] Kumar A., Chowdhuri A. R., Laha D., Mahto T. K., Karmakar P., & Sahu S. K. (2017). Green synthesis of carbon dots from *Ocimum sanctum* for effective fluorescent sensing of Pb^{2+} ions and live cell imaging, *Sensors and Actuators B: Chemical*. 242679–242686, https://doi.org/https://doi.org/10.1016/j.snb.2016.11.109.

[109] Zhou C., He X., Ya D., Zhong J., & Deng B. (2017). One step hydrothermal synthesis of nitrogen-doped graphitic quantum dots as a fluorescent sensing strategy for highly sensitive detection of metacycline in mice plasma, *Sensors and Actuators B: Chemical*. 249256–249264, https://doi.org/https://doi.org/10.1016/j.snb.2017.04.092.

[110] Ensafi A. A., Nasr-Esfahani P., & Rezaei B. (2018b). Synthesis of molecularly imprinted polymer on carbon quantum dots as an optical sensor for selective fluorescent determination of promethazine hydrochloride, *Sensors and Actuators B: Chemical*. 257889–257896, https://doi.org/https://doi.org/10.1016/j.snb.2017.11.050.

[111] Kaur R. P., Gupta V., Christopher A. F., & Bansal P. (2015). Potential pathways of pesticide action on erectile function – A contributory factor in male infertility, *Asian Pacific Journal of Reproduction*. 4322–4330, https://doi.org/https://doi.org/10.1016/j.apjr.2015.07.012.

[112] Poshteh Shirani M., Rezaei B., & Ensafi A. A. (2019). A novel optical sensor based on carbon dots embedded molecularly imprinted silica for selective acetamiprid detection, *Spectrochimica Acta Part A: Molecular and Biomolecular Spectroscopy*. 21036–21043, https://doi.org/https://doi.org/10.1016/j.saa.2018.08.030.

[113] Wing K. D., Sacher M., Kagaya Y., Tsurubuchi Y., Mulderig L., Connair M., & Schnee M. (2000). Bioactivation and mode of action of the oxadiazine indoxacarb in insects, *Crop Protection*. 19537–19545, https://doi.org/https://doi.org/10.1016/S0261-2194(00)00070-3.

[114] Taheri Mirghaed A., Ghelichpour M., Mirzargar S. S., Joshaghani H., & Ebrahimzadeh Mousavi H. (2018). Toxic effects of indoxacarb on gill and kidney histopathology and biochemical indicators in common carp (*Cyprinus carpio*), *Aquaculture Research*. 491616–491627, https://doi.org/https://doi.org/10.1111/are.13617.

[115] Shirani M. P., Rezaei B., Ensafi A. A., & Ramezani M. (2021). Development of an eco-friendly fluorescence nanosensor based on molecularly imprinted polymer on silica-carbon quantum dot for the rapid indoxacarb detection, *Food Chemistry*. 339127920, https://doi.org/https://doi.org/10.1016/j.foodchem.2020.127920.
[116] Jalili R., & Khataee A. (2020). Application of molecularly imprinted polymers and dual-emission carbon dots hybrid for ratiometric determination of chloramphenicol in milk, *Food and Chemical Toxicology*. 146111806, https://doi.org/https://doi.org/10.1016/j.fct.2020.111806.
[117] Pan M., Sun J., Wang Y., Yang J., Wang Z., Li L., & Wang S. (2024). Carbon-dots encapsulated luminescent metal–organic frameworks @ surface molecularly imprinted polymer: A facile fluorescent probe for the determination of chloramphenicol, *Food Chemistry*. 442138461, https://doi.org/https://doi.org/10.1016/j.foodchem.2024.138461.
[118] Amiri A., Faridbod F., & Zoughi S. (2021). An optical nanosensor fabricated by carbon dots embedded in silica molecularly imprinted polymer for sensitive detection of ceftazidime antibiotic, *Journal of Photochemistry and Photobiology A: Chemistry*. 408113111, https://doi.org/https://doi.org/10.1016/j.jphotochem.2020.113111.
[119] Yuphintharakun N., Nurerk P., Chullasat K., Kanatharana P., Davis F., Sooksawat D., & Bunkoed O. (2018). A nanocomposite optosensor containing carboxylic functionalized multiwall carbon nanotubes and quantum dots incorporated into a molecularly imprinted polymer for highly selective and sensitive detection of ciprofloxacin, *Spectrochimica Acta Part A: Molecular and Biomolecular Spectroscopy*. 201382–201391, https://doi.org/https://doi.org/10.1016/j.saa.2018.05.034.
[120] Hu Y., Yang D., Yang C., Feng N., Shao Z., Zhang L., Wang X., Weng L., Luo Z., & Wang L. (2018). A Novel "off-on" fluorescent probe based on carbon nitride nanoribbons for the detection of citrate anion and live cell imaging, *Sensors*. 18, https://doi.org/10.3390/s18041163.
[121] Zhou Z., Shen Y., Li Y., Liu A., Liu S., & Zhang Y. (2015). Chemical cleavage of layered carbon nitride with enhanced photoluminescent performances and photoconduction, *ACS Nano*. 912480–912487, https://doi.org/10.1021/acsnano.5b05924.
[122] Chen J., Chen Q., Chen J., & Qiu H. (2016). Magnetic carbon nitride nanocomposites as enhanced peroxidase mimetics for use in colorimetric bioassays, and their application to the determination of H_2O_2 and glucose, *Microchimica Acta*. 1833191–1833199, https://doi.org/10.1007/s00604-016-1972-7.
[123] Xu S., Ding J., & Chen L. (2018b). A fluorescent material for the detection of chlortetracycline based on molecularly imprinted silica–graphitic carbon nitride composite, *Analytical and Bioanalytical Chemistry*. 4107103–4107112, https://doi.org/10.1007/s00216-018-1310-5.
[124] Xu S., Chen L., & Ma L. (2018a). Fluorometric determination of quercetin by using graphitic carbon nitride nanoparticles modified with a molecularly imprinted polymer, *Microchimica Acta*. 185492, https://doi.org/10.1007/s00604-018-3016-y.
[125] Zhou T., Halder A., & Sun Y. (2018). Fluorescent nanosensor based on molecularly imprinted polymers coated on graphene quantum dots for fast detection of antibiotics, *Biosensors*. 882. https://doi.org/10.3390/bios8030082
[126] Sun Q., Fu L., Yin C., Wu M., Liu H., Niu N., & Chen L. (2022). Construction of biomass carbon dots@molecularly imprinted polymer fluorescent sensor array for accurate identification of 5-nitroimidazole antibiotics, *Sensors and Actuators B Chemical*. 373132716, https://doi.org/https://doi.org/10.1016/j.snb.2022.132716.
[127] Ren Y., & Fan Z. (2023). Synthesis of fluorescent probe based on molecularly imprinted polymers on nitrogen-doped carbon dots for determination of tobramycin in milk, *Food Chemistry*. 416135792, https://doi.org/https://doi.org/10.1016/j.foodchem.2023.135792.
[128] Hu X., Zhao Y., Dong J., Liu C., Qi Y., Fang G., & Wang S. (2021). A strong blue fluorescent nanoprobe based on Mg/N co-doped carbon dots coupled with molecularly imprinted polymer for ultrasensitive and highly selective detection of tetracycline in animal-derived foods, *Sensors and Actuators B: Chemical*. 338129809, https://doi.org/https://doi.org/10.1016/j.snb.2021.129809.

[129] Hakiem A. F. A., Urriza-Arsuaga I., & Urraca J. L. (2023). Development of a screening method for fluoroquinolones in meat samples using molecularly imprinted carbon dots, *Biosensors*. 13972. https://doi.org/10.3390/bios13110972

[130] Wang L., Wen L., Zhao L., Chao J., Tao F., Wang F., & Li C. (2022). Development of fluorescence sensor and test paper based on molecularly imprinted carbon quantum dots for spiked detection of domoic acid in shellfish and lake water, *Analytica Chimica Acta*. 1197339515. https://doi.org/10.1016/j.aca.2022.339515

[131] Zhang D., Tang J., & Liu H. (2019b). Rapid determination of lambda-cyhalothrin using a fluorescent probe based on ionic-liquid-sensitized carbon dots coated with molecularly imprinted polymers, *Analytical and Bioanalytical Chemistry*. 4115309–4115316. https://doi.org/10.1007/s00216-019-01912-0

Arunima Reghunadhan, Viswanathan Pillai Aromal, Linu Tresa K. G., Yadhu Krishna M. A., Satheesh Babu, Sari P. Sasidharan, Prasanth Raghavan, and Sindhu Karthika Ammini*

Chapter 10
Carbon-imprinted polymers for forensic applications: a game-changing tool in the crime scene

Abstract: In recent years, the field of forensic science has seen remarkable advancements through the integration of novel materials and technologies. In the realm of forensic science, the development of cutting-edge technologies has played a pivotal role in aiding investigators to solve complex cases. One such advancement that has captured the attention of forensic experts worldwide is the use of carbon-imprinted polymers (CIPs). These polymers, designed to selectively recognize and bind to specific molecules, offer significant potential for the detection and analysis of forensic evidence. This chapter explores the synthesis, properties, and forensic applications of CIPs, highlighting their advantages and future prospects.

Keywords: Molecular imprinting, carbon imprint, carbon, polymers, forensic, crime

***Corresponding author: Sindhu Karthika Ammini**, Department of Paediatrics, Kinder Multispecialty Hospital, Kalamassery, Kochi, Kerala, India
Arunima Reghunadhan, Materials Science and Nano Engineering Lab (MSNE-Lab), Department of Polymer Science and Rubber Technology, Cochin University of Science and Technology, Kochi; Bernal Institute, University of Limerick, Castletroy, Limerick, Republic of Ireland
Viswanathan Pillai Aromal, Materials Science and Nano Engineering Lab (MSNE-Lab), Department of Polymer Science and Rubber Technology, Cochin University of Science and Technology, Kochi; Government Medical College-Pariyaram, Kerala University of Health Sciences, Kannur, Kerala, India
Linu Tresa K. G., Materials Science and Nano Engineering Lab (MSNE-Lab), Department of Polymer Science and Rubber Technology, Cochin University of Science and Technology, Kochi
Yadhu Krishna M. A., Materials Science and Nano Engineering Lab (MSNE-Lab), Department of Polymer Science and Rubber Technology, Cochin University of Science and Technology, Kochi; Department of Ship Technology, Cochin University of Science and Technology, Kochi
Satheesh Babu, Department of Ship Technology, Cochin University of Science and Technology, Kochi
Sari P. Sasidharan, Department of Polymer Science and Rubber Technology, Cochin University of Science and Technology, Kochi, India
Prasanth Raghavan, Materials Science and Nano Engineering Lab (MSNE-Lab), Department of Polymer Science and Rubber Technology, Cochin University of Science and Technology, Kochi; Department of Materials Engineering and Convergence Technology, Gyeongsang National University (GNU), Jinju-daero, Republic of Korea; International and Inter University Centre for Nanoscience and Nanotechnology (IIUCNN), Mahatma Gandhi University, Kottayam, Kerala, India

https://doi.org/10.1515/9783111503325-010

10.1 Introduction

Forensic analysis is a multidisciplinary field that applies scientific principles and techniques to investigate crimes and gather evidence for legal proceedings. It involves the meticulous examination of physical, digital, and biological evidence to reconstruct events, identify suspects, and support criminal investigations [1]. Forensic analysis is essential in modern criminal justice systems due to its ability to provide objective and scientific evidence that can corroborate or refute testimonies and investigative findings [2]. It plays a critical role in solving crimes by enabling the accurate identification of perpetrators, exonerating the innocent, and uncovering the truth behind criminal activities. Techniques used in forensic analysis range from DNA profiling, fingerprint analysis, and bloodstain pattern analysis to digital forensics, which involves the recovery and investigation of data from electronic devices [3]. Chemical analysis plays a pivotal role in forensic science by enabling the identification and quantification of substances found at crime scenes. Through techniques such as chromatography, spectroscopy, and mass spectrometry (MS), forensic chemists can analyze a wide range of materials, including drugs, explosives, toxins, and trace evidence like fibers and paints. This analysis can determine the composition of unknown substances, link evidence to specific sources, and provide insights into the circumstances of a crime [4–6]. For example, chemical analysis can detect the presence of poisons in biological samples, identify accelerants used in arson, and match paint samples to a specific vehicle. The precise and reliable results obtained from chemical analysis are crucial for constructing accurate narratives of criminal events, supporting investigations, and providing compelling evidence. Forensic analysis involves examining evidence to support investigations and legal proceedings. The procedure for forensic analysis generally follows various steps as summarized in Figure 10.1.

Figure 10.1: Methodologies involved in the forensic analysis process.

Polymers have inevitable roles in forensic analysis. Synthetic polymers, such as plastics, fibers, and paints, are commonly encountered at crime scenes and can be analyzed to link suspects or objects to specific locations or events. Polymers also play a key role in forensic tools, such as fingerprint powders and DNA extraction materials, enhancing the precision and reliability of forensic investigations [7–11]. The use of polymers in diagnostics has been broadened by the molecular printing process. The process of creating artificial polymer matrices with particular target molecule recognition sites is called molecular imprinting. In order to create cavities that are complementary to the target molecule in terms of size, shape, and functional groups, a polymer is formed around the template molecule and subsequently removed. These imprinted polymers can selectively rebind the target molecule, making them useful in a variety of applications as described in Figure 10.2 [12–14].

Figure 10.2: Selected important biological applications of molecularly imprinted polymers.

In a simple sense, it can be explained as the making of a suitable lock for a key. Accuracy in the preparation of molecular recognition matrices is the key advantage of this technology [15]. The binding between the target and the polymer can be by covalent or non-covalent interactions, both having their own advantages. Covalent imprinting involves the formation of covalent bonds between the template molecule and functional monomers during the polymerization process. After polymerization, these covalent bonds are cleaved to remove the template, leaving behind binding sites that can reform the covalent bonds with the target molecule. Non-covalent imprinting involves the use of weaker, non-covalent interactions (e.g., hydrogen bonding, van der Waals forces, ionic interactions, and hydrophobic effects) between the template molecule and the functional monomers [16].

Molecularly imprinted polymers (MIPs) can be categorized depending on the interaction, synthetic strategy, the end product structure and the type of template. Depending on the end polymeric structure, we can have monolithic (bulk polymers that are ground into particles), spherical (uniformly shaped particles, often produced by suspension or precipitation polymerization), surface imprinted (the template is imprinted on the surface of a support material), and nanostructured (with nanometer-scale features, such as nanoparticles, nanofibers, or nanotubes) [17, 18]. The different preparation methods of MIPs are summarized in Figure 10.3. While the introduction of nanostructures can enhance uniformity, it is not feasible to create perfect contact zones or connect functionality through ligand molecules to individual nanoparticles uniformly [19].

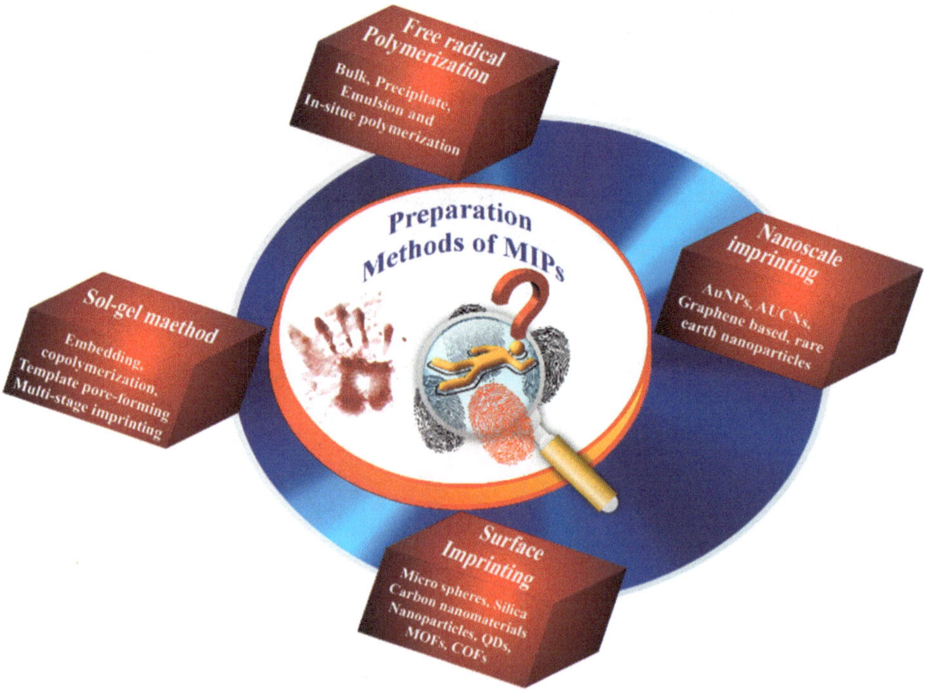

Figure 10.3: The popular preparation methods for molecularly imprinted polymers.

The polymerization process can be initiated by heat, light, or chemical initiators. After the polymerization is complete, the template is removed with the help of suitable solvents and during the removal, cavities, pores or binding sites are formed in the polymeric structure that are complementary to the template molecule in terms of size, shape, and functional groups. Figure 10.4 displays the schematic representation of molecular imprinting process for the formation of MIPs.

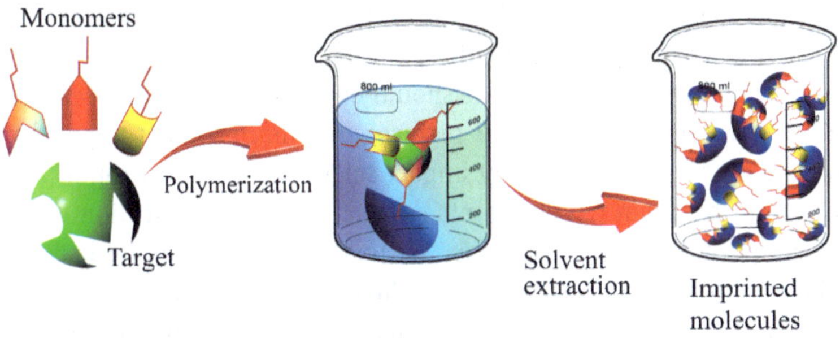

Figure 10.4: Schematic representation of molecular imprinting process.

From the schematic, one may feel that the imprinting process is simple and easy to carryout. But there are many technical issues on molecular imprinting process. Let us discuss some of the issues in the following paragraph. Generally, the challenge with classic molecular imprinting is the biological macromolecules' low solubility in organic solvents. Additionally, detrimental to the process are poor mass transfer kinetics due to the massive protein structures seen in snake or bee venom. Proteins and other parasites may cross-react with MIPs in a manner similar to that of antibodies, leading to unpleasant detection as well as decreased binding ability [20]. Furthermore, one of the main difficulties in terms of complexity and flexibility is imprinting big molecules [21]. Many distinct binding sites (polyclonal characters) with various binding properties will arise from conformational variations in molecules [22]. Additionally, removing large-sized templates is especially challenging when there is excessive crosslinking [23]. Insufficient template extraction leads to "template leakage" or "template bleeding," which causes a false positive signal and poor chromatographic separation in sensing [18]. Along with to the significant risk that template leakage presents for forensic and clinical analysis, non-specific interactions between the heterogeneous binding sites reduce MIP's efficiency [22]. Higher order non-specific binding will take place in bulk MIP if the template-monomer complex is routed randomly in the polymer network, leading to template-template contact. In reality, the template-template interaction can actually improve homogeneity in some situations, in addition to the semi-covalent method and stoichiometric non-covalent imprinting [24]. It is inevitable that there will be nonspecific interactions between the template and interfering molecules because of the many affinity zones that the huge surface area of MIPs provides. As a result, the surface area ought to be optimal rather than high [25].

Because common monomers used in the molecular imprinting process are insoluble and require hydrogen bonding to form pre-polymer complexes, water molecules may compete with the template or monomer, resulting in decreased affinity [21, 26].

The synthesis of MIPs lacks a standard preparation process, which makes individual imprinting necessitates procedure adaption for various scenarios [26, 27]. Once all requirements are satisfied, scaling up the synthesis in the autonomous reactors will be simple and cost-effective. Inconsistency issue should be solved in template removal procedures [28, 29]. Surface, epitope-mediated surface, and micro-contact imprinting methods can address variables that adversely impact selectivity and adsorption [29–32]. Consideration should be given to constraints such as lengthy preparation times, trouble regulating the polymer film's thickness, the possibility of template leakage, and difficult template removal from a thick polymer layer [33]. A further contributing aspect is the poor elution profile, which causes the accumulation of template residue and reduces selectivity. Bad elution profiles will therefore be shown as negative errors, which can be offset by using better elution agents or by using computer modeling [34]. Dummy templates and surface imprinting can also be applied if the original template is dangerous/unstable or too expensive. In the event that the original template is too costly, unstable, or harmful, dummy templates and surface imprinting may alternatively be utilized [35, 36]. However, these two options may also have certain disadvantages, such as making it difficult to get appropriate fake templates and making it impossible to completely remove templates, particularly when dealing with big molecules during surface imprinting [37]. MIPs will be in high demand as components of marketed kits that do detections in a matter of seconds if they are coupled with cutting-edge technologies like artificial intelligence, which will lead to the intended practical solutions.

In summary, MIP remains far superior to classical techniques in terms of sensitivity and selectivity. It will also be easier to handle because to less expensive production setup and straightforward application processes. Finding suitable functional monomers that fit the target molecules – especially those without functional groups – remains difficult, and using hazardous chemicals is frequently unavoidable. Less sensitive data are obtained when imprinting a single atom or a molecule with almost the same 3D structure, departing from the main purpose of imprinting. To advance and improve imprinting, researchers ought to concentrate on innovative techniques.

10.2 Carbon-imprinted polymers

CIPs are a specialized type of MIPs designed to recognize and selectively bind carbon-containing compounds. The imprinting process involves the creation of polymer matrices with specific recognition sites for carbon-based molecules, such as organic compounds, gases, or pollutants. CIPs are tailored to selectively adsorb carbon-containing molecules from mixtures. These polymers leverage the unique properties of carbon, such as high surface area, excellent thermal and chemical stability, and electrical conductivity, to create highly selective and efficient recognition sites for target molecules. The synthesis of CIPs involves the integration of carbon materials, like carbon nanotubes (CNTs), graphene, or activated carbon, with functional monomers (Figure 10.5) and cross-linkers (Figure 10.6) in the presence of the template molecule. The carbon atoms in these materials can adopt several hybridization states (sp^1, sp^2, and sp^3), resulting in various allotropic structures. A broad class of metallic and nonmetallic elements, as well as other carbon atoms, can form strong covalent bonds with carbon atoms among these allotropes, leading to the formation of an immense variety of compounds ranging in size from small molecules to lengthy chains to solids. These materials have played significant roles in our civilization and have attracted a lot of attention from academics looking to explore previously unexplored possibilities due to their variety in morphology, structure, dimensionality, and characteristics.

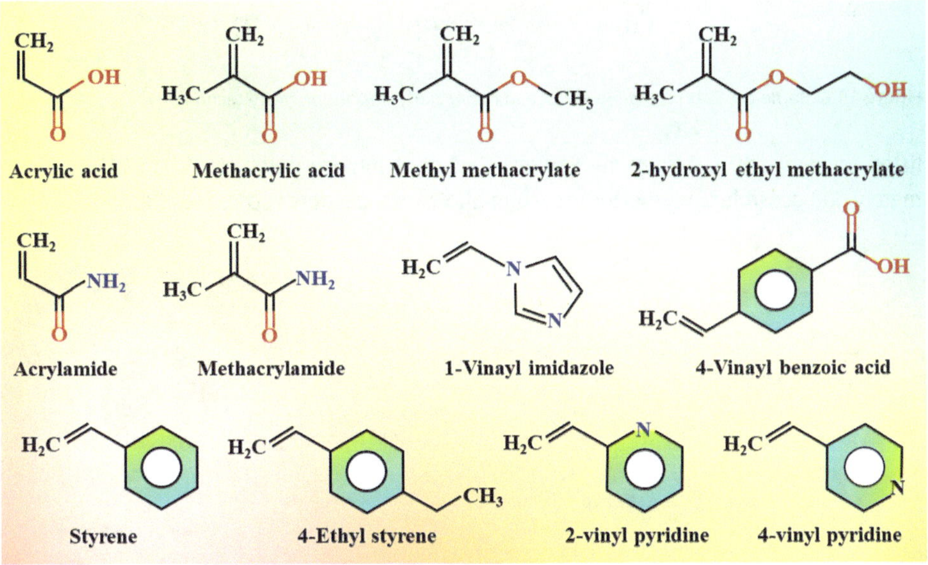

Figure 10.5: Some common monomers used in molecularly imprinting (MIP) technology.

There is a vast array of characteristics and dimensionality, from zero to three dimensions, for carbon materials and their derivatives. For example, the dimensionalities of zero-dimension fullerene, one-dimensional CNTs, two-dimensional graphene, and three-dimensional graphite differ, but they all belong to the same family and contain the same carbon element arranged in a hexagonal lattice. Similarly, the properties of ultrahard, insulator diamond, and soft, highly electrically conductive graphite differ. Carbon is well-known for its usual polymorphic materials and for the several polymorphs of carbon

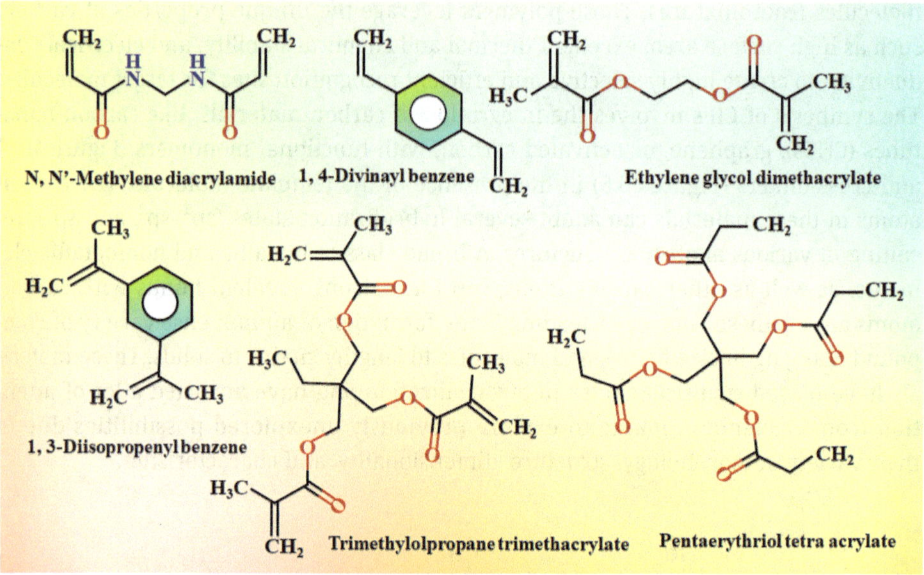

Figure 10.6: Some common cross-linker used in molecularly imprinting (MIP) technology.

listed in Figure 10.7. As per the Samara Carbon Allotrope database (SACADA), approximately 500 possible three-periodic carbon allotropes are now recognized [38, 39].

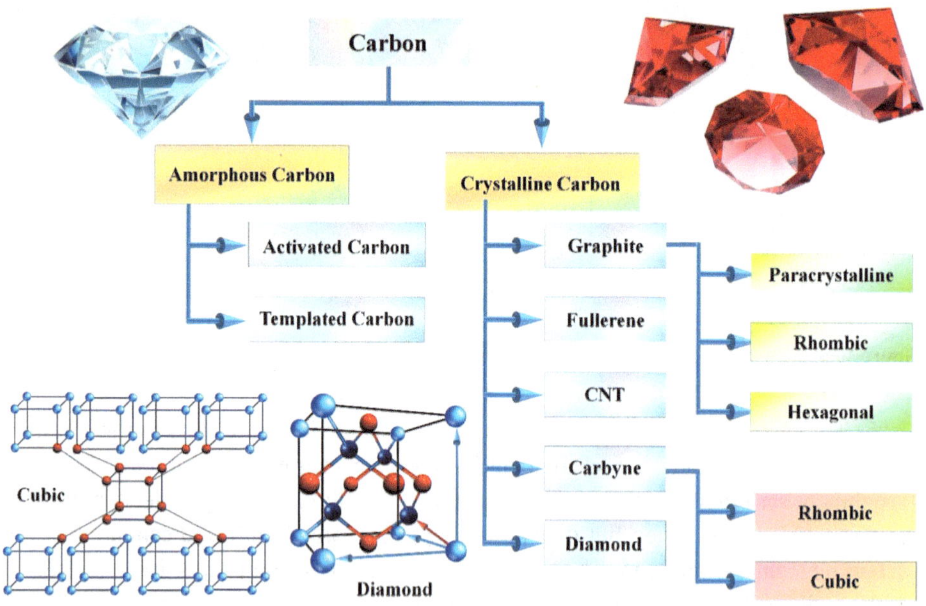

Figure 10.7: Classification of carbon materials and different carbon allotropes.

Carbon-based materials are being extensively investigated for the creation of CIPs for a range of applications, including forensic science, because they have demonstrated endless opportunities to be combined with other materials (especially nanomaterials) to enhance their properties.

In forensic applications, CNTs imprinting can be used to analyze residue from crime scenes. Graphene with similar properties to CNTs also plays important role in forensic analysis (Figure 10.8). The intrinsic characteristics enhances the applicability of graphene an ideal material for developing advanced forensic tools and sensors and the graphene-based sensors are used to detect and analyze trace amounts of substances. Graphene can also be applied in latest fingerprinting devices and electrochemical sensors.

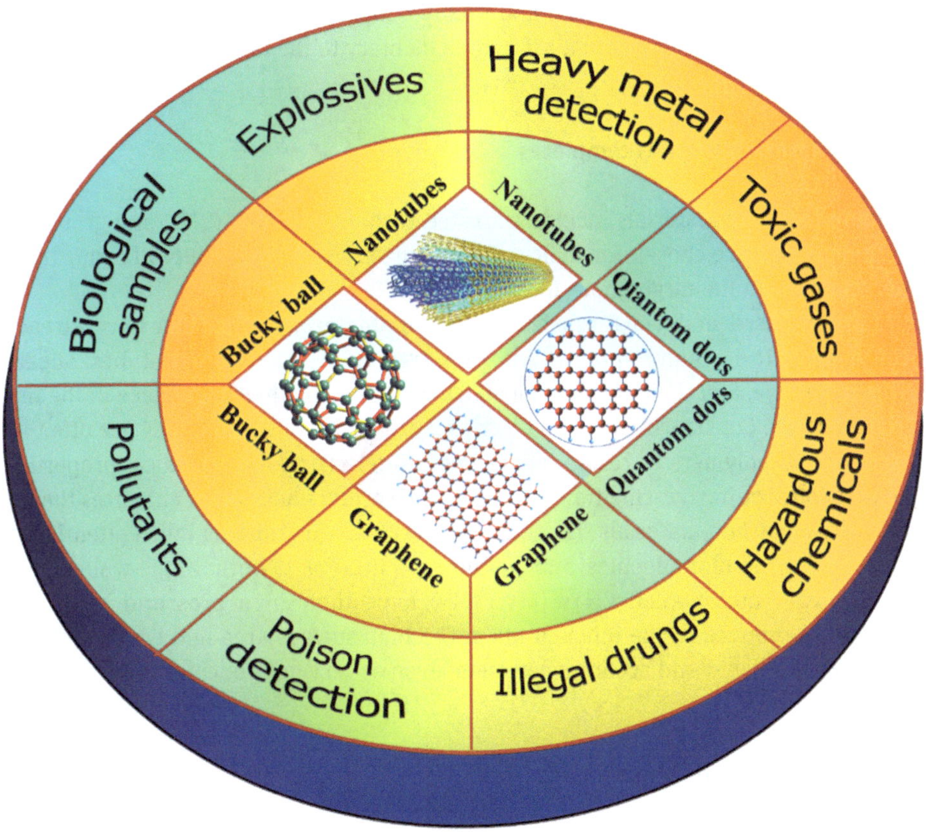

Figure 10.8: Role of carbon-based materials in forensic analysis.

In addition to these the zero-dimensional carbon dots are nowadays incorporated due to their superior optical properties such as photostability and fluorescence. Also, they have less toxicity and are employed to analyze biological samples such as blood, saliva, and other bodily fluids. They are effectively incorporated in the finger print analysis. In the following sections a detailed description of the role of carbon-based nanomaterials and MIPs in forensic analysis will be given.

10.2.1 Synthesis of carbon-imprinted polymers

The synthesis of CIPs involves the use of carbon-based templates to create imprinted sites within a polymer matrix. These sites are designed to match the shape, size, and functional groups of the target molecules. The general procedure includes template selection (choosing a carbon-based template that resembles the target molecule), polymerization (mixing the template with monomers and cross-linkers, followed by polymerization to form a solid matrix) and finally template removal (extracting the template to leave behind cavities that are complementary to the target molecule).

10.2.1.1 Materials used as templates

Common template materials include activated carbon, graphite, CNTs, and graphene. These materials are chosen for their stability, surface area, and ability to form strong interactions with the target molecules.

Carbon-based materials, especially the nanomaterials of carbon have already been a favorite in many applications. The derivatives such as nanotubes, sheets, wires, graphene, and the quantum dots (Figure 10.9) made their own place in the material research. Considering the good intrinsic properties of carbon, they are often incorporated to polymers to result enhancement in electrical and chemical properties. The possibility of introducing functionalities and second phase often enhances the applicability of carbon materials. The MIPs with carbon are grabbing much attention in the detection of biomolecules and pollutants [40]. CNTs, both single-walled CNT (SWCNT) and multiwalled CNTs (MWCNTs) have their own pros and cons. The MWCNT is preferred to the other in MIPs. The high surface area and conductivity of CNTs allow for rapid and accurate detection, even at very low concentrations.

10.2.1.2 Polymerization techniques

Polymerization techniques for MIPs are essential for creating these highly specific synthetic receptors. Various polymerization techniques are employed, including bulk polymerization, suspension polymerization, and emulsion polymerization (Figure 10.10). The choice of technique depends on the desired properties of the final CIP, such as porosity, surface area, and mechanical strength. Bulk polymerization is a straightforward technique where monomers and cross-linkers polymerize in the presence of a template molecule, followed by grinding and sieving to obtain the MIP particles. Suspension poly-

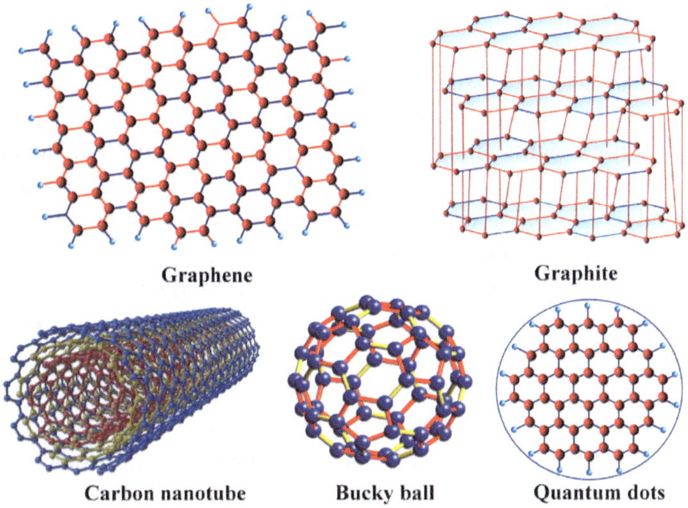

Figure 10.9: The structural representation of nano-sized carbon materials.

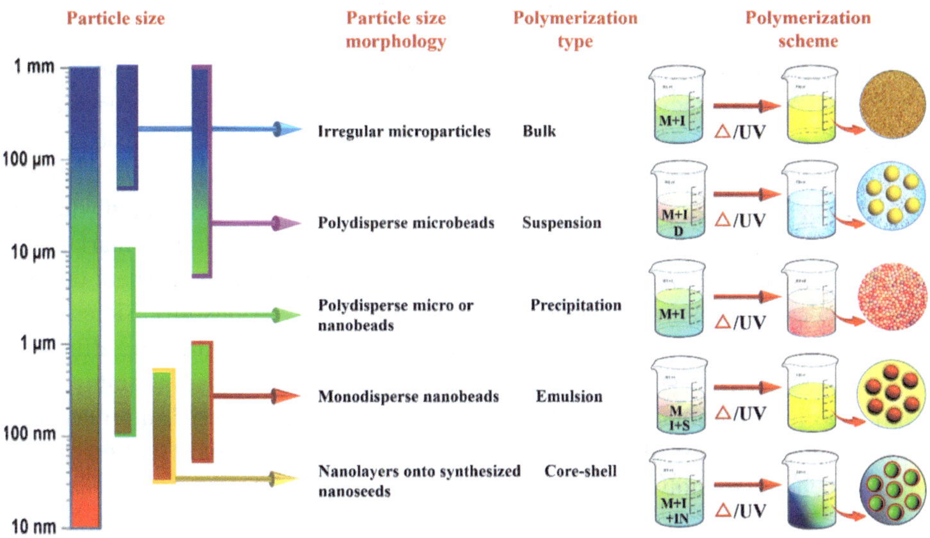

Figure 10.10: Different techniques used in the production of carbon-imprinted polymers (CIPs).

merization involves dispersing the monomer-template mixture in a continuous phase to form spherical MIP beads. Emulsion polymerization uses surfactants to create micelles, leading to uniform nanoparticle MIPs. Sol-gel polymerization incorporates organic-inorganic hybrid materials, providing high thermal stability. Finally, surface imprinting techniques create MIPs on the surface of substrates, allowing easy template removal and high binding site accessibility.

10.2.1.3 Characterization of CIPs

The synthesized CIPs are characterized using techniques such as spectroscopy, scanning electron microscopy, and surface area analysis. These methods confirm the successful imprinting and provide insights into the polymer's morphology and functional properties. Nitrogen adsorption-desorption isotherms are measured using the Brunauer-Emmett-Teller method to determine the specific surface area, pore volume, and pore size distribution. This helps in understanding the material's capacity for adsorption and interaction with target molecules. Adsorption isotherms and binding studies are performed to evaluate the affinity and specificity of CIPs for target molecules. These experiments help determine the binding capacity, selectivity, and imprinting efficiency of the polymers. The spectral techniques reveal the presence of specific bonding environments and chemical modifications introduced during synthesis. Most of the identification and detection using CIPs are based on the electrical properties and hence, the conductivity is also studied.

10.2.1.4 Properties of carbon-imprinted polymers

Considering the forensic analysis, the conventional methods of analysis and identification has several disadvantages which are overcame by the latest techniques such as molecular imprinting. The carbon materials are widely used in the forensic analysis to detect very trace amount of chemicals in biological samples and environmental samples. The primary advantage of CIPs is their high selectivity and sensitivity toward target molecules. The imprinted sites are tailored to fit specific forensic markers, allowing for precise detection even in complex samples. CIPs exhibit excellent thermal and chemical stability, making them suitable for use in various forensic environments. They can withstand harsh conditions without losing their binding affinity or structural integrity. One of the significant benefits of CIPs is their reusability. After capturing the target molecules, CIPs can be regenerated and reused multiple times without significant loss of performance. This feature makes them cost-effective and environmentally friendly.

10.3 Forensic applications of carbon-imprinted polymers

Carbon-based materials have been a favorite candidate in forensic analysis as discussed earlier. They are designed to selectively bind molecules based on their carbon content, offering various forensic applications. Innovative ways to identify DNA from fingerprints, explosives, and gunshot residues when combined with carbon nanomaterials like fullerenes (C60), CNTs, graphene (G) and its derivatives, graphene oxide (GO), and reduced graphene oxide (rGO) can expedite the investigative process. By leveraging their selective binding capabilities, CIPs enhance the sensitivity and specificity of forensic analyses, contributing to more accurate and reliable evidence process. CIPs are very useful in the detection of evidence since they can selectively capture and identify organic compounds, such as drugs or explosives, from crime scenes, aiding in the detection of trace evidence. CIPs can be used to enhance the collection and analysis of fingerprint residues by selectively binding to carbon-based residues left behind. They assist in profiling chemical substances found at crime scenes, helping to match substances to known standards or to determine their origin. They can be used to identify and quantify drugs or toxins in biological samples, such as blood or urine, supporting toxicological investigations. CIPs are employed in analyzing environmental samples, such as soil or water, to detect and quantify contaminants related to criminal activities.

10.3.1 Detection of explosives

CIPs are highly effective in detecting trace amounts of explosives. By imprinting the polymer with explosive compounds, such as TNT or RDX, the resulting CIPs can selectively bind to these substances, enabling their detection in field conditions or laboratory settings [41]. MIPs can be effectively used in selective detection of explosives by adding them as selective sorbents. Most commonly available explosive materials are nitro aromatic compounds or NACs. They have an aromatic system with conjugated pi electrons and hence they can easily interact with other electron rich monomers by pi-pi manner or charge transfer. To study the presence of explosives, a widely used strategy is to create a dummy template [42]. The majority of the synthetic polymers were employed in conjunction with transducers for sensing. Because NACs can block the fluorescence emission of surrounding fluorophores or undergo stepwise reduction in an electrochemical reaction, the bulk of these sensors were either electrochemical or fluorescence sensors. There has been a report on a highly sensitive GCE modified with a combination of MWCNTs and RDX-imprinted polymer nanoparticles. The preparation strategy of RDX-imprinted polymer nanoparticles is illustrated in Figure 10.11a. Sensitive (LOD: 20×10^{-12} mol L^{-1}) and selective RDX determination in water samples was made possible by the coupling of the distinct electrical properties of MWCNTs with the selectivity and porosity of MIP nanospheres. The surface morphology of the nano-MIP/MWCNTs nanocomposite is displayed in Figure 10.11b. It was also investigated how the extraction time affected the DPV signal. MIP/MWCNT-GC electrode was submerged in 0.5 μmol L^{-1} RDX solution for varied periods of time in order to achieve this goal. Following its removal from the extraction solution, the electrode was briefly washed before being used in the electrochemical determination process. Research demonstrated that the simultaneous presence of MIP and MWCNTs on the electrode surface significantly boosts the electrode signal's magnitude to RDX (Figure 10.11c)[43]. MIP nanospheres/MWCNTs composite which acts as the sensing layer on the GC electrode has a special structure that increases the MIP contacting area with the target molecule and also helps the RDX molecules to involve in the electrochemical reactions mediated by MWCNTs at the electrode surface.

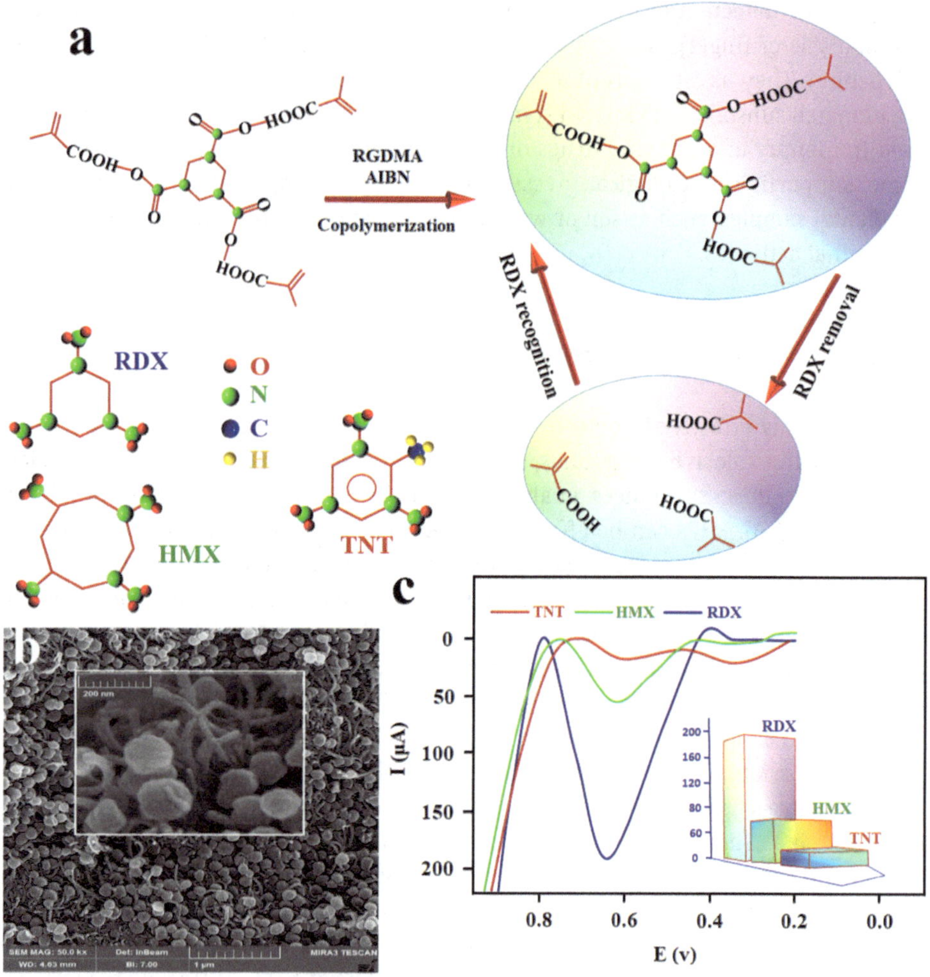

Figure 10.11: (a) Schematic representation of the MIP preparation methodology, (b) scanning electron microscopy images of prepared nano-MIP/MWCNTs nanocomposite, casted on GC electrode and (c) comparison of differential pulse voltammetry responses of the MIP/ MWCNTs-GCE to RDX, TNT, and HMX in order to show the selectivity of the electrode to RDX.

The detection of nitrocellulose is possible with MIP nanocomposites on MWCNTs. The surface polymerization technique used nitrocellulose as a template, methacrylic acid as monomer, 2,2′-azobis (2-methylpropionitrile) as an initiator ethylene glycol dimethacrylate as a cross-linker. A glassy carbon (GC) electrode was used to create the MWCNTs-MIP sensor, which was then evaluated using differential pulse voltammetry (DPV) and cyclic voltammetry (CV). Using DPV with a 3.45×10^{-10} mg mL^{-1} detection limit, the detection of NC at the MWCNTs-MIP/GC electrode in the range of 0 to 7×10^{-5} mg mL^{-1} was observed [44]. In order to fluorescently label an MIP made using 2,4-dinitrotoluene (DNT) as a template, amino-functionalized carbon dots (AC-dots) were prepared. MIPs show 2 to 2.5 imprinting efficiencies, indicating higher adsorption than NIPs. The MIP's cavities selectively absorb DNT, which then interacts with the surface's AC dots to quench the fluorescence [45].

10.3.2 Detection of illegal drugs and chemicals

In forensic toxicology, CIPs are used to identify and quantify drugs of abuse in biological samples. In forensic and toxicological applications, CIPs facilitate the detection of trace levels of illicit drugs and their metabolites, enhancing the accuracy and reliability of analytical results. Polymers imprinted with drug molecules like cocaine, heroin, or amphetamines can selectively extract these substances from complex matrices like blood, urine, or hair. Sedatives and barbiturates are commonly employed illegally and the over dosage can lead to serious health issues. The techniques that used MIPs and CIPs have found to be very successful in the detection of these chemicals. The superiority of CIPs and MIPs lies on the fact that the drug molecules are normally present in a very low detectable level by other techniques and during the detection by the conventional techniques there may be interfaces of some complex samples with the matrix. Another disadvantage of the conventional techniques that was resolved by the CIPs are the alteration of the existing drug molecules and the usage of additives that produces analogues of the same drug molecule. The selectivity of the MIPs to a certain template is exploited in the analysis of the analogous drug materials.

The usage of illegal drugs and stimulants are often reported with athletes. The commonly used illegal drugs are categorized into seven namely: analgesics, tranquilizers, stimulants, anabolic steroids, peptide hormones, thiazide diuretics and aldosterone drugs, masking agents and β-blockers. By selecting these drugs as templates, the detection with CIPs are quite easy and accurate [46, 47]. Graphene based electrodes and assemblies are often selected in the forensic drug detection and analysis. Reports are supporting graphene and rGO as eligible candidates in MIP-based detections. The detection of the presence of drug molecules in food materials can be identified with carbon nanomaterial-imprinted polymers. Researchers tried MWCNT@MIP composites for the identification of amoxycillin in milk and honey samples. Functionalized MWCNT incorporated MIPs are employed in the detection of Norfloxacin. Similarly, the presence of metronidazole is detected using the MWCNT@MIPs. Dopamine is electropolymerized in the presence of metronidazole (MNZ) template onto the surface of

MWCNT to create the MNZ-MIP/MWCNT/GCE electrode, which exhibits significant linearity, 0.049 μg L^{-1}, and good stability and reusability [48]. Many graphene-based MIP sensors have been reported for the detection of antibiotics among other types of analytes. Because of the large number of π-electrons on its planar surface, graphene has a strong electroconductive characteristic. Furthermore, graphene has superior electrocatalytic qualities. A MIP-based sensor on GCE was created using AuNPs/RGO/SWCNT nanocomposites, where *o*-phenylenediamine was electrochemically polymerized onto the nanocomposites' surface while a pefloxacin template was present [45].

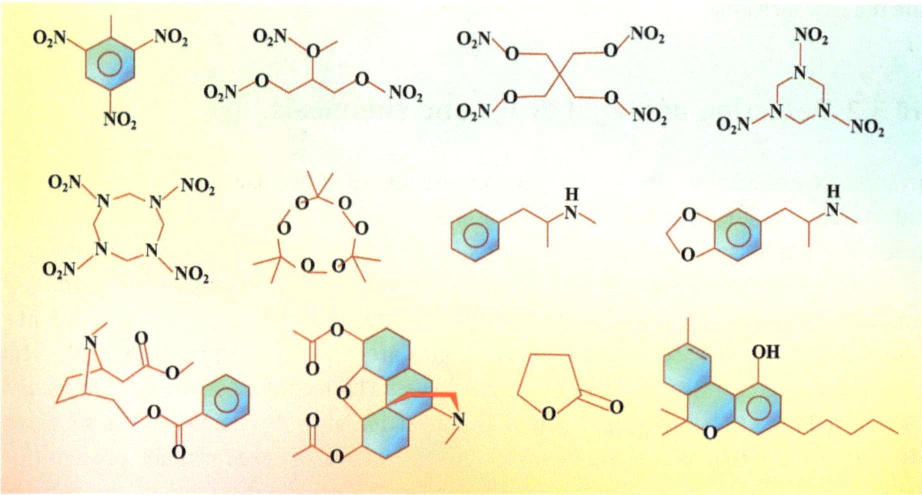

Figure 10.12: Structure of commonly used illicit drugs and explosives (Chemical structures of selected illicit drugs and explosives). (1) Trinitrotoluene (TNT); (2) nitroglycerine; (3) pentaerythritol tetranitrate (PETN); (4) cyclotrimethylene triamine (RDX); (5) cyclotetramethylene tetranitramine (HMX); (6) triacetone triperoxide (TATP); (7) methamphetamine; (8) methylenedioxmethamphetamine (MDMA, ecstasy); (9) cocaine; (10) heroin; (11) γ-butyrolactone; and (12) Δ9-tetrahydrocannabinol.

Peroxide-based explosives (PBEs) can be detected sensitively and selectively on the spot or in the field. Electrochemical sensors that uses well-dispersed MWCNTs/polyethyleneimine (MWCNTs/PEI) for the direct detection of organic peroxide explosives, such as triacetone triperoxide (TATP) and hexamethylene triperoxide diamine (HMTD)GC/MWCNTs/PEI electrode, was developed which is a modified GC electrode The sensor electrode responded to intact TATP solutions in neutral medium, which is a clear benefit over prior approaches of a similar nature because it meant that TATP did not interact with acids or bases that would have converted it into H_2O_2 [49]. Figure 10.12 displays the structure of commonly used illicit drugs and explosives.

10.3.3 Food adulteration/poisoning

CNT-imprinted polymers can be employed to detect the presence of poison or any adulterant in food materials. It is possible to easily imprint a variety of analytes, from microorganisms and enzymes to pollutants, poisons, and adulterants. This makes it easier to remove analytes from food matrices, such as dairy, fruits, baked goods, and drinks, and then do additional analysis. The possibility of poisoning through food materials can thus be easily detected with the MIPs. A report suggests the usage of 3D printed CNT @ MIP for the effective detection of profenofos, a banned pesticide in food materials. MIP was grafted to the CNT using methacrylic acid (MAA) as monomer, ethylene glycol dimethacrylate (EGDMA) as the cross linker, and 2,2′-azobisisobutyronitrile (AIBN) as the initiator (Figure 10.13).

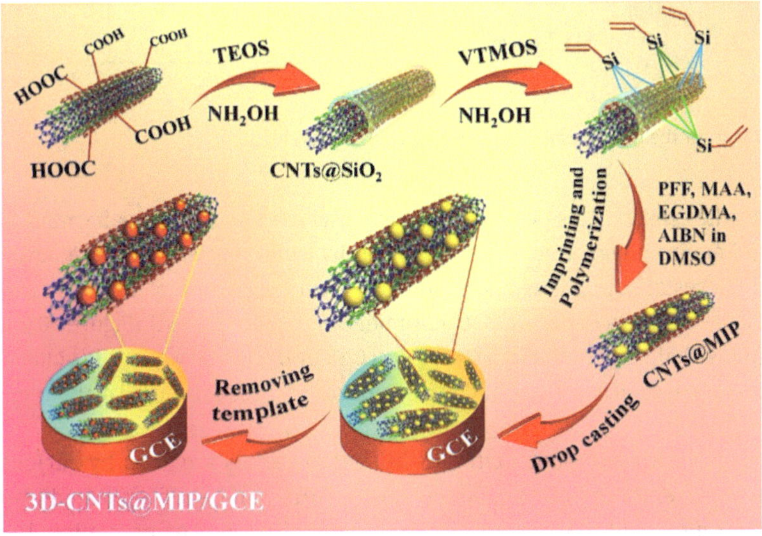

Figure 10.13: Schematic for the fabrication of the molecularly imprinted 3D-CNTs@MIP/GCE (carbon-imprinted polymer).

Sensor was developed by coating the surface of glass carbon electrode with the imprinted CNTs and then removing the templates by solvent extraction (Figure 10.14). The association of MIP with CNT increased the surface area and hence enhanced imprinting sites. This enabled the sensor material faster recognition of profenofos by voltammetry and amperometry. The 3D-imprinted sensor provided a linear response over the range 0.01–200 μM (r^2 = 0.995) with lower detection limit of 0.002 μM (3σ) [50]

A very interesting study conducted by Mugo and coworkers revealed the potential of conducting polymers in the detection of pesticide molecules by means of the carbon imprinting mechanism. The detection of imidacloprid was carried out using the PANI-

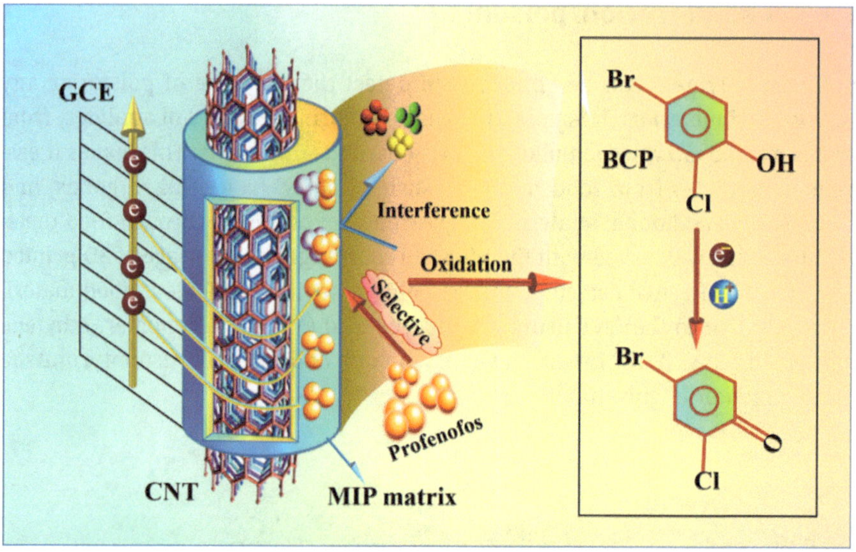

Figure 10.14: Schematic for the mechanism of identification of the pesticide.

modified microneedle sensors. This pesticide is widely used worldwide and is harmful to human health and environment. They were fabricated by layer-by-layer technique using CNT/cellulose nanocrystal (CNC) composite and the receptor film was the pesticide-imprinted PANI layer co-polymerized with CNC (imidazole functionalized) (Figure 10.15). The detection limit with the microneedle sensors was 0.35 or 0.06 μM using cyclic voltammetry and DPV, respectively. One of the added advantages of these sensors was the reusability and reliability [51].

The detection of toxic hazardous materials in food materials are done by various strategies. Some requires incorporating the powdered MIPs can be incorporated into the membranes by means of plasticizers and then embodied to sensing device. We can use imprinted beads in a two-phase system of chloroform and silicon oil [52]. In these cases, electrochemical detection as previously mentioned is carried out. Presence of pesticides such as diazinon, cypermethrin, cyromazine, malathion, carbendazim, etc. are detected by the methods of imprinting[53–55]. In order to increase the performance of MIP-based sensors, functional monomers will be selected with the help of theoretical calculations. The performance of MIPs will be redefined by using chemicals that can produce several contacts at once and by diversifying their cross-linker repertoire. Even though there have not been many attempts at multiple recognitions thus far, this is a problem that has to be investigated [25].

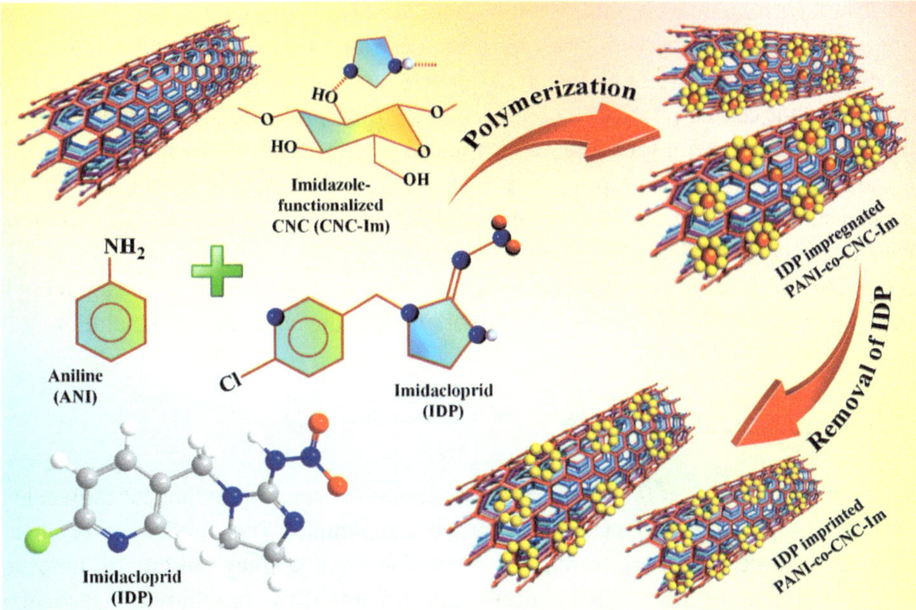

Figure 10.15: Schematic for the fabrication of the IDP-imprinted PANI/CNC@CNT/CNC MN sensor (carbon-imprinted polymer).

10.3.4 Environmental forensics

CIPs are also applied in environmental forensics to detect pollutants and contaminants. Presence of heavy metals plays crucial role in some investigations. Carbon imprinting can be a tool for this to detect even trace amounts in liquid and gaseous samples. Studies have shown that CIPs exhibit high selectivity toward specific heavy metal ions due to the molecular recognition sites created during the imprinting process. This selectivity allows for accurate and precise detection of heavy metals in complex sample matrices. For instance, CIPs imprinted with polycyclic aromatic hydrocarbons can selectively capture these pollutants from soil or water samples, aiding in environmental crime investigations. CIPs can be used for sample preconcentration, enabling the enrichment of trace amounts of heavy metals from complex crime scene samples. This preconcentration step enhances the sensitivity of detection methods and improves the overall detection limits. They can be integrated into solid-phase extraction techniques for the isolation and purification of heavy metal ions from forensic samples. This extraction process facilitates the separation of target analytes from interfering substances, leading to more accurate detection results.

Carbon electrode-based sensing devices can detect samples of very low concentration with high accuracy. For the detection, suitable carbon-based electrodes will be fabricated with monomers, and carbon forms like CNTs, graphene, or CDs. The detection is made possible by voltammetry. A fairly straightforward method was used to create Pb^{2+} ion-imprinted polymeric nanoparticles using itaconic acid, a functional monomer that can be purchased commercially. In the presence of common potential interfering heavy metal ions, the application of the novel IIP as the modifying agent of the MWCNT/CP electrode produced a very selective voltammetric sensor for Pb^{2+} ion measurement. It has been demonstrated that adding MWCNTs and IIP nanoparticles to the carbon paste electrode's composition improves the electrode's analytical signal. The electrode's selectivity increased as a result of washing it after removing it from the Pb^{2+} ion solution and before immersing it in the electroanalysis solution.

10.3.5 Biological evidence

The high specificity of CIPs makes them suitable for isolating biological molecules, such as DNA, proteins, and enzymes, from forensic samples. This application is particularly useful in cases where traditional extraction methods may fail due to the presence of inhibitors or low sample concentrations. These CIPs are tailored to recognize and bind specific biomolecules, such as proteins, nucleic acids, or small metabolites, in complex biological matrices. CIPs can be tailored to selectively bind drugs of abuse, such as cocaine, heroin, and methamphetamine, from blood or urine samples. This allows for rapid and accurate identification of these substances, crucial for toxicological analysis. CIPs can be designed to recognize and capture toxins or toxic compounds in biological samples, including tissues or body fluids. This selective binding capability allows for the efficient extraction and detection of toxins, facilitating the identification of poisoning cases or exposure to harmful substances. CIPs can be designed to capture specific alcohol biomarkers like ethyl glucuronide from urine or blood samples. These biomarkers provide evidence of alcohol consumption even after the ethanol itself has been metabolized. Forensic cases sometimes require the analysis of hormones, such as steroids or stress hormones, in biological fluids to establish patterns of behavior or physiological states. CIPs can be designed to selectively capture these hormones [40]. MIPs have significant role in the biomedical field for the detection of biological samples such as serum albumin, hemoglobin, urine, and ferritin. In forensic serology, CIPs can be utilized for the detection and analysis of blood group antigens or other blood-related markers in biological samples. These polymers can aid in the identification of bloodstains, saliva traces, or other bodily fluids, contributing to the establishment of crucial forensic evidence in criminal cases. CIPs can be employed for the selective extraction and purification of specific proteins or biomarkers from biological samples. By imprinting the polymer with target protein molecules, researchers can isolate and concentrate proteins of interest for further analysis, such as proteomic studies or biomarker identification in forensic investigations. Also, by incorporating CIPs into DNA purification kits, forensic laboratories can enhance the efficiency and yield of DNA extraction, enabling accurate genetic profiling and identification of individuals based on their DNA profiles.

10.4 Future prospects and challenges

10.4.1 Advancements in CIP technology

Advancements in CIP technology have significantly enhanced their performance and broadened their applications. Ongoing research aims to enhance the performance of CIPs through the development of new templates, improved polymerization techniques, and advanced characterization methods. Innovations in nanotechnology and material science are expected to play a crucial role in these advancements. Development of more specific and diverse carbon templates has increased the selectivity of CIPs for a broader range of target molecules. Innovations in polymerization methods, such as surface imprinting and controlled radical polymerization, have led to improved binding affinity and specificity.

10.4.2 Integration with analytical techniques

The integration of advanced analytical techniques for forensic applications is essential for keeping pace with the evolving pharmaceutical landscape. Different analytical techniques are widely used in forensic science to extract information from trace evidence and other sources. Hence the integration of CIPs with advanced analytical techniques, such as MS and chromatography, holds great promise for forensic applications. This combination can lead to highly sensitive and specific detection methods, further improving the reliability of forensic analyses. Spectroscopic techniques such as infrared spectroscopy, ultraviolet-visible spectroscopy, vibrational spectroscopy, nuclear magnetic resonance spectroscopy, MS, and gas and liquid chromatography are commonly used in forensic science to identify unknown substances found at crime scenes or on evidence. The integration includes evidence types such as fibers, paint, pesticides, drugs, gunshot residues, chemical warfare agents, and explosives. The data collection, subsequent analysis, what information has been obtained and what this means in the context of a case is could stand as strong scientific evidence which is equally important like an eye witness. Integrating with analytical chemistry, data science and artificial intelligence in line with trace evidence can problem solve the nature of much of forensic analysis.

10.4.2.1 Overcoming challenges

Despite their advantages, CIPs face challenges such as specific recognition sites, selectivity, and stability, template leakage, nonspecific binding, and the need for rigorous validation in forensic contexts. In addition, they suffer from a series of drawbacks, including loss of conductivity, electrocatalytic activity, and cost, which limit their use in forensic analysis. Addressing these issues through continued research and development is essential for the widespread adoption of CIPs in forensic science. There are not many studies are conducted to utilize computational approaches for determining the composition of imprinting materials. If put some efforts in this direction would make possible to implement computationally well-designed systems to be used as effective, time saving and useful tools in the area of molecular imprinting.

10.5 Conclusion

CIPs represent a significant breakthrough in forensic science, offering unparalleled selectivity and sensitivity for detecting and analyzing forensic evidence. Their versatility and robustness make them invaluable tools for various forensic applications, from explosive detection to drug analysis and environmental forensics. The CIPs can be effectively utilized in sensors that detect even the ppm level of food adulterants, toxic chemicals, and banned drug materials. One of the major break-through in the forensic analysis using the CIPs is the detection of explosives. Sensors based on carbon electrodes have been successfully employed for the explosive detection. Graphite and carbon quantum dots are utilized in electrode materials and fluorescent sensors to detect the heavy metals even in trace amounts. As research progresses, CIPs are poised to become an integral part of forensic investigations, contributing to more accurate and efficient crime-solving efforts.

References

[1] Díez-Pascual A. M. (2023). Perspectives of polymers in forensic analysis, *Macromol*, 3, 108–119, https://doi.org/10.3390/macromol3020008.

[2] Causin V. (2015). Causin V. Ed. *Polymers on the crime scene BT – Polymers on the crime scene: forensic analysis of polymeric trace evidence*, Springer International Publishing, Cham, 105–166, https://doi.org/10.1007/978-3-319-15494-7_4.

[3] Causin V. (2010). Polymers on the crime scene: How can analytical chemistry help to exploit the information from these mute witnesses?, *Analytical Methods*. 2, 792–804, https://doi.org/10.1039/c0ay00115e.

[4] Farah S., Kunduru K. R., Tsach T., Bentolila A., & Domb A. J. (2015). Forensic comparison of synthetic fibers, *Polymers for Advanced Technologies*. 26, 785–796, https://doi.org/https://doi.org/10.1002/pat.3540.

[5] Akhtar S., Razaq A., Faras K. S., Ashiq M. I., & Tahir M. A. (2022). Identification of the polymer-bounded drugs on the fabric surface: A challenge to the forensic drug analysts, *Journal of Forensic Sciences*. 67, 1267–1273, https://doi.org/https://doi.org/10.1111/1556-4029.14980.

[6] De Bruin-hoegée M., Schoorl J., Zoon P., Van der Schans M. J., Noort D., & Van Asten A. C. (2023). A novel standard for forensic elemental profiling of polymers by LA-ICP-TOF-MS, *Forensic Chemistry*. 35, 100515, https://doi.org/https://doi.org/10.1016/j.forc.2023.100515.

[7] Causin V. (2015). Polymers on the Crime Scene. In: Polymers on the Crime Scene. Springer, Cham. https://doi.org/10.1007/978-3-319-15494-7_4.

[8] Bhatia T. (2022). Novel nanomaterials in forensic investigations: A review, *Materials Today: Proceedings*. 50, 1071–1079.

[9] Stuart B. H. (2012). Forensic Analytical Techniques. John Wiley & Sons, UK.

[10] Lewis P. R. (2016). Forensic Polymer Engineering: Why Polymer Products Fail in Service. Woodhead, Publishing Sawston, United Kingdom.

[11] It is a research gate article published. But no volume and journal information is given

[12] Wulff G. (2013). Fourty years of molecular imprinting in synthetic polymers: Origin, features and perspectives, *Microchimica Acta*. 180, 1359–1370.

[13] Haupt, K., Linares, A. V., Bompart, M., & Bui, B. T. (2012). Molecularly imprinted polymers, *Topics in current chemistry*, *325*, 1–28. https://doi.org/10.1007/128_2011_307.

[14] Dong C., Shi H., Han Y., Yang Y., Wang R., & Men J. (2021). Molecularly imprinted polymers by the surface imprinting technique, *European Polymer Journal*. 145, 110231.

[15] Cheong W. J., Yang S. H., & Ali F. (2013). Molecular imprinted polymers for separation science: A review of reviews, *Journal of Separation Science*. 36, 609–628, https://doi.org/10.1002/jssc.201200784.

[16] Chen L., Xu S., & Li J. (2011). Recent advances in molecular imprinting technology: Current status, challenges and highlighted applications, *Chemical Society Reviews*. 40, 2922–2942, https://doi.org/10.1039/C0CS00084A.

[17] Dvorakova G., Haschick R., Chiad K., Klapper M., Müllen K., & Biffis A. (2010). Molecularly imprinted nanospheres by nonaqueous emulsion polymerization, *Macromolecular Rapid Communications*. 31, 2035–2040, https://doi.org/https://doi.org/10.1002/marc.201000406.

[18] Advincula R. C. (2011). Engineering molecularly imprinted polymer (MIP) materials: Developments and challenges for sensing and separation technologies, *Korean Journal of Chemical Engineering*. 28, 1313–1321, https://doi.org/10.1007/s11814-011-0133-2.

[19] Refaat D., Aggour M. G., Farghali A. A., Mahajan R., Wiklander J. G., Nicholls I. A., & Piletsky S. A. (2019). Strategies for molecular imprinting and the evolution of MIP nanoparticles as plastic antibodies-synthesis and applications, *International Journal of Molecular Sciences*. 20, https://doi.org/10.3390/ijms20246304.

[20] Dorkó Z., Szakolczai A., Verbić T., & Horvai G. (2015). Binding capacity of molecularly imprinted polymers and their nonimprinted analogs, *Journal of Separation Science*. 38, 4240–4247, https://doi.org/10.1002/jssc.201500874.

[21] Turner N. W., Jeans C. W., Brain K. R., Allender C. J., Hlady V., & Britt D. W. (2006). From 3D to 2D: A review of the molecular imprinting of proteins, *Biotechnology Progress*. 22, 1474–1489, https://doi.org/10.1021/bp060122g.

[22] Cieplak, M.; Węgłowski, R.; Iskierko, Z.; Węgłowska, D.; Sharma, P.S.; Noworyta, K.R.; D'Souza, F.; Kutner, W. Protein Determination with Molecularly Imprinted Polymer Recognition Combined with Birefringence Liquid Crystal Detection. Sensors 2020, 20, 4692, MDPI, https://doi.org/10.3390/s20174692.

[23] N.M. Bergmann N. A. (2008). Peppas, Molecularly imprinted polymers with specific recognition for macromolecules and proteins, *Progress in Polymer Science*. 33, 271–288, https://doi.org/https://doi.org/10.1016/j.progpolymsci.2007.09.004.

[24] Karlsson B. C. G., Mahony J. O., Karlsson J. G., Bengtsson H., Eriksson L. A., & Nicholls I. A. (2009). Structure and dynamics of monomer–template complexation: An explanation for molecularly imprinted polymer recognition site heterogeneity, *Journal of the American Chemical Society*. 131, 13297–13304, https://doi.org/10.1021/ja902087t.

[25] Nicholls I. A. (1995). Thermodynamic considerations for the design of and ligand recognition by molecularly imprinted polymers, *Chemistry Letters*. 24, 1035–1036, https://doi.org/10.1246/cl.1995.1035.

[26] Vasapollo G., Del Sole R., Mergola L., Lazzoi M. R., Scardino A., Scorrano S., & Mele G. (2011). Molecularly imprinted polymers: Present and future prospective, *International Journal of Molecular Sciences*. 12, 5908–5945, https://doi.org/10.3390/ijms12095908.

[27] Szenczi A., Kardos J., Medgyesi G. A., & Závodszky P. (2006). The effect of solvent environment on the conformation and stability of human polyclonal IgG in solution, *Journal of the International Association of Biological Standardization*. 34, 5–14, https://doi.org/10.1016/j.biologicals.2005.06.007.

[28] Poma A., Guerreiro A., Caygill S., Moczko E., & Piletsky S. (2014). Automatic reactor for solid-phase synthesis of molecularly imprinted polymeric nanoparticles (MIP NPs) in water, *RSC Advances*. 4, 4203–4206, https://doi.org/10.1039/C3RA46838K.

[29] Vaneckova T., Bezdekova J., Han G., Adam V., & Vaculovicova M. (2020). Application of molecularly imprinted polymers as artificial receptors for imaging, *Acta Biomaterialia*. 101, 444–458, https://doi.org/10.1016/j.actbio.2019.11.007.

[30] Perçin I., Idil N., Bakhshpour M., Yılmaz E., Mattiasson B., & Denizli A. (2017). Microcontact imprinted plasmonic nanosensors: Powerful tools in the detection of *Salmonella paratyphi*, *Sensors*. 17, https://doi.org/10.3390/s17061375.

[31] Nishino H., Huang C.-S., & Shea K. J. (2006). Selective protein capture by epitope imprinting, *Angewandte Chemie International Edition in English*. 45, 2392–2396, https://doi.org/10.1002/anie.200503760.

[32] Nicholls I. A., & Rosengren J. P. (2001). Molecular imprinting of surfaces, *Bioseparation*. 10, 301–305, https://doi.org/10.1023/A:1021541631063.

[33] Yan H., & Row K. H. (2006). Characteristic and synthetic approach of molecularly imprinted polymer, *International Journal of Molecular Sciences*. 7, 155–178, https://doi.org/10.3390/i7050155.

[34] Uzun L., & Turner A. P. F. (2016). Molecularly-imprinted polymer sensors: Realising their potential, *Biosensors and Bioelectronics*. 76, 131–144, https://doi.org/https://doi.org/10.1016/j.bios.2015.07.013.

[35] Ma X., Ji W., Chen L., Wang X., Liu J., & Wang X. (2015). Molecularly imprinted polymers with synthetic dummy templates for the preparation of capsaicin and dihydrocapsaicin from chili peppers, *Journal of Separation Science*. 38, 100–107, https://doi.org/10.1002/jssc.201400911.

[36] Xu S., Lu H., Li J., Song X., Wang A., Chen L., & Han S. (2013). Dummy molecularly imprinted polymers-capped CdTe quantum dots for the fluorescent sensing of 2,4,6-trinitrotoluene, *ACS Applied Materials & Interfaces*. 5, 8146–8154, https://doi.org/10.1021/am4022076.

[37] Zhang W., Qin L., He X.-W., Li W.-Y., & Zhang Y.-K. (2009). Novel surface modified molecularly imprinted polymer using acryloyl-beta-cyclodextrin and acrylamide as monomers for selective recognition of lysozyme in aqueous solution, *Journal of Chromatography A*. 1216, 4560–4567, https://doi.org/10.1016/j.chroma.2009.03.056.

[38] Giubileo F., Di Bartolomeo A., Iemmo L., Luongo G., & Urban F. (2018). Field emission from carbon nanostructures, *Applied Sciences*. 8, 1–21, https://doi.org/10.3390/app8040526.

[39] Hoffmann R., Kabanov A. A., Golov A. A., & Proserpio D. M. (2016). Homo citans and carbon allotropes: For an ethics of citation, *Angewandte Chemie International Edition*. 55, 10962–10976, https://doi.org/10.1002/anie.201600655.

[40] Kamyab H., Chelliapan S., Tavakkoli O., Mesbah M., Bhutto J. K., Khademi T., Kirpichnikova I., Ahmad A., & ALJohani A. A. (2022). , A review on carbon-based molecularly-imprinted polymers (CBMIP) for detection of hazardous pollutants in aqueous solutions, *Chemosphere*. 308, 136471, https://doi.org/https://doi.org/10.1016/j.chemosphere.2022.136471.
[41] McCluskey A., Holdsworth C. I., & Bowyer M. C. (2007). Molecularly imprinted polymers (MIPs): Sensing, an explosive new opportunity?, *Organic & Biomolecular Chemistry*. 5, 3233–3244, https://doi.org/10.1039/B708660A.
[42] Zarejousheghani M., Lorenz W., Vanninen P., Alizadeh T., Cämmerer M., & Borsdorf H. (2019). molecularly imprinted polymer materials as selective recognition sorbents for explosives: A review, *Polymers*, MDPI Basel, 11, https://doi.org/10.3390/polym11050888.
[43] Alizadeh T., Atashi F., & Ganjali M. R. (2019). Molecularly imprinted polymer nano-sphere/multi-walled carbon nanotube coated glassy carbon electrode as an ultra-sensitive voltammetric sensor for picomolar level determination of RDX, *Talanta*. 194, 415–421, https://doi.org/10.1016/j.talanta.2018.10.040.
[44] Meng X., Xiao Z., & Scott S. K. (2019). Preparation and application of electrochemical sensor based on molecularly imprinted polymer coated multi-walled carbon nanotubes for nitrocellulose detection, propellants, explos, *Pyrotech*. 44, 1337–1346, https://doi.org/https://doi.org/10.1002/prep.201900055.
[45] Shi X., Zuo Y., Jia X., Wu X., Jing N., Wen B., & Mi X. (2020). A novel molecularly imprinted sensor based on gold nanoparticles/reduced graphene oxide/single-walled carbon nanotubes nanocomposite for the detection of pefloxacin, *International Journal of Electrochemical Science*. 15, 9683–9697, https://doi.org/https://doi.org/10.20964/2020.10.38.
[46] Xiao D., Jiang Y., & Bi Y. (2018). Molecularly imprinted polymers for the detection of illegal drugs and additives: A review, Microchimica Acta (2018) 185: 247 https://doi.org/10.1007/s00604-018-2735-4.
[47] Djavanshir D., Farajzadeh M., Sorouraddin S., & Baheri T. (2011). Synthesis and application of high selective monolithic fibers based on molecularly imprinted polymer for SPME of trace methamphetamine, *Chromatographia*. 73, 975–983, https://doi.org/10.1007/s10337-011-1984-8.
[48] Singhal A., Sadique M. A., Kumar N., Yadav S., Ranjan P., Parihar A., Khan R., & Kaushik A. K. (2022). Multifunctional carbon nanomaterials decorated molecularly imprinted hybrid polymers for efficient electrochemical antibiotics sensing, *Journal of Environmental Chemical Engineering*. 10, https://doi.org/10.1016/j.jece.2022.107703.
[49] Arman A., Sağlam Ş., Üzer A., & Apak R. (2021). Direct electrochemical determination of peroxide-type explosives using well-dispersed multi-walled carbon nanotubes/polyethyleneimine-modified glassy carbon electrodes, *Analytical Chemistry*. 93, 11451–11460, https://doi.org/10.1021/acs.analchem.1c01397.
[50] Amatatongchai M., Sroysee W., Sodkrathok P., Kesangam N., Chairam S., & Jarujamrus P. (2019). Novel three-Dimensional molecularly imprinted polymer-coated carbon nanotubes (3D-CNTs@MIP) for selective detection of profenofos in food, *Analytica Chimica Acta*. 1076, 64–72, https://doi.org/https://doi.org/10.1016/j.aca.2019.04.075.
[51] Mugo S. M., Lu W., & Robertson S. V. (2022). Molecularly imprinted polymer-modified microneedle sensor for the detection of imidacloprid pesticides in food samples, *Sensors*. 22, https://doi.org/10.3390/s22218492.
[52] Arreguin-Campos R., Jiménez-Monroy K. L., Diliën H., Cleij T. J., Van Grinsven B., & Eersels K. (2021). Imprinted polymers as synthetic receptors in sensors for food safety, *Biosensors*. 11, https://doi.org/10.3390/bios11020046.
[53] Li Y., Zhang L., Dang Y., Chen Z., Zhang R., Li Y., & Ye B.-C. (2019). A robust electrochemical sensing of molecularly imprinted polymer prepared by using bifunctional monomer and its application in detection of cypermethrin, *Biosensors and Bioelectronics*. 127, 207–214.

[54] Motaharian A., Motaharian F., Abnous K., Hosseini M. R. M., & Hassanzadeh-Khayyat M. (2016). Molecularly imprinted polymer nanoparticles-based electrochemical sensor for determination of diazinon pesticide in well water and apple fruit samples, *Analytical and Bioanalytical Chemistry*. 408, 6769–6779.

[55] Aghoutane Y., Diouf A., Österlund L., Bouchikhi B., & El Bari N. (2020). Development of a molecularly imprinted polymer electrochemical sensor and its application for sensitive detection and determination of malathion in olive fruits and oils, *Bioelectrochemistry*. 132, 107404, https://doi.org/ https://doi.org/10.1016/j.bioelechem.2019.107404.

Part 4: **Case studies of molecular imprinting-based smart carbon nanomaterials**

Jemmyson Romário de Jesus and Tatianny de Araujo Andrade

Chapter 11
Binding kinetics and thermodynamics of carbon-based molecularly imprinted polymers

Abstract: This chapter delves into the binding kinetics and thermodynamics of carbon-based molecularly imprinted polymers (MIPs), exploring the critical role these studies play in understanding and optimizing MIPs. It begins by discussing the fundamental principles of molecular recognition and the dynamic interactions between MIPs and target molecules. The chapter highlights the importance of kinetic and thermodynamic analyses, which, when combined with mathematical modeling, provide a mechanistic understanding of binding processes. Important experimental techniques such as Fourier transform infrared spectroscopy , surface plasmon resonance , isothermal titration calorimetry , atomic force microscopy , and nuclear magnetic resonance are discussed in detail, showing their application in obtaining kinetic and thermodynamic profiles of MIPs. The integration of these experimental methods with computational simulations allows a comprehensive evaluation of carbon-based MIPs, providing important information on binding efficiency, selectivity, and the thermodynamic parameters governing analyte-polymer interactions. These parameters are crucial for optimizing the performance of MIPs in a variety of applications, including sensors, controlled release systems, and purification processes.

Keywords: Binding kinetics , thermodynamic analysis , molecular recognition, computational simulations

11.1 Introduction

Molecular imprinting has generated significant interest due to its relevance in diverse applications. Described as a "lock-and-key" process, this methodology is an essential

Acknowledgments: The authors thank Fundação de Amparo à Pesquisa do Estado de Minas Gerais (FAPEMIG, grants APQ-01786-22, APQ-05429-23, and RED-00144-22), Conselho Nacional de Desenvolvimento Científico e Tecnológico (CNPq, grand number 05828/2022-5), and Coordenação de Aperfeiçoamento de Pessoal de Nível Superior for financial support.

Jemmyson Romário de Jesus, Tatianny de Araujo Andrade, Research Laboratory in Bionanomaterials, LPbio, Department of Chemistry, Federal University of Viçosa, 36570-900 Viçosa, Minas Gerais, Brazil

https://doi.org/10.1515/9783111503325-011

route to directly creating polymers that mimic antibodies [1]. The molecular printing process generally involves an automatic assembly of functional monomers around a molecular model (Figure 11.1). Then, polymerization occurs in the presence of a cross-linking agent to repair this organized structure. After polymerization, the template is removed from the polymer, leaving binding sites that are complementary in terms of structure and function. These binding sites give the polymer a molecular memory, allowing it to specifically model the molecule [2].

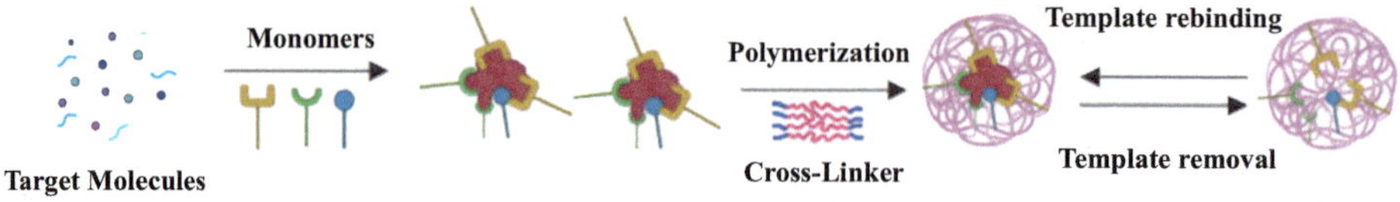

Figure 11.1: General scheme of molecularly imprinted polymer (MIP) production.

Similar to interactions between natural biomolecules such as antibody-antigen, receptor-ligand, and enzyme-substrate, the molecularly imprinted polymers (MIPs) can recognize their model molecules with high specificity. This specificity is based on the "lock-and-key" principle. The architecture of the imprints plays a crucial role in directing this specificity, as specific sorption and molecular recognition result from complementary conformations. The organized arrangement of functional groups within the imprint forms a binding structure that allows the polymer to selectively bind to the template molecule. Furthermore, the print format can act as a molecular sieve, allowing only molecules identical to or smaller than the template to access its interior. However, although this theory offers logical explanations for several general applications, the conclusions are often qualitative, modal or theoretical. Furthermore, there are limitations in specific cases, where the recognition behavior does not seem to depend exclusively on the specific impression [3]. Dynamic interaction behaviors between MIP and target molecules can interfere with molecular recognition. In this context, kinetic and thermodynamic studies have been widely explored in the context of MIPs with target molecules. Such knowledge is crucial to optimize the design and functionality of MIPs in diverse applications, such as pollutant removal [4], sensors [5, 6] and controlled drug delivery systems [7, 8].

In this chapter, the thermodynamic and kinetic considerations related to the dynamic interactions between MIPs and target molecules are discussed. These studies not only quantify the interactions but also provide an in-depth understanding of the fundamental principles of molecular recognition. Furthermore, the main techniques used to determine these parameters are highlighted, including: (i) Fourier transform infrared spectroscopy (FT-IR); (ii) surface plasmon resonance (SPR); (iii) isothermal titration calorimetry (ITC); (iv) atomic force microscopy (AFM); (v) nuclear magnetic resonance (NMR) spectroscopy; and (vi) computational simulations .

11.2 Kinetic studies and main mathematical models

The evolution of mass transport steps and chemical reactions provides a solid basis for improving MIP performance, ensuring effective operation under desired conditions. Understanding kinetic models provides insights into how quickly and efficiently MIPs can capture and release target molecules. Several kinetic models have been established to understand the dynamics of molecular recognition and the rate-limiting step. These include: (i) pseudo-first-order model; (ii) pseudo-second-order model; (iv) external mass transfer model (iv) Weber and Morris sorption kinetic model; (v) Boyd kinetic model; (vi) Bhattacharya and Venkobachar model, and (vii) Elovich model. Below, each model is briefly presented.

11.2.1 Pseudo-first-order model

The pseudo-first-order kinetic model is one of the simplest and most widely used models to describe the adsorption process in various systems, including MIPs. This model assumes that the occupation rate of adsorption sites is directly proportional to the number of unoccupied sites [9, 10]. The model is often applied to systems where the adsorption process is relatively straightforward and involves a single, rate-limiting step. The pseudo-first-order kinetic equation is given by Lagergren (1898) [11], and can be expressed as:

$$\frac{dq}{dt} = k_1(q_e - q) \tag{11.1}$$

where q_t is the amount of adsorbate adsorbed at time t (min) (mg g^{-1}), q_e is the amount of adsorbate adsorbed at equilibrium (mg g^{-1}), and k_1 is the rate constant of pseudo-first-order adsorption (min^{-1}).

To facilitate the integration and practical application of this model, the equation can be rearranged and integrated with boundary conditions $t = 0$ min and $q_t = 0$ mg g^{-1}, resulting in the linear form:

$$\ln(q_e - q) = \ln q_e - k_1 t \tag{11.2}$$

This linear equation allows the determination of the adsorption rate constant k_1 and the equilibrium adsorption capacity q_e by plotting ln (q_e – q_t) versus t (min). The slope of the resulting line is $-k_1$, and the intercept is ln q_e.

The pseudo-first-order model is widely used due to its simplicity and the straightforward nature of its application. It is particularly useful in systems where the initial concentration of the adsorbate is low, and the adsorption process is dominated by physisorption. However, this model has limitations. It may not accurately describe adsorption kinetics in systems where

- the adsorption involves complex interactions or multiple steps;

- chemisorption is significant;
- there are variations in the adsorbate or adsorbent properties over time.

In such cases, more complex kinetic models like the pseudo-second-order model or other multi-step models may provide better fits and more accurate descriptions of the adsorption processes.

11.2.2 Pseudo-second-order model

The pseudo-second-order kinetic model is a predominant model used to describe the adsorption kinetics of MIPs. This model is especially relevant in scenarios where the adsorption process is more complex than the pseudo-first-order model. It assumes that the rate limiting step involves chemisorption, which includes valence forces through the sharing or exchange of electrons between the adsorbent and the adsorbate [9].

The pseudo-second-order kinetic equation is expressed as

$$\frac{dq_t}{dt} = k_2(q_e - q_t)^2 \tag{11.3}$$

where q_t is the amount of adsorbate adsorbed at time t (min) (mg g^{-1}), q_e is the amount of adsorbate adsorbed at equilibrium (mg g^{-1}), and k_2 is the rate constant of pseudo-secondorder adsorption (g mg^{-1} min^{-1}).

Integrating the equation with the boundary conditions $t = 0$ min and $q_t = 0$ mg g^{-1} gives

$$\frac{t}{q_t} = \frac{1}{k_2 q_e^2} + \frac{t}{q_e} \tag{11.4}$$

This linear form allows for the determination of q_e and k_2 by plotting t/q_t against t (min). The slope of the plot is $1/q_e$ and the intercept is $1/(k_2\, q_e^2)$.

In the context of MIPs, the pseudo-second-order model is particularly useful due to the nature of interactions between the polymer and the target molecules. MIPs are designed to have specific binding sites that match the molecular structure of the target analytes, often leading to strong and selective interactions. By analyzing the kinetic parameters (q_e and k_2) it is possible to optimize the design of MIPs. Higher q_e values indicate better capacity for the target molecule, while higher k_2 values suggest faster adsorption rates, both of which are desirable characteristics in practical applications, including sensor design, drug delivery, and environmental cleanup [12].

In general, the pseudo-second-order model often provides a more accurate description of the adsorption kinetics compared to the pseudo-first-order model, especially in systems involving chemisorption. Addition, it offers insights into the mechanisms of adsorption, including the involvement of chemical bonds and the potential for electron exchange between the MIP and the target molecule [13]. However, this model assumes that the adsorption process is controlled exclusively by the chemisorption mechanism, which may not be the case in all scenarios. Other factors like pore diffusion or physical adsorption might also play significant roles.

11.2.3 Principles of the external diffusion model

The external diffusion model, also known as the film diffusion model, is another important kinetic model used to describe mass transfer processes in adsorption systems, including those involving MIPs. This model focuses on the transport of adsorbate molecules from the bulk solution to the surface of the adsorbent, which is a crucial step in the overall adsorption process. It is particularly relevant when the mass transfer resistance within the boundary layer is significant compared to other steps in the sorption process, such as pore diffusion or surface adsorption [14].

The external diffusion model can be described by the following equation:

$$\ln(1-F) = -k_{fd}t \tag{11.5}$$

where F is the fractional attainment of equilibrium ($F = q_t/q_e$), q_t is the amount of adsorbate adsorbed at time t (mg g^{-1}), q_e is the amount of adsorbate adsorbed at equilibrium (mg g^{-1}), and k_{fd} is the rate constant for external diffusion (min^{-1}).

Alternatively, the model can be represented in a linearized form, which can be used to determine the rate constant:

$$\ln\left(\frac{q_e - q_t}{q_e}\right) = -k_d t \tag{11.6}$$

By plotting $\ln((q_e - q_t)/q_e)$ versus t (min), the rate constant k_d can be determined from the slope of the resulting line.

In the context of MIPs, the external diffusion model is useful for understanding and optimizing the initial stages of the adsorption process, where mass transfer from the bulk solution to the surface of the MIP particles is crucial. By applying the external diffusion model, it is possible to optimize various parameters, such as stirring speed, temperature, and concentration gradients, to increase the mass transfer rate and consequently the overall adsorption efficiency of MIPs [15].

The external diffusion model provides a straightforward approach to evaluating mass transfer kinetics, making it easy to apply and interpret. This model is useful for studying the initial stages of adsorption, where external diffusion is often the dominant process [10]. However, this model assumes that external diffusion is the exclusive rate-limiting step, which may not always be the case. Other processes, such as pore diffusion and surface adsorption, can also play significant roles [13]. In addition, the model depends on an accurate estimate of the boundary layer thickness, which can be difficult to determine experimentally.

11.2.4 Weber and Morris kinetic model (internal diffusion model)

The Weber and Morris kinetic sorption model, also known as the intraparticle diffusion model, assumes that the adsorption process is controlled by the diffusion of adsorbate molecules into the interior of the adsorbent particles. It is particularly useful for systems where external mass transfer (film diffusion) is not the only rate-limiting step, and the internal structure of the adsorbent significantly influences the overall adsorption rate [16].

The Weber and Morris kinetic sorption model is typically expressed as [17]:

$$q = k_{\mathrm{WM}\,\sqrt{t}+C} \tag{11.7}$$

where q_t is the amount of adsorbate adsorbed at time t (min) (mg g^{-1}), k_{WM} is the intraparticle diffusion rate constant (mg g^{-1} min), and C is the intercept (mg g^{-1}), which reflects the boundary layer effect.

A plot of q_t versus $\sqrt{t}$ should yield a straight line if intraparticle diffusion is the rate-limiting step. The slope of this line gives the intraparticle diffusion rate constant k_{WM}, and the intercept C provides insights into the thickness of the boundary layer [14].

For MIPs, the Weber and Morris model allows understanding how target molecules diffuse through the polymer matrix to reach the specific binding sites. The model helps differentiate between different diffusion mechanisms, such as external diffusion, intraparticle diffusion, and surface adsorption. By examining the linearity of the q_t versus $\sqrt{t}$ plot, it is possible to determine whether the adsorption process is primarily controlled by internal diffusion [18].

The Weber and Morris model can guide the synthesis of MIPs with optimized pore structures and surface properties. For instance, adjusting the porosity and cross-linking density can enhance the intraparticle diffusion rate, thereby improving the adsorption efficiency. Applying the Weber and Morris model allows for a detailed evaluation of the performance of MIPs in various environments. This includes assessing how well the MIPs capture and release target molecules under different conditions, such as varying pH, temperature, and ionic strength. In addition, it provides a comprehensive analysis of the intraparticle diffusion process, offering important information into the internal transport mechanisms within MIPs [19, 20].

However, the model assumes that intraparticle diffusion is the primary rate-limiting step, which may not always hold for complex systems with multiple mechanisms.

11.2.5 Other kinetic models

11.2.5.1 Boyd kinetic model

Given the dual nature of intraparticle diffusion, which encompasses both film and pore diffusion, accurately determining the rate-controlling step in the adsorption process is essential for optimizing MIP performance. The Boyd kinetic equation, is expressed mathematically to analyze the kinetic data of adsorption processes. By applying this model, it is possible to identify whether the rate-controlling step is the diffusion of the solute through the boundary layer (film diffusion) or the diffusion within the pores of the MIP (pore diffusion). This distinction is important because it influences the design and synthesis of MIPs, ensuring that they are tailored to facilitate the most efficient adsorption pathways [14, 21].

The Boyd kinetic equation is expressed as follows:

$$F = \frac{6}{\pi^2} \exp\left(-Bt\right) \tag{11.8}$$

and

$$F = \frac{q}{q_e} \tag{11.9}$$

where q_e is the amount of solute adsorbed at equilibrium (mg g^{-1}), q is the amount of solute adsorbed at any time t (min), F represents the fraction of solute adsorbed at any time t (min), and B_t is the mathematical function of F.

The previous equation can be rearranged as

$$B_t = -0.4977 - \ln(1 - F) \tag{11.10}$$

The plot of [−0.4977 − ln (1 − F)] versus t (min) can be used to assess the linearity of the experimental data. The degree of linearity in these plots helps distinguish between adsorption rates controlled by external transport and those controlled by intraparticle transport.

A straight line passing through the origin indicates that adsorption processes are governed by particle diffusion mechanisms; otherwise, they are governed by kinetic diffusion.

The diffusion coefficient D_i (m^2 g^{-1}) can be calculated from the B values using the following relationship:

$$B = \frac{\pi^2 D_i}{r^2} \tag{11.11}$$

where D_i is the effective diffusion coefficient of solute in the adsorbent phase, and r is the radius of adsorbent particles.

11.2.5.2 Bhattacharya and Venkobachar kinetic model

The Bhattacharya and Venkobachar model provide allows for studying the adsorption kinetics of MIP, distinguishing between external mass transfer and intraparticle diffusion. This model was developed by Bhattacharya and Venkobachar in 1984, and is typically represented by the following equation [22]:

$$\log[1 - U(T)] = -\left(\frac{k}{2.303}\right)t \tag{11.12}$$

where $U(T) = [(C_0 - C_t)/(C_0 - C_e)]$, k is the Bhattacharya and Venkobachar rate constant (min^{-1}), which can be calculated from the slope of the plot $\log[1 - U(T)]$ versus t (min), C_0 and C_e are the initial and equilibrium concentration of target molecules, respectively, and C_t is the concentration of pollutant at any time t.

When applied to MIPs, the Bhattacharya and Venkobachar model helps in evaluating the efficiency of MIPs in capturing target molecules. MIPs are designed to have specific binding sites that are complementary to the target molecules, mimicking natural recognition processes. The model assists in quantifying the adsorption kinetics and determining the dominant mechanisms involved. In addition, it is effective in identifying whether external mass transfer or intraparticle diffusion controls the adsorption rate.

To use the Bhattacharya and Venkobachar model, experimental data on the concentration of solute over time is required. The log-transformed concentration ratios are plotted against time to assess the linearity and determine the rate constant *(k)*. The slope of the plot indicates the rate-controlling mechanism, helping to refine the design and application of MIPs in practical scenarios.

11.2.5.3 Elovich kinetic model

Elovich kinetic is another model widely used to describe adsorption kinetics in heterogeneous systems. Originally developed for catalytic reactions on solid surfaces, this model is particularly useful for analyzing processes where the adsorption rate decreases exponentially with increasing surface coverage [22, 23]. Applying the Elovich model to molecule MIP offers valuable insights into the efficiency and mechanisms of adsorption of specific analytes. Elovich kinetic model is described by the following equation:

$$qt = \frac{1}{\beta}\ln(\alpha\beta) + \frac{1}{\beta}\ln(t) \tag{11.13}$$

where q_t is the amount of adsorbate (e.g., a substance being adsorbed) on the adsorbent surface at time t (min). α is the initial adsorption rate (mg g^{-1} min^{-1}). It represents the rate of adsorption at the very beginning of the process, where the surface of the adsorbent is most available. β is the desorption constant (g mg^{-1}). It is related to the extent of surface coverage and the activation energy for chemisorption. A higher β-value indicates a slower decrease in the adsorption rate over time. t represents the time at which the amount of adsorbate on the adsorbent surface is measured.

The Elovich model is particularly useful when dealing with adsorption processes that do not fit simple kinetic models like first-order or second-order kinetics, especially in cases involving chemisorption where the adsorption energy varies with surface coverage.

11.3 Thermodynamic studies

Thermodynamic studies are also crucial to understanding the interactions between the analyte and the MIPs recognition sites. These studies involve the analysis of parameters such as Gibbs free energy (ΔG), enthalpy (ΔH), and entropy (ΔS) of adsorption [24]. By analyzing thermodynamic parameters, it is possible to gain a detailed understanding of the molecular interactions between recognition sites and the analyte. Furthermore, based on thermodynamic information, it is possible to adjust the synthesis and modification conditions of MIPs to improve their selectivity and efficiency [25].

Gibbs free energy indicates the spontaneity of the adsorption process. Negative values of ΔG suggest that adsorption is spontaneous. Adsorption enthalpy provides information about the exothermic or endothermic nature of the process. Negative values of ΔH indicate exothermic interactions, while positive values indicate endothermic interactions. Adsorption entropy, in turn, gives an idea of the disorder associated with the adsorption process. An increase in entropy (positive ΔS values) suggests that adsorption is associated with an increase in disorder, possibly due to the release of solvent molecules from the MIP cavities during analyte adsorption [26].

Thermodynamic analysis can be carried out through adsorption experiments at different temperatures, allowing the construction of an adsorption isotherm and the application of van't Hoff equations to determine the thermodynamic parameters [27]. The van't Hoff equation is given by

$$\ln K = -\frac{\Delta H}{RT} + \frac{\Delta S}{R} \tag{11.14}$$

where K is the equilibrium constant, R is the universal gas constant (8.314 J mol^{-1} K^{-1}) and T is the absolute temperature (Kelvin). From this equation, it is possible to determine the ΔH and ΔS of adsorption through the slope and intersection of the straight line obtained by plotting $\ln K$ versus $1/T$.

The Gibbs free energy (ΔG), enthalpy (ΔH), and entropy (ΔS) are related by the equation:

$$\Delta G = \Delta H - T\Delta S \tag{11.15}$$

11.3.1 Main mathematical models involving adsorption isotherms

Mathematical models are essential to describe and predict adsorption behavior in MIPs. Among the most applied models are: (i) Langmuir model ; (ii) Freundlich model ; and (iii) Dubinin-Radushkevich (D-R) model.

11.3.1.1 Langmuir model

The Langmuir model assumes that adsorption occurs at specific sites on a homogeneous surface, without interaction between the adsorbed molecules [26, 28]. The Langmuir equation is given by

$$q_e = \frac{q_m K_L C_e}{1 + K_L C_e} \tag{11.16}$$

where q_e is the amount of analyte adsorbed in the equilibrium (mg g^{-1}), q_m is the maximum adsorption capacity (mg g^{-1}), K_L is the Langmuir constant (L mg^{-1}), and C_e is the concentration of the analyte at equilibrium (mg L^{-1}).

The Langmuir equation can model the coverage of molecules on a solid surface, such as an MIP up to the equilibrium concentration of the adsorbate in solution at a constant temperature. Langmuir isotherms are based on three fundamental assumptions: (*i*) the adsorbent surface is in contact with a solution containing the adsorbate, which is strongly attracted to the surface; (*ii*) the surface has a finite and defined number of specific sites where solute molecules can be adsorbed; and (*iii*) adsorption results in the formation of only one layer of molecules on the surface, characterizing a monolayer adsorption [14, 29]. The Langmuir isotherm parameters can be obtained from its linearized form using the equation:

$$\frac{C_e}{q_e} = \frac{1}{K_L Q_0} + \frac{C_e}{Q_0} \tag{11.17}$$

A plot of C_e/q_e versus C_e should indicate a straight line of slope $1/Q_0$ and an intercept of $1/(K_L Q_0)$.

11.3.1.2 Freundlich model

The Freundlich model is empirical and assumes a heterogeneous surface with different types of adsorption sites [30]. The Freundlich equation is expressed as

$$\ln\ q_e = \ln K_F + \left(\frac{1}{n}\right) \ln C_e \tag{11.18}$$

where K_F is the Freundlich constant (mg g^{-1}) characteristic of the adsorption system, indicating adsorption capacity, and n indicates the adsorption intensity.

The parameter n is an empirical value that varies with the degree of heterogeneity and is related to the distribution of ions bound on the adsorbent surface. Generally, $n > 1$ indicates that the adsorbate is favorably adsorbed onto the adsorbent, with higher n values signifying stronger adsorption intensity [31].

The Freundlich model often describes data better at low concentrations.

11.3.1.3 Dubinin-Radushkevich (D-R) model

The Dubinin-Radushkevich (D-R) isotherm model assumes that the characteristics of the sorption curves are related to the porosity of the adsorbent [32]. The linear form of this model is expressed by

$$\ln(q_e) = \ln(q_m) - \beta\varepsilon^2 \tag{11.19}$$

where q_e is the amount of adsorbate adsorbed per unit dosage of the adsorbent (mg g^{-1}), q_m is the theoretical monolayer saturation capacity (mg g^{-1}), β is the activity coefficient (mol^2 kJ^{-2}) related to the mean sorption energy, and ε is the Polanyi potential described as

$$\varepsilon = RT\ln\left[1 + \left(\frac{1}{C_e}\right)\right] \tag{11.20}$$

where R is the gas constant in kJ mol^{-1} K^{-1}, T is the temperature in Kelvin, and C_e is the concentration of the analyte at equilibrium (mg L^{-1}).

From the plot of ln q_e versus ε^2 the values of β and q_m are determined by the slope and intercept of the linear plot.

11.4 Analytical methods applied to investigate binding kinetics and thermodynamic parameters involving carbon-based MIPs

To investigate the binding kinetics and thermodynamics of carbon-based MIPs, several experimental techniques and analytical methods can be used, among them, the following stand out: (i) FT-IR; (ii) SPR; (iii) ITC; (iv) AFM; (v) NMR; and (vi) computational simulations. Below, each technique is discussed, presenting advantages and limitations.

11.4.1 Fourier transform ifrared spectroscopy (FT-IR) in MIPs

Fourier transform infrared spectroscopy is one of the main techniques that allows monitoring the formation of interaction complexes between an analyte and the active sites of carbon-based MIPs [33]. It offers a non-destructive approach to analyze the structural changes and binding mechanisms that occur during the adsorption process. In general, FT-IR allows the precise identification of functional groups present in both the MIP and the analyte, making it possible to determine which groups are involved in the formation of interaction complexes [34]. In addition, the technique can be used to monitor in real time the interaction between the analyte and the MIP, observing the changes in the absorption bands that correspond to different functional groups [35]. For example, changes in the intensity or shift of carbonyl (C=O), hydroxyl (O–H), or amine (N–H) stretching bands may indicate the formation of hydrogen bonds or other intermolecular interactions [33-35]. From this information, it is possible to elucidate the binding mechanisms between the analyte and the MIP by comparing the spectra before and after adsorption. Changes in the characteristic bands may suggest the type of interaction (e.g., hydrogen bonds, van der Waals interactions, and π-π interactions) occurring during adsorption [36].

Sullivan et al. [37] employed FT-IR to evaluate a series of molecularly imprinted thin-sheet hydrogel polymers using a series of acrylamide-based monomers selective for the target protein myoglobin (Mb). Direct production of the thin-sheet MIP offers an alternative biorecognition surface that is robust, stable, and uniform, and has the potential to be adapted for biosensing applications. The MIP containing the functional monomer *N*-hydroxymethylacrylamide (NHMAm) produced optimal specific rebinding of the target protein (Mb) with rebinding of 84.9 (±0.7) and imprinting factors and selectivity of 1.41 and 1.55, respectively. The combination of FT-IR and computational methods allowed the evaluation of the effects of hydrogen bonding within the protein-MIP complex. The predictions from quantum mechanical calculations of the monomer carbonyl peak shift were confirmed in FT-IR spectra, with three of the MIPs, acrylamide, *N*-(hydroxymethyl) acrylamide, and *N*(hydroxyethyl)acrylamide, showing peak shifts of 4, 11, and 8 cm^{-1}, respectively [37].

Although FT-IR is an effective technique for monitoring the formation of complexes involving MIPs and target molecules, it is sensitive to the presence of water or organic solvents, since water presents intense absorption bands in the infrared region, especially in the range of 3,200–3,600 cm^{-1} (O–H stretching) [34]. This can make data interpretation and identification of specific interactions difficult. Furthermore, FT-IR may be less sensitive to weak interactions, such as Van der Waals interactions, which may not produce significant changes in the absorption bands [36]. This limits the ability to detect all types of interactions between the analyte and the MIP.

11.4.2 Surface plasmon resonance (SPR)

Another technique widely used to study the binding kinetics between molecules is SPR [38]. When applied to carbon-based MIPs, SPR provides detailed insights into the adsorption and desorption rates of analytes at the MIP active sites, allowing for an in-depth understanding of the interaction mechanisms and kinetic properties.

SPR is based on the detection of changes in the refractive index at the interface between a metal (usually gold) and a dielectric medium (such as a solution) [39]. When a beam of light is coupled to a metal film, electrons on the metal surface can oscillate in response to the electric field of the light, generating a phenomenon known as surface plasmon resonance. The resonance condition is highly sensitive to changes in the refractive index near the surface, which allows the detection of molecular interactions in real time without the need for fluorescent markers or labels [39]. To apply SPR to carbon-based MIPs, it is necessary to immobilize the MIPs on the surface of the SPR sensor [40]. This can be done by functionalizing the sensor surface with groups that promote MIP anchoring, such as thiols or silanes, which chemically bond to carbon-based materials such as graphene or carbon nanotubes. During the molecular imprinting process, MIPs are synthesized in the presence of the analyte or template molecule. After template removal, molecular recognition sites are left in the MIP structure, ready to re-recognize the analyte during SPR experiments [40].

SPR allows monitoring of binding kinetics in real time, measuring the change in resonance angle as the analyte binds or detaches from the active sites of the MIP [38]. The temporal sensitivity curves obtained allow calculation of the association (k_a) and dissociation (k_d) rates, essential for determining K_D of the system [38, 39, 41]. In this technique, different kinetic models can be used, such as the Langmuir model or more complex models that consider cooperative interactions or multiple binding sites, to fit the experimental data and extract kinetic parameters. For instance, Safran et al. [42] used SPR to selectively and sensitively detect Penicillin G (PENG) antibiotics in aqueous solutions and milk samples using MIP. For this purpose, a PENG-imprinted poly(2-hydroxyethyl methacrylate-*N*-methacryloyl-(L)cysteine methyl ester-gold nanoparticles-*N*-methacryloyl-L-phenylalanine methyl ester (MIP-AuNPs) nanosensor was prepared. Control experiments were performed using the nanosensor without the addition of AuNPs (MIP) to examine the effect of the incorporated gold nanoparticles (AuNPs) to enhance the surface plasmon resonance signal response. Furthermore, to evaluate the imprinting efficiency, the non-imprinted nanosensor (NIP-AuNPs) was designed using the same polymerization recipe except the addition of PENG molecule. As a result, the MIP-AuNPs nanosensor was 9.87 times more selective for the target PENG molecule than amoxicillin and 16.78 times than ampicillin. Furthermore, antibiotics in real milk sample were determined using the nanosensor.

Among the main advantages of the SPR technique, its sensitivity in detecting small changes in the interaction between the analyte and the MIPs stands out, even at very low concentrations [39, 41]. This is especially useful for systems where the analyte is a molecule of biological interest, such as a disease biomarker. In addition, the technique allows monitoring interactions in real time, without the need for labels or markers, preserving the integrity of the analyte-MIP interaction and allowing a more faithful study of the kinetic properties [40]. However, the reproducible immobilization of carbon-based MIPs on the surface of the SPR sensor can be challenging. Small variations in the surface preparation or in the MIP functionalization can lead to significant variations in the experimental results. In addition, the SPR technique detects changes in the refractive index close to the surface, which means that interferences from other species present in the solution or changes in the composition of the medium can influence the results. This can be particularly problematic in complex systems such as biological samples, especially if binding sites exhibit significant heterogeneity or if non-specific interactions occur [40].

11.4.3 Isothermal titration calorimetry (ITC)

The ITC technique is the main strategy applied to study binding kinetics and obtain thermodynamic data in systems involving interactions between molecules, including the recognition of MIPs to target molecules [43, 44]. ITC directly measures the heat changes associated with molecular interactions, allowing a detailed analysis of both the forces governing binding and the dynamics of release of molecules from the active sites of MIPs. For instance, Wang et al. [45] used ITC to evaluate the binding kinetics of bovine serum albumin MIPs (BSA-MIPs) incorporating orientation-controllable binding ligands via sequential thiol-ene click chemistry and thiol-ene-amine conjugation. This study demonstrated that the optimal thiol-ene-amine conjugates and binding ligands played a crucial role in determining the rebinding ability and selectivity of the MIPs. Among the tested MIPs, those containing pyrrolidinyl ligands exhibited the highest adsorption capacity of 352 ± 22 mg g^{-1} and an impressive imprinting factor of 4.72. These promising results were further explored using computational simulations. Molecular docking studies revealed preferential binding free energies and hydrogen bonding interactions between BSA residues and the thiol-ene-amine conjugates. Furthermore, the pyrrolidinyl ligand motif promoted entropy-favorable affinity through hydrophobic interactions with the BSA template, as evidenced by ITC thermodynamics. As a result of these favorable interactions, the MIPs exhibited exceptional adsorption specificity for BSA over competing proteins, suggesting that MIPs with orientation-controllable conjugates can serve as effective alternatives for use with real biological samples [45].

One of the main advantages of ITC is the ability to provide a complete characterization of the interactions between the analyte and the active sites of the MIP, without the need for labels or markers [46]. By measuring the heat evolved or absorbed during the titration of the analyte in the MIP, ITC allows the direct determination K_D, ΔH, ΔS, and ΔG associated with the binding process. ITC is also extremely useful for studying the release kinetics of MIPs [43]. During the titration process, the technique can monitor how the release of the analyte from the imprinted sites alters the energy balance of the system. This is particularly relevant for applications where precise control of analyte release is essential, such as in controlled drug delivery systems [44]. Another important aspect is that ITC allows the analysis of interactions under near-physiological conditions, such as temperature and pH, which is crucial for the application of MIPs in biological environments. The technique is applicable to a wide range of systems and can provide valuable information on how different factors, such as MIP composition or the presence of interferents, affect binding and release efficiency [46].

However, despite its important advantages, the ITC has some limitations [43]. One of the main ones is the need for relatively large amounts of material for measurements, which can be challenging for MIPs, especially if the material synthesis is complex or expensive. Furthermore, the sensitivity of ITC may be insufficient to detect low-affinity interactions, which may limit its application to systems where the interactions between the analyte and the MIP are not strong [46]. In addition, ITC provides an average of the interactions across the system, which may mask heterogeneities in the MIP binding sites [46]. In systems where the MIP surface is heterogeneous, with binding sites of different affinities, ITC may not capture the full complexity of the interaction process, making data interpretation difficult.

11.4.4 Atomic force microscopy (AFM)

Binding kinetics in systems involving MIPs, including carbon-based ones, can also be evaluated using AFM [47]. This technique is characterized by its versatility and allows surface characterization at the nanometric scale, providing important information about the interactions between the analyte and the MIP active sites [48].

AFM has the ability to provide nanometer-resolution images of the MIP surface, allowing detailed characterization of its morphology, roughness, and topography [47]. This is crucial to understand how molecular recognition sites are distributed on the surface of carbon-based MIPs and how the structure of the MIP influences the interaction with the analyte. Furthermore, AFM can be used to directly measure interaction forces between the MIP and the analyte [48]. This is accomplished through force spectroscopy experiments, where a probe functionalized with the analyte, or a specific functional group is brought closer to and removed from the MIP surface. The attractive and repulsive forces that occur during this process provide quantitative information about the binding strength, affinity, and dissociation energy of the complexes formed. This ability to quantify interactions at the molecular level is particularly advantageous for studying binding kinetics in MIPs, allowing the analysis of parameters that would be difficult to obtain by other techniques. Çakir et al. reported the development of a novel quartz crystal microbalance (QCM) sensor utilizing MIP nanofilms for the detection of amitrole. In this study, AFM was employed as a key technique to characterize the binding interactions between the MIP and the target molecules. The results demonstrated that the imprinted nanofilms exhibited greater sensitivity to the target molecule compared to the non-imprinted counterparts. Additionally, the adsorption kinetics were analyzed at various concentrations of the target molecule. Competitive adsorption experiments were conducted to evaluate the selectivity of the imprinted nanofilms against other pesticides.

AFM technique also allows experiments to be performed in different environments, such as in solutions or under controlled environmental conditions, which is useful for studying how binding kinetics vary with changing conditions, such as pH, ionic strength, or the presence of other solutes [49]. The flexibility of the technique makes it suitable for a wide range of applications, from the detection of biomolecules to the study of small molecule interactions in MIPs [49].

However, despite its advantages, AFM has some important limitations. One of the main ones is the complexity and time required to set up and perform experiments, especially compared to other binding kinetics techniques, such as SPR [47, 49]. Proper preparation of the probe and MIP surface is crucial to ensure accurate results, and any variation in preparation can introduce artifacts that complicate data interpretation. Furthermore, interpreting force spectroscopy data in AFM can be challenging. Measurements are sensitive to a variety of factors, such as probe geometry, approach and retraction speed, and environmental conditions [47]. This can make it difficult to obtain reproducible and quantitative data, especially in complex systems where multiple intermolecular interactions are present. Another significant limitation is that AFM often provides information about local interactions, which may not be representative of the overall behavior of the MIP [47–49]. This can be problematic when studying MIPs with significant heterogeneity in their binding sites, where properties measured at one point may not reflect behavior at other locations on the MIP surface. The need to scan multiple areas to obtain an overview can be time-consuming and can introduce additional variability into the data.

11.4.5 Nuclear magnetic resonance spectroscopy (NMR)

NMR spectroscopy is another analytical technique that allows the evaluation of binding kinetics and thermodynamic data in systems involving carbon-based MIPs [51]. NMR allows detailed analysis of the interactions between analytes and the active sites of MIPs, providing information on both the structure and molecular dynamics, essential for understanding the binding and release processes [52]. Through titration experiments, where the analyte is gradually added to the MIP, NMR can measure changes in chemical shifts or coupling constants, which are directly related to binding affinity and intermolecular forces. Analysis of these data can provide information on KD, as well as on ΔH and ΔS involved in the binding process [53]. However, NMR has low sensitivity compared to other techniques, such as ITC or SPR. This can make it difficult to detect low-affinity interactions or interactions at very low concentrations, which are common in some MIP systems. Another limitation is the complexity of the data obtained by NMR [53-55]. Interpreting NMR spectra, especially in systems where multiple interactions can occur simultaneously, can be challenging and require advanced modeling to extract reliable kinetic and thermodynamic parameters [55]. Furthermore, NMR is a technique that requires relatively large sample amounts and can be sensitive to material purity, which can be a challenge in studies involving synthetically complex or difficult-to-purify MIPs. This can limit the technique ability to provide a complete view of the system.

11.4.6 Computational simulations

Computational simulations have become an increasingly important tool in the study of kinetics and thermodynamics involving molecular interactions between MIPs and target molecules [55]. These simulations allow modeling and predicting the behavior of MIPs at the molecular level, providing detailed insights into the interaction mechanisms between the polymer active sites and the analytes [56]. This includes the analysis of different MIP conformations, variations in the analyte structure, and changes in environmental conditions such as temperature and pH [55–57]. Molecular dynamics simulations, for example, allow real-time observation of the analyte binding and release process, providing information on the kinetics of the system [56]. In addition, methods such as density functional theory can be used to calculate thermodynamic parameters, such as binding energies and free energy profiles, which are essential to understand the stability and selectivity of MIPs [56]. However, computer simulations also have limitations [58]. The accuracy of the results strongly depends on the quality of the models and parameters used, which may not fully capture the complexity of real systems [58]. Furthermore, simulations can be computationally intensive and require significant resources, especially for large systems or when high accuracy is sought [56].

11.5 Conclusion

Kinetic and thermodynamic studies, together with the application of mathematical models, are essential to provide insights into the interactions between recognition sites and analytes, allowing a mechanistic understanding of the binding between MIPs and target molecules. Several techniques are employed to obtain these kinetic and thermodynamic parameters including FT-IR, SPR, ITC, AFM, and NMR. Combining these experimental techniques with computational simulations allows a comprehensive analysis of carbon-based MIPs, offering detailed information into binding efficiency, selectivity, and the thermodynamic parameters governing analyte-polymer interactions. These data are crucial to optimize MIP performance in applications such as sensors, controlled release systems, and purification processes.

Abbreviations

AFM	Atomic force microscopy
AuNPs	Gold nanoparticles
BSA	Bovine serum albumin
BSA-MIPs	Bovine serum albumin molecularly imprinted polymers
FT-IR	Fourier transform infrared spectroscopy
ITC	Isothermal titration calorimetry
Mb	Myoglobin
MIPs	Molecularly imprinted polymers
NHMAm	*N*-Hydroxymethylacrylamide
NIP-AuNPs	Non-imprinted nanosensor
NMR	Nuclear magnetic resonance
PENG	Penicillin G
QCM	Quartz crystal microbalance
SPR	Surface plasmon resonance
ΔG	Gibbs free energy
ΔH	Enthalpy
ΔS	Entropy

References

[1] Bel B J. J. (2019). Molecularly imprinted polymers, *Chemical Reviews*. 119(1), 94–119.

[2] Tan H., Wang Y., Zhao Z., Guan Y., & Zhang Y. (2024). Molecularly imprinted polymer for selective insulin capture and separation, *Polymer*. Guildf, 307, 127290. Elsevier, https://doi.org/10.1016/j.polymer.2024.127290.

[3] Himshweta V. N., Trehan N., & Singh M. (2024). Molecularly imprinted polymers in the analysis of chlorogenic acid: A review. *Analytical Biochemistry*, 694. 115616.

[4] Tang M., Wan J., Wang Y., Ye G., Yan Z., Ma Y., & Sun J. (2023). Insights into molecular imprinting polydopamine in-situ activating peroxydisulfate for targeted removal of refractory organic pollutants: Overlooked N site. *Applied Catalysis B*, 334. 122852.

[5] Alizadeh T., Atashi F., & Ganjali M. R. (2019). Molecularly imprinted polymer nano-sphere/multi-walled carbon nanotube coated glassy carbon electrode as an ultra-sensitive voltammetric sensor for picomolar level determination of RDX. *Talanta*, 194. 415–421.

[6] Wei P., Zhu Z., Song R., Li Z., & Chen C. (2019). An ion-imprinted sensor based on chitosan-graphene oxide composite polymer modified glassy carbon electrode for environmental sensing application. *Electrochimica Acta*, 317. 93–101.

[7] Mokhtari P., & Ghaedi M. (2019). Water compatible molecularly imprinted polymer for controlled release of riboflavin as drug delivery system. *European Polymer Journal*, 118. 614–618.

[8] Khaneghah S. S., Sohrabi N., & Mohammadi R. (2023). Synthesis and characterization of magnetic molecularly imprinted polymer of polydopamine/graphene oxide as drug carrier for rivastigmine. *Journal of Molecular Liquids*, 391. 123238.

[9] Ezzati R. (2023). A new insight into the pseudo-second-order model and the physical meaning of its rate constant in adsorption, *Journal of Dispersion Science and Technology*. 46, 1–8.

[10] Wang J., & Guo X. (2020). Adsorption kinetic models: Physical meanings, applications, and solving methods. *Journal Hazard Materials*, 390. 122156.

[11] Lagergreen S. (1907). Zur Theorie der sogenannten Adsorption gelöster Stoffe. *Zeitschrift Für Chemie Und Industrie der Kolloide*, 2. 15–15.

[12] Xu L., Zhao Z.-X., Huang Y.-A., & Zhu Q.-J. (2020). Preparation of chitosan molecularly imprinted polymers and the recognition mechanism for adsorption of alpha-lipoic acid. *Molecules*, 25. 312.

[13] Pooresmaeil M., & Namazi H. (2020). Application of polysaccharide-based hydrogels for water treatments, *Hydrogels Based on Natural Polymers*. Elsevier, Amsterdam, Netherlands, 411–455.

[14] Gautam R. K., & Chattopadhyaya M. C. (2016). Kinetics and equilibrium isotherm modeling: Graphene-based nanomaterials for the removal of heavy metals from water, *Nanomaterials for Wastewater Remediation*. Elsevier, Amsterdam, Netherlands, 79–109.

[15] Surikumaran H., Mohamad S., & Sarih N. (2014). Molecular imprinted polymer of methacrylic acid functionalised β-cyclodextrin for selective removal of 2,4-Dichlorophenol. *International Journal of Molecular Sciences*, 15. 6111–6136.

[16] Campos N. F., Barbosa C. M., Rodríguez-Díaz J. M., & Duarte M. M. (2018). Removal of naphthenic acids using activated charcoal: Kinetic and equilibrium studies. *Adsorption Science & Technology*, 36. 1405–1421.

[17] Weber W. J., & Morris J. C. (1963). Kinetics of adsorption on carbon from solution. *Journal of the Sanitary Engineering Division*, 89. 31–59.

[18] Yoro K. O., Amosa M. K., Sekoai P. T., Mulopo J., & Daramola M. O. (2020). Diffusion mechanism and effect of mass transfer limitation during the adsorption of CO_2 by polyaspartamide in a packed-bed unit. *International Journal of Sustainable Engineering*, 13. 54–67.

[19] Foroughirad S., Haddadi-Asl V., Khosravi A., & Salami-Kalajahi M. (2021). Effect of porogenic solvent in synthesis of mesoporous and microporous molecularly imprinted polymer based on magnetic halloysite nanotubes. *Materials Today Communications*, 26. 101780.

[20] Öter Ç., & Zorer Ö. S. (2021). Molecularly imprinted polymer synthesis and selective solid phase extraction applications for the detection of ziram, a dithiocarbamate fungicide. *Chemical Engineering Journal Advances*, 7. 100118.

[21] William Kajjumba G., Emik S., Öngen A., Özcan H K., & Aydın S. (2019). Modelling of adsorption kinetic processes – errors, theory and application, *Advanced Sorption Process Applications*. Ongen, Atakan, IntechOpen Limited, London, United Kingdom, 1–19.

[22] Musah M., Azeh Y., Mathew J., Umar M., Abdulhamid Z., & Muhammad A. (2022). Adsorption kinetics and isotherm models: A review. *Caliphate Journal of Science and Technology*, 4. 20–26.

[23] Yakub E., Agarry S. E., Omoruwou F., & Owabor C. N. (2020). Comparative study of the batch adsorption kinetics and mass transfer in phenol-sand and phenol-clay adsorption systems. *Particulate Science and Technology*, 38. 801–811.

[24] Denderz N., Lehotay J., Čižmárik J., Cibulková Z., & Šimon P. (2012). Thermodynamic study of molecularly imprinted polymer used as the stationary phase in high performance liquid chromatography, *Journal of Chromatography A*. 1235, 77–83.

[25] Yu M., Wang S., Wang N., Wu Y., Li L., & Lu X. (2024). Magnetic dummy-template molecularly imprinted polymer with an ultrathin shell for selective enrichment of geosmin in water sample. *Journal of Industrial and Engineering Chemistry*, 136. 349–358.

[26] Azizi-Khereshki N., Mousavi H. Z., Dogaheh M. G., Farsadrooh M., Alizadeh N., & Mohammadi A. (2023). Synthesis of molecularly imprinted polymer as a nanosorbent for dispersive magnetic

micro solid-phase extraction and determination of valsartan in biological samples by UV–Vis Spectrophotometry: Isotherm, kinetics, and thermodynamic studies. *Spectrochimica Acta Part A, Molecular and Biomolecular Spectroscopy*, 296. 122656.

[27] Feng M., Yu S., Wu P., Wang Z., Liu S., & Fu J. (2021). Rapid, high-efficient and selective removal of cationic dyes from wastewater using hollow polydopamine microcapsules: Isotherm, kinetics, thermodynamics and mechanism. *Applied Surface Science*, 542. 148633.

[28] Hooshyar S. P., Panahi H. A., Moniri E., & Farsadrooh M. (2021). Tailoring a new hyperbranched PEGylated dendrimer nano-polymer as a super-adsorbent for magnetic solid-phase extraction and determination of letrozole in biological and pharmaceutical samples. *Journal of Molecular Liquids*, 338. 116772.

[29] Alafnan S., Awotunde A., Glatz G., Adjei S., Alrumaih I., & Gowida A. (2021). Langmuir adsorption isotherm in unconventional resources: Applicability and limitations. *Journal of Petroleum Science and Engineering*, 207. 109172.

[30] Siddiqui S. I., Zohra F., & Chaudhry S. A. (2019). Nigella sativa seed based nanohybrid composite-Fe2O3–SnO2/BC: A novel material for enhanced adsorptive removal of methylene blue from water, *Environmental Research*. 178,108667.

[31] Mozaffari Majd M., Kordzadeh-Kermani V., Ghalandari V., Askari A., & Sillanpää M. (2022). Adsorption isotherm models: A comprehensive and systematic review (2010-2020). *Science of the Total Environment*, 812. 151334.

[32] Hu Q., & Zhang Z. (2019). Application of Dubinin–Radushkevich isotherm model at the solid/solution interface: A theoretical analysis. *Journal of Molecular Liquids*, 277. 646–648.

[33] Tiernan H., Byrne B., & Kazarian S. G. (2020). ATR-FTIR spectroscopy and spectroscopic imaging for the analysis of biopharmaceuticals. *Spectrochimica Acta Part A: Molecular and Biomolecular Spectroscopy*, 241. 118636.

[34] Fadlelmoula A., Pinho D., Carvalho V. H., Catarino S. O., & Minas G. (2022). Fourier Transform Infrared (FTIR) spectroscopy to analyse human blood over the last 20 years: A review towards lab-on-a-chip devices, *Micromachines*. 13(2), 187,13.

[35] Tiquia-Arashiro S., Li X., Pokhrel K., Kassem A., Abbas L., Coutinho O., Kasperek D., Najaf H., & Opara S. (2023). Applications of Fourier Transform-Infrared spectroscopy in microbial cell biology and environmental microbiology: Advances, challenges, and future perspectives. *Frontiers in Microbiology*, 14. 1304081.

[36] Gong Y., Chen X., & Wu W. (2024). Application of Fourier transform infrared (FTIR) spectroscopy in sample preparation: Material characterization and mechanism investigation. *Advances in Sample Preparation*, 11. 100122.

[37] Sullivan M. V., Dennison S. R., Hayes J. M., & Reddy S. M. (2021). Evaluation of acrylamide-based molecularly imprinted polymer thin-sheets for specific protein capture – A myoglobin model. *Biomedical Physics & Engineering Express*, 7. 45025.

[38] Sparks R. P., Jenkins J. L., & Fratti R. (2019,). Use of surface plasmon resonance (SPR) to determine binding affinities and kinetic parameters between components important in fusion machinery, *Methods in Molecular Biology*. 1860, 199–210.

[39] Patching S. G. (2014,). Surface plasmon resonance spectroscopy for characterisation of membrane protein-ligand interactions and its potential for drug discovery, *Biochimica Et Biophysica Acta: Biomembranes*. 1838, 43–55.

[40] Alberti G., Zanoni C., Spina S., Magnaghi L. R., & Biesuz R. (2023). Trends in Molecularly Imprinted Polymers (MIPs)-based plasmonic sensors, *Chemosensors* 2023. 11(2), 144.

[41] Drescher D. G., Drescher M. J., & Ramakrishnan N. A. (2009). Surface plasmon resonance (SPR) analysis of binding interactions of proteins in inner-ear sensory epithelia. *Methods in Molecular Biology*, 493. 323–343.

[42] Safran V., Göktürk I., Bakhshpour M., Yılmaz F., & Denizli A. (2021). Development of molecularly imprinted polymer-based optical sensor for the sensitive penicillin G detection in milk. *Chemistry Select*, 6. 11865–11875.
[43] Bastos M., & Velazquez-Campoy A. (2021). Isothermal titration calorimetry (ITC): A standard operating procedure (SOP). *European Biophysics Journal*, 50. 363–371.
[44] Lima Cavalcanti I. D., Xavier Junior F. H., Santos Magalhães N. S., & Lira Nogueira M. C. D. B. (2023). Isothermal titration calorimetry (ITC) as a promising tool in pharmaceutical nanotechnology. *International Journal of Pharmaceutics*, 641. 123063.
[45] Wang M., Zhou J., Zhang G., Liu Q., & Zhang Q. (2022). Pyrrolidinyl ligand motif-assisted bovine serum albumin molecularly imprinted polymers with high specificity. *Journal of Colloid and Interface Science*, 609. 102–113.
[46] Matilla M. A., Martín-Mora D., & Krell T. (2020). The use of isothermal titration calorimetry to unravel chemotactic signalling mechanisms, *Environmental Microbiology*. 22(8), 3005–3019.
[47] Last J. A., Russell P., Nealey P. F., & Murphy C. J. (2010). The applications of atomic force microscopy to vision science, *Investigative Ophthalmology & Visual Science*. 51(12), 6083–6094.
[48] Yip C. M. (2001). Atomic force microscopy of macromolecular interactions, *Current Opinion in Structural Biology*. 11(5), 567–572.
[49] Wei M., Zhang Y., Wang Y., Liu X., Li X., & Zheng X. (2024). Employing Atomic Force Microscopy (AFM) for microscale investigation of interfaces and interactions in membrane fouling processes: New perspectives and prospects, *Membranes (Basel)*. 14(2), 35.
[50] Roy K. S., Mazumder A., Goud D. R., & Dubey D. K. (2018). A simplistic designing of molecularly imprinted polymers for derivative of nerve agents marker using 31P{1H}NMR, *European Polymer Journal 2018*. 98, 105–115.
[51] Dubey A., Takeuchi K., Reibarkh M., & Arthanari H. (2020). The role of NMR in leveraging dynamics and entropy in drug design, *Journal of Biomolecular NMR*. 74(10–11), 479–498.
[52] Dalvit C., Gmür I., Rößler P., & Gossert A. D. (2023). Affinity measurement of strong ligands with NMR spectroscopy: Limitations and ways to overcome them, *Progress in Nuclear Magnetic Resonance Spectroscopy*. 138–139, 52–69.
[53] Ben-Tal Y., Boaler P. J., Dale H. J. A., Dooley R. E., Fohn N. A., Gao Y., García-Domínguez A., Grant K. M., Hall A. M. R., Hayes H. L. D., Kucharski M. M., Wei R., & Lloyd-Jones G. C. (2022). Mechanistic analysis by NMR spectroscopy: A users guide. *Progress in Nuclear Magnetic Resonance Spectroscopy*, 129. 28–106.
[54] Krishnan V. V. (2019). Molecular thermodynamics using nuclear magnetic resonance (NMR) spectroscopy. *Inventions*, 4. 1–15.
[55] Kong Y., Wang N., Ni X., Yu Q., Liu H., Huang W., & Xu W. (2016). (2016) Molecular dynamics simulations of molecularly imprinted polymer approaches to the preparation of selective materials to remove norfloxacin. *Journal Applied Polymer Science*, 133. 1–11.
[56] Decherchi S., & Cavalli A. (2020). Thermodynamics and kinetics of drug-target binding by molecular simulation, *Chemical Reviews*. 120(23), 12788–12833.
[57] Alauddin M., Parvez M. M., & Matin M. A. (2023). A computational study of microhydrated N-Acetyl-Phenylalaninylamide (NAPA): Kinetics and thermodynamics, *Computational Molecular Bioscience 2023*. 13, 63–74.
[58] Peluso P., & Chankvetadze B. (2024). Recent developments in molecular modeling tools and applications related to pharmaceutical and biomedical research, *Journal of Pharmaceutical and Biomedical Analysis 2024*. 238, 115836.

Part 5: **Various aspects and future prospects of molecular imprinting-based smart carbon nanomaterials**

Jonathan Tersur Orasugh*, Lesego Tabea Temane,
and Suprakash Sinha Ray*

Chapter 12
Green aspects of carbon-based molecularly imprinted polymers

Abstract: Carbon-based (C-based) molecularly imprinted polymers (MIPs) have attracted wide relevance due to their eco-friendliness and extensive application scope. Regarding this, the greener features of C-based MIPs, from synthesis routes to environmentally friendly characteristics and applications, are discussed in this chapter. Utilizing carbon sources that are renewable in the synthesis of C-based MIPs and environment-friendly solvents that could minimize the ecological footprint associated with conventional polymerization techniques are discussed. Intrinsic properties of carbon materials, including tunable porosity, high surface area, and excellent chemical stability, increase the performance and sustainability of MIPs in environmental remediation, sensing, and drug delivery. The green attributes of the C-based MIPs are emphasized in this chapter by their use in developing sustainable materials for future environmental and biomedical niches as well as their challenges.

Keywords: Carbon-based (C-based), molecularly imprinted polymers (MIPs), green synthesis, eco-friendly, sustainable applications

12.1 Introduction

Over the last decade, molecularly imprinted polymers (MIPs) have emerged as robust tools in material science due to their ability to selectively recognize, similar to natural receptors. The incorporation of carbon (C)-based materials into MIP systems has recently opened new perspectives for enhancing performance and broadening their

***Corresponding author: Jonathan Tersur Orasugh**, Department of Chemical Sciences, University of Johannesburg, Doorfontein, Johannesburg 2028, South Africa; Centre for Nanostructures and Advanced Materials, DSI-CSIR Nanotechnology Innovation Centre, Council for Scientific and Industrial Research, Pretoria 0001, South Africa, e-mails: calunivjto@gmail.com
***Corresponding author: Suprakash Sinha Ray**, Department of Chemical Sciences, University of Johannesburg, Doorfontein, Johannesburg 2028, South Africa; Centre for Nanostructures and Advanced Materials, DSI-CSIR Nanotechnology Innovation Centre, Council for Scientific and Industrial Research, Pretoria 0001, South Africa, e-mail: rsuprakas@csir.co.za
Lesego Tabea Temane, Department of Chemical Sciences, University of Johannesburg, Doorfontein, Johannesburg 2028, South Africa; Centre for Nanostructures and Advanced Materials, DSI-CSIR Nanotechnology Innovation Centre, Council for Scientific and Industrial Research, Pretoria 0001, South Africa.

https://doi.org/10.1515/9783111503325-012

range of applications [1, 2]. Meanwhile, significant emphasis has been placed on developing methods for synthesizing and applying such materials using eco-friendly or greener techniques. In this trend-setting area of material science, the search for greener, more sustainable approaches has become essential; hence, C-based MIPs turn out to be a promising class of materials owing to their inherently structural versatility as well as their eco-friendliness [1].

This chapter intends to discuss the synthesis of C-based MIPs by incorporating aspects of green chemistry. Researchers have developed innovative pathways to create MIPs with a minimum environmental footprint using such sustainable carbon sources as biomass-derived precursors or renewable polymers [1–3]. Such synthetic strategies not only reduce dependence on fossil-based resources but also result in a decrease in carbon emission, thus conforming to the principles of sustainability and circular economy [1, 3, 4]. Besides, due to the structural intricacies of C-based MIPs, they exhibit exceptional properties applied in wide fields [1]. From environmental remediation, and sensing technologies to drug delivery systems (DDSs), the versatility of these materials is endless. Their ability to bind target molecules in complex matrices will not only bring up their activity but also drastically reduce the waste generated, fostering a greener approach toward materials design [1, 3]. In parallel, the environmental impact of C-based MIPs should be considered with great care. Although their green synthesis pathways give a promising future, the end-of-life aspects require a holistic approach [1, 3], through detailed lifecycle analysis and ecotoxicological studies, whereby researchers try to determine the environmental fate of these materials to ensure that their use is in line with the general aim of sustainability.

C-based materials like carbon nanotubes (CNTs), graphene, and/or carbon dots attracted great interest owing to their extraordinary properties, which include mechanical strength, a large surface area, and superior electrical conductivity [1]. After their incorporation into MIP systems, these carbon materials enhance binding capacity, and selectivity, and enable better signal transduction in sensing applications [1]. However, most conventional syntheses and applications of C-based MIPs entail energy-intensive procedures, the use of hazardous chemicals, and nonbiodegradable components, raising many fears over their impact on the environment.

Anastas and Warner introduced the concept of "green chemistry" at the end of the 1990s: a conceptual guide that helped develop chemical processes and products more harmoniously [5]. Their ideas have recently been applied to C-based MIPs since performance in research studies started finding an optimal balance with environmental concerns. This chapter discusses various approaches adopted to develop eco-friendlier C-based MIPs, from synthesis to application. The first focus area is thus the development of green synthesis approaches for C-based MIPs. Traditional syntheses usually require organic solvents and toxic cross-linking agents, which raise environmental and health concerns. Very recently, some methodologies that were assessed include utilizing eco-friendly solvents like supercritical CO_2 [1, 3, 5], and/or synthesis in aqueous (water-based) routes [1, 3, 4]. Such techniques not only diminish the ecological footprint of synthesis but also present MIPs with improved properties and performances.

Again, the selection of monomers and cross-linkers plays a decisive role in determining the green credentials and attributes of C-based MIPs. Biobased and biodegradable materials have been increasingly used as alternatives to the traditional petroleum-derived polymers. For instance, Luo et al. [6] revealed the application of monomers generated from cellulose in combination with graphene oxide to create biodegradable MIPs for environmental remediation. Also, the incorporation of cross-linkers such as genipin can weaken C-based MIP toxicity and enable them to behave as biocompatible materials [7]. In addition, the energy consumption in the course of synthesis and application of MIPs is another important factor for green C-based MIPs development. Hence, energy-saving methodologies have emerged, such as microwave-assisted and photopolymerization techniques, to complement conventional thermal polymerization [8]. All these methods contribute not only to energy savings but also to reduced reaction time and better control over the nature of the synthesized MIP structure. The C-based MIP application in environmental monitoring and remediation shows its possibilities for fulfilling the needs of sustainable development. These materials are used for the effective selective removal of contaminants from wastewater and air, heavy metals, and organic pollutants, including emerging ones such as microplastics [1, 3, 4]. The high selectivity and adsorption capacity of C-based MIPs, combined with their potential for regeneration and reuse, make them attractive alternatives to traditional adsorbents.

In green analytical chemistry, C-based MIPs have been employed to create very sensitive and selective environmental monitoring sensors [1, 3, 4]. These sensors mostly require a minimum of sample preparation and can be performed on-site, thereby minimizing laboratory procedures that may be very complicated and generate chemical waste [1, 3, 4]. Integrating C-based MIPs into smartphone-based detection platforms has even further expanded their prospects toward applications in broad environmental monitoring [9].

While research in the area is continuing, there are several challenges and opportunities that exist and are solely discussed, but not limited to, in this chapter. For instance, the scalability of the green synthesis methods in C-based MIPs needs attention if such methods are ever to see industrial applications [1, 3, 4]. Moreover, the long-term environmental fate and possible ecotoxicological effects of these materials need further investigation to establish their sustainability over their life cycle.

This chapter reviews the status, recent developments, and prospects of green C-based MIPs. It provides insight, from synthesis and application to further environmental implications, into how such materials could potentially contribute toward a more sustainable future. While emphasizing the green aspects of C-based MIPs, it is crystal clear that besides being environmentally friendly, the introduction of the principles of sustainability into materials design and application will contribute to performance and new functionalities in development.

12.1.1 The need for green and sustainable synthesis of MIPs

The imperative for green and sustainable synthesis of C-based MIPs arises from the escalating concerns over environmental pollution and the imperative to mitigate the ecological footprint of synthetic processes [1, 3, 4]. Conventional methods of synthesizing polymers usually involve toxic solvents, energy-consuming processes, and unregenerative resources, which all increase environmental destruction [1, 3, 4]. On the other hand, in green synthesis methods, the approaches focus on nontoxic precursors, renewable feedstocks, and nontoxic reaction conditions to reduce any negative impact on the environment [1, 3, 4].

Green synthesis of the C-based MIPs follows the principles of green chemistry, hence reduction of wastes, energy-saving, and avoidance of hazardous chemicals [1, 3–5]. The precursors that can be reckoned as sustainable are bio-based monomers, natural polymers, and carbon sources from biomass, which negate the use of raw materials arising from petrochemical-derived feedstock; hence, their carbon footprint has a positive reduction effect and raises resource efficiency. Besides, employing green solvents, like water, or other benign organic solvents further reduces VOC emissions and lessens the environmental impacts due to poor solvent disposal methods [1, 3–5]. Furthermore, green synthesis routes can also bring opportunities for MIP production intensification and energy efficiency by means of carbon footprint reduction. This enables one to perform very rapid and effective polymerization under soft conditions, thereby reducing energy usage and reaction time by using methods such as microwave-assisted synthesis, sonochemistry, and photochemical reactions [1, 3, 4]. Besides, the incorporation of catalytic processes or enzymatic catalysts enables the use of milder reaction conditions and higher selectivity of reactions, hence further improving the sustainability profile of C-based MIP synthesis [1, 3, 4].

The imperative for the green and sustainable synthesis of carbon-based MIPs goes well beyond environmental aspects to include many economic and social benefits. Upon embracing green chemistry principles, researchers will find ways to develop cost-efficient and scalable routes of synthesis, thus promoting the commercial viability of MIP-based technologies [1, 3–5]. This leads to more sustainable manufacturing processes that are innovative, technologically advanced, economically active, and provide a scientific contribution to environmental care [1, 3–5].

The synthesis of C-based MIPs should, therefore, be carried out in an environmentally friendly way with consideration of environmental protection, resource efficiency, and societal well-being [1, 3, 4]. With the green chemistry concept and advanced synthesis methodologies, various ways might be opened toward the preparation of environmentally benign MIPs with sundry applications in environmental remediation, sensing, and resource recovery.

12.1.2 Sustainable and/or green sources of raw materials for MIP synthesis

Most MIP syntheses depend on feedstocks derived from petrochemical sources, and their preparation contributes to environmental degradation and the depletion of natural resources [4, 10]. In the last decade, however, bio-based feedstock alternatives for MIP synthesis have been researched due to the opportunities provided by green chemistry technologies.

12.1.2.1 Bio-based monomers

Bio-based monomers have emerged as a promising green alternative in the synthesis of MIPs and, in conjunction with green chemistry concepts, can address environmental issues associated with petroleum-based ingredients [1, 3, 4, 10]. For this reason, such a shift toward bio-based monomers represents a significant step in further greening MIPs while maintaining or even enhancing their characteristics.

Bio-based monomers come from sustainable and renewable resources such as plants, animals, and microorganisms [1, 3, 4, 10, 11]. Natural functional monomers, having a hydrophilic exterior and a hydrophobic cavity, like aptamers, gelatin, chitosan, cellulose, sodium alginate, and cyclic oligosaccharides, are bio-based monomers that readily undergo polymerization under moderate conditions [10]. They possess a number of advantages over the use of synthetic monomers, including being biodegradable and less toxic, with a lower carbon footprint. When bio-based monomers are used in MIP synthesis, they enhance the general sustainability of the material life cycle, which starts with raw material extraction and continues through eventual disposal or degradation. Among the most studied bio-based monomers for MIP synthesis is chitosan, a polysaccharide obtained from the exoskeleton of crustaceans [1, 3, 4, 10]. It can be said that MIPs with a chitosan matrix exhibit excellent performance for various applications, especially environmental remediation and biosensing. Deng et al. [12], in fact, used a chitosan-based MIP for the removal of red X-3B dye in wastewater with great adsorption and reusable capacity. Because of its rich functional group abundance, molecular imprinting of chitosan is quite effective, followed by target recognition, as reflected in Figure 12.1a.

Other promising bio-based monomers include cellulose (especially in its nano form) and its derivatives [13]. Cellulose, being the most abundant biopolymer on Earth, would be a very sustainable source for the synthesis of MIPs [14]. Yoon et al. [13], described cellulose nanocrystal-templated TiO_2 as a new bio-based template for preparing MIPs that showed enhanced mechanical properties and improved thermal stability. The resulting MIPs exhibited excellent photodegradation of rhodamine B, methylene blue, and methyl orange while maintaining the biodegradability of cellulose-based materials intrinsically.

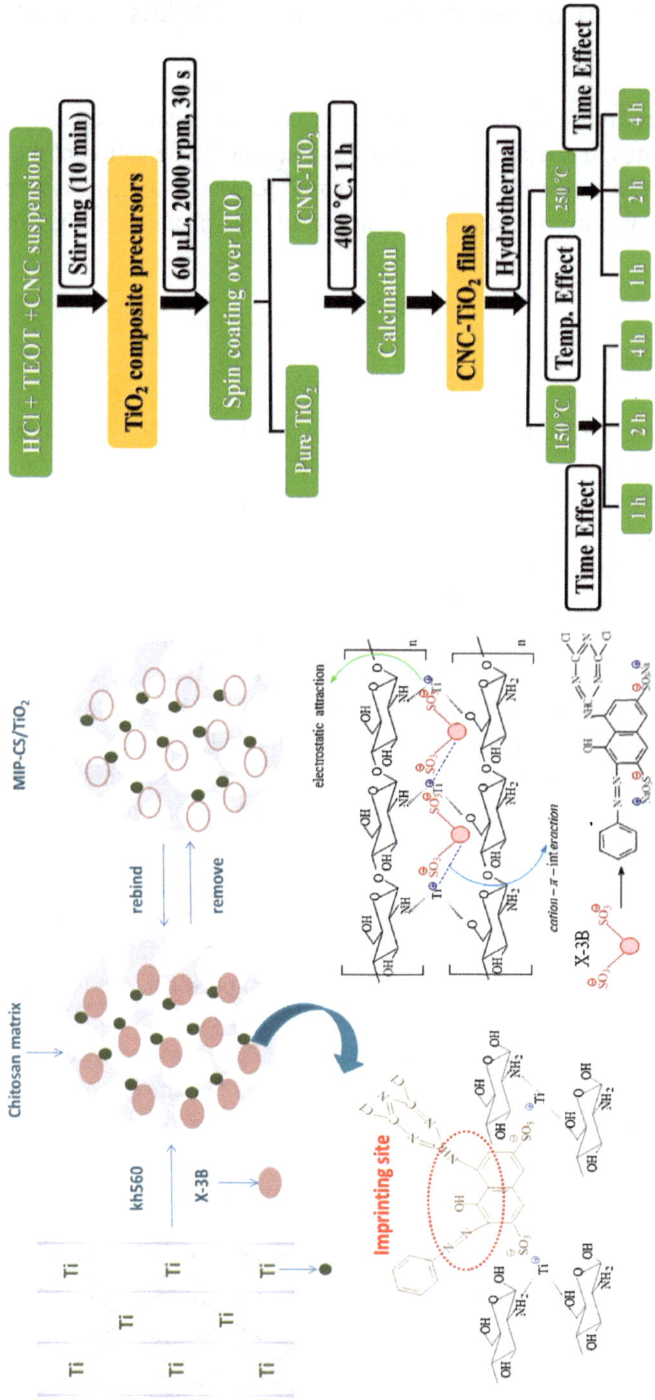

Figure 12.1: (a) Schematic representation of the synthesis procedure and adsorption mechanism. Reproduced with permission from Deng et al. [12], Copyright 2017, Elsevier Science Ltd. (b) Process diagram for the synthesis of TiO_2 films. Reproduced with permission from Yoon et al. [13], Copyright 2023, Elsevier Science Ltd.

The films were made by spin-coating a CNC template onto an ITO substrate using a mixed sol and then calcining it for one hour at 400 °C, as shown in Figure 12.1b. The produced films were then heated to 150 and 250 °C and hydrothermally treated for varying lengths of time (1, 2, and 4 h) in a pure aqueous solution. Besides, as a bio-based monomer in MIP synthesis, the complex biopolymer lignin from plant cell walls has also been attracting attention [1].

While the application of bio-based monomers in the synthesis of MIPs allows for many real benefits, drawbacks still persist. These include the assured quality and reproducibility properties of bio-based monomers, as they usually strongly depend on sources and extraction methods. Moreover, the optimization of the imprinting process in a bio-based monomer might differ from the approach usually followed when using traditional synthetic monomers.

Bio-based monomers provide a sustainable, green source for raw materials in MIP synthesis and offer a route to more environmentally friendly, possibly better functional materials. Further innovations in the development of bio-based MIPs will be paramount to advancing the general concept of sustainability in materials science and chemistry as research in this area continues to evolve.

12.1.2.2 Natural polymers

Natural polymers extracted from renewable biological materials represent an emerging class of "green" alternatives to classical synthetic polymers in MIP synthesis and are based on polysaccharides, such as cellulose and chitosan, or proteins like gelatin and albumin [1, 3, 4, 10]. Natural polymers tend to be biocompatible, biodegradable, and versatile, making them candidates of choice to be used as the basis for C-MIP design for different applications.

12.1.2.3 *Biomass-derived carbon sources*

Preparation using biomass-derived carbon sources such as lignin, cellulose, and biomass pyrolysis/hydrothermal product-based approaches offers an efficient alternative method for incorporating an environmentally sustainable approach into the improvement of the properties of MIPs [15]. Various examples of sourcing biomaterial-based sustainable C-materials are shown in Table 12.1. Carbon material originating from biomass is typically characterized by several peculiarities that distinguish its structural and surface characteristics: for example, a very big surface area and porosity can increase MIPs' target analyte selection ability along with its adsorption capacity. Biochar, derived from biomass pyrolysis, combines high surface area with rich surface chemistry and is thus suitable for developing MIPs with higher adsorption capacity and selectivity. Biomass-sourced activated carbon undergoes activation processes for increased porosity and surface area, which significantly enhances the performance of MIPs concerning target molecule capture. These materials reduce dependence on fossil-based carbon sources and add value related to waste valorization and carbon sequestration.

Table 12.1: Green source-derived carbon for MIPs-C systems.

Source	Carbon synthesis approach	Solvent used (if applicable)	Process temperature and time	References
Sweet potato peels (SPP)	Hydrothermal method	Di-water	200 °C for 3 h	[16]
Cedrus	Hydrothermal method	Di-water	180 °C for 12 h	[17]
Citric acid (CA)	Hydrothermal method	Di-water/ethylenediamine	200 °C for 3 h	[18]
Glucose	Hydrothermal method	Di-water/phosphoric acid	90 °C for 0.5 h	[19]
Starch	Hydrothermal method	Di-water/phosphoric acid	160 °C for 1 h	[20]
Mango peels	Hydrothermal method	Di-water/phosphoric acid	200 °C for 4 h	[21]
CA	Reflux	Branched polyethylenimine	170 °C for 2 h	[22]
Urea	Recrystallization + calcination	Urea	Recrystallization (urea solution (0.5 g/mL) heated for 24 h at 80 °C); calcination (crystallized urea heated at 550–1,200 °C for 6 h)	[23]
Grape juice	Hydrothermal method	Water	180 °C for 8 h	[24]
CA/urea	Hydrothermal method	DMF	180 °C for 8 h	[25]

Such biomass-derived carbon materials can be functionalized by introducing specific chemical groups that promote the process of imprinting and, hence, increase the capability for recognition by MIPs. Functionalization with carboxyl, hydroxyl, and/or amino groups, for example, favors interaction between MIPs and target analytes [3, 4, 10, 15, 22, 26]. Biomass-derived C-based MIPs have, therefore, been increasingly tried in the monitoring of environments, biomedical diagnosis, and safety of food.

12.1.2.4 Green solvents

Green solvents also play a crucial role in developing C-based MIPs to meet the basic principles of green chemistry, limiting environmental impact as much as possible [27]. Aqueous catalysis and other chemical reactions benefit from the use of water and other green solvents [4, 27, 28], supercritical fluids (such as supercritical CO_2) [4, 29], ionic liquids (ILs) [4], deep eutectic solvents (DES) [4, 27, 30], and fluorous media [27, 31]. These are just a few of the many applications of GC that include these benefits. Mechanochemistry [32], magnetic field-assisted synthesis [33], hydro/solvothermal reactions [4, 16–18], microwave- and ultrasound-assisted procedures [4, 34], and UV irradiation [1, 4, 10] are further environmentally friendly techniques. These solvents are derived from renewable resources and possess low toxicity; many are biodegradable, making them an attractive substitute for traditional organic solvents.

In the synthesis of MIPs, ILs and DES represent two of the most promising green solvents [4, 27]. Due to the tunability of their properties, ILs can be tailored for better solubilization of monomers and templates in order to facilitate imprinting processes, thus yielding improved efficiency and selectivity of MIPs. Besides this, ILs normally possess negligible vapor pressure and therefore result in lower volatile organic compound (VOC) emissions in the synthesis of polymers [27, 30].

DES, prepared from natural compounds such as choline chloride and urea, is an environmentally friendly and beneficial medium for the preparation of MIPs. DESs are associated with certain benefits, including high thermal stability, low cost, and simplicity of preparation. Due to their specific solvation characteristics, DESs might allow for well-defined recognition sites within the MIP matrix and, hence, improve their performance in sensing and separation applications [27].

The synthesis of C-based MIPs by using green solvents supports not only sustainability but also improves the functional properties of the resulting polymers. This approach falls within the growing demand for environmentally benign materials in different industrial and research applications.

12.1.2.5 Waste-derived feedstocks

The utilization of waste-derived feedstocks, such as agricultural residues, food waste, or industrial by-products for MIP synthesis, represents a sustainable approach toward the valorization of waste materials and the reduction of environmental pollution. Waste-derived feedstocks can act as inexpensive and abundant sources of raw materials for MIP synthesis, encouraging resource efficiency and supporting the circular economy. Waste feedstock such as SPP [16], *Cedrus* [17], waste tea [35], and mango peels [21] have been adopted as carbon for the preparation of C-MIPs. One excellent instance of waste feedstock utilization is presented in the effort by Sun et al. [21], where the inventors utilized mango skin as green feedstock for the carbon quantum dot (CQD) preparation. They used the method outlined in Figure 12.2 for the synthesis of CQDs@MIPs with fluorescent properties using sol-gel hydrolysis. First, via the hydrothermal process, mango peels were used as carbon sources to create the CQDs. Then, the MIPs were further coated by modifying the CQDs with APTES. Upon hydrolysis of the alkoxy groups of APTES, Si-O-Si condensation occurs, combining with CQDs to develop a cross-linked hybrid structure through silsesquioxane. Finally, with luminous material CQDs, template mesotrione, functional monomer APTES, cross-linker TEOS, and catalyst $NH_3 \cdot H_2O$, CQDs@MIPs were developed. Once mesotrione was added back to the CQDs@MIPs, there was a dampening of fluorescence. The impact of the template molecule, an effective electron acceptor, primarily occurred due to electron-hole recombination at the CQDs@MIPs interface. Recombination between electrons and holes after they took off the template molecule leads to the recovery of fluorescence [21].

By leveraging sustainable and green sources of raw materials, researchers can develop environmentally benign MIPs with a reduced environmental footprint and enhanced sustainability. These eco-friendly approaches not only mitigate the environmental impact of MIP synthesis but also contribute to the development of sustainable solutions for environmental remediation, sensing, and other applications.

12.2 Green chemistry and its relevance to green aspects of C-based MIPs

Green chemistry, often called sustainable chemistry, is a scientific methodology that aims to develop chemical processes and products by decreasing environmental impact while using less resources with a stress on the health and safety of humans [4, 36]. It can also be characterized by a set of principles for developing benign and sustainable practices in the life cycle of chemical products and processes. Anastas and Warner's list of the 12 green chemistry principles provides a guideline toward these goals presented in Figure 12.3 and also discussed below in short [4, 36].

Waste prevention: Waste should always be avoided rather than treated or cleaned up later. Green MIP synthesis methods aim at waste minimization by employing competent reaction pathways, reducing the usage of hazardous reagents, and maximizing the yield of desired products [4].

Atom economy: Synthetic procedures should be designed to maximize the inclusion of all elements into the finished product. This is because green synthesis of C-based MIPs focuses on maximizing the use of raw materials while reducing the generation of by-products or non-functionalized polymers, thus improving atom economy [4].

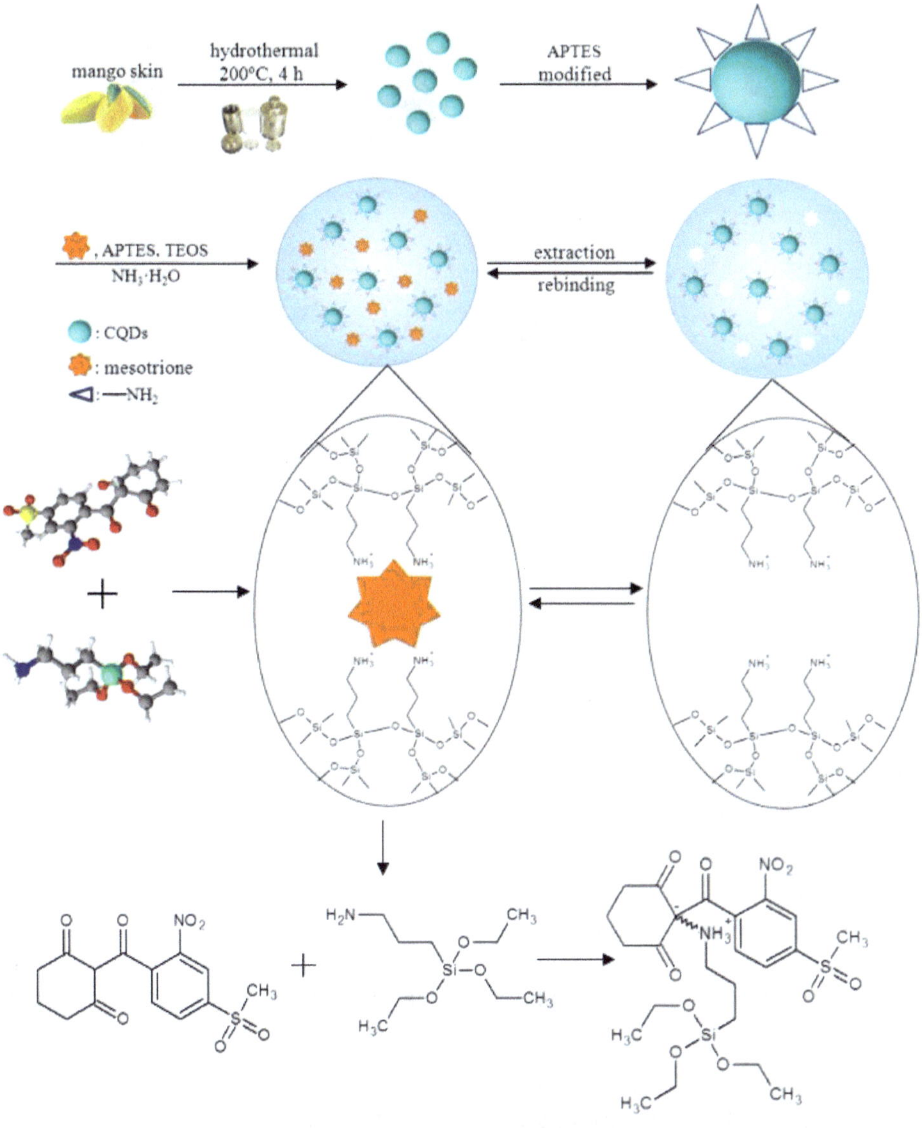

Figure 12.2: Illustration of the preparation of CQDs@MIPs. Reproduced with permission from Sun et al. [21], Copyright 2019, Springer Nature.

Less hazardous chemical syntheses (avoidance of harmful synthetic techniques): Synthetic processes, which produce compounds with low to no toxicity to the environment and human health, should be employed whenever feasible. This means green synthesis routes for C-based MIPs prioritize the use of nontoxic or benign solvents,

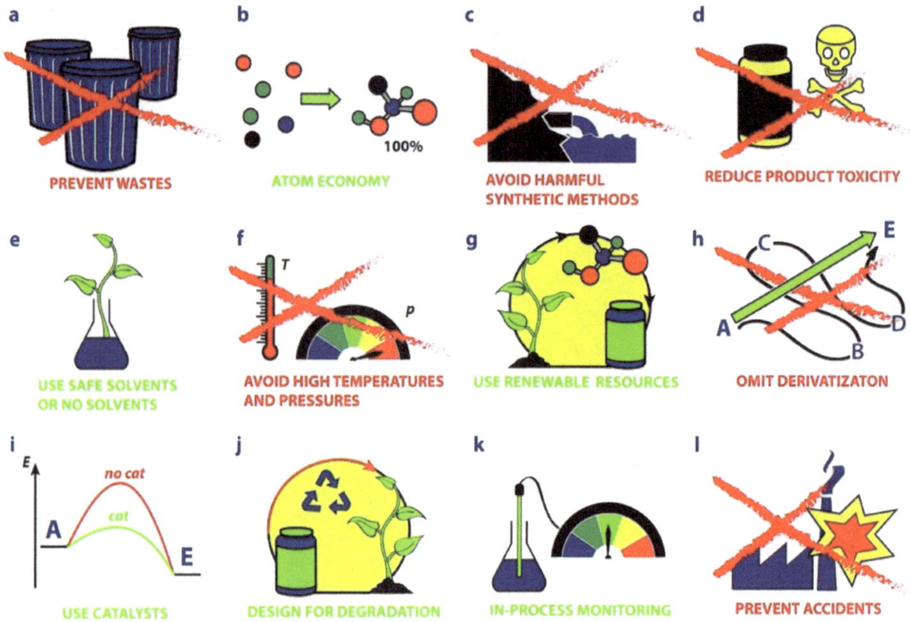

Figure 12.3: The 12 principles of green chemistry. Reproduced with permission from Unterlass et al. [36]. Copyright 2016, Wiley.

monomers, and cross-linkers to minimize the environmental and health risks associated with chemical synthesis [4].

Designing safer chemicals (toxicity reduction): It is important to reduce the toxicity of chemical products without compromising their functionality. In other words, green MIP design entails the selection of monomers and functional groups that are biocompatible and environmentally benign to ensure that the produced polymers will not present any hazards in real applications for biomedical or environmental purposes [4].

Safer solvents and auxiliaries (no solvent(s) or safe solvent utilization): Wherever possible, auxiliary substances such as solvents or separation agents should be avoided or used in a manner that is innocuous. This is because the green synthesis of C-based MIPs explores the usage of solvent-free or solvent-light methods and substitutes traditional organic solvents with eco-friendly alternatives like supercritical CO_2 or water that can minimize environmental impact [4].

Design for energy efficiency (avoidance of extreme temperatures and/or pressure): The energy demand of the chemical process should be assessed and reduced regarding environmental impact and economic considerations. In order to minimize energy usage, green approaches for MIP synthesis use mild reaction conditions and energy-efficient methods of synthesis, for example, microwave-assisted synthesis and renewable feedstock. It may also entail avoiding high temperatures and/or pressures [4].

Use of renewable feedstocks (utilization of sustainable/renewable resources): Whenever possible, renewable raw materials and feedstocks should be employed. The green synthesis of C-based MIPs develops the use of biomass-derived, renewable feedstocks for monomers, cross-linkers, and carbon sources, contributing to resource efficiency and lessening dependency on limited fossil fuels [4].

Reduce/avoid derivatives (derivatization omission): When feasible, unnecessary derivatization, which includes blocking groups, protection/deprotection, and short-term changes to physical or chemical processes should be minimized or avoided, as it requires extra reagents and can produce waste. Green MIP synthesis methods aim to streamline reaction pathways and minimize the use of protecting groups or derivatization steps, thereby reducing waste production and reagent consumption.

Catalysis (utilization of catalysts): Catalytic reagents are preferred over stoichiometric ones. Green MIP synthesis focuses on the development of catalytic processes or enzyme-mediated reactions, which offer improved reaction selectivity with shorter reaction times, thereby reducing the consumption of stoichiometric reagents, ensuring resource efficiency, and minimizing waste.

Design for degradation (reduced toxicity): Chemical products should be designed to break down into innocuous degradation products at the end of their useful lives and not persist in the environment. This is because green MIP design considers the biodegradability and environmental fate of polymers, aiming to develop materials that degrade into nontoxic by-products, thus minimizing environmental persistence and pollution.

Real-time analysis for pollution prevention (in-process monitoring): Real-time, in-process monitoring and control of any synthesized compounds far in advance of the emergence of potentially dangerous intermediates is not yet fully implemented. In any green method developed for the synthesis of MIPs, in situ process monitoring methods, spectroscopic techniques, or chromatography are incorporated to study the reaction progress, optimizing such reaction conditions toward prevent the possible formation of hazardous intermediates or by-products.

Inherently safer chemistry for accident prevention: In order to reduce the risk of chemical accidents such as leaks, explosions, and fires, the compounds and their forms employed in a chemical process should be carefully chosen. Green MIP synthesis favors inherently safer chemical processes and materials that reduce risks of accidents, hazardous releases, and environmental contamination associated with conventional modes of synthesis [4].

C-based MIPs may be designed, synthesized, and applied in accordance with the principles of green chemistry, using more environmentally benign molecular recognition materials that have very low environmental impact and superior performance. On the one hand, considering the developmental and application principles of C-based MIPs based on green chemistry will undoubtedly contribute to the achievement of its objectives: sustainability, environmental stewardship, and human health and safety.

12.3 C-based MIPs and their green synthesis methods

12.3.1 C-based MIPs

C-based MIPs represent an important class of synthetic polymers that, have garnered a lot of attention lately because of their high selectivity and sensitivity for use in sensing, separation, and catalysis [2, 37]. One should remember that MIPs are cross-linked polymers designed to possess specific detection sites tailored to a target molecule in a way similar to natural antibodies or enzymes. The carbon-based material graphene, CNTs, and activated carbon with MIP improve its potential due to extraordinary electrical conductivity with high mechanical strength and greater surface area of the base C-based materials as presented in Table 12.2 with reference to available literature. Incorporation in MIP resulted in great performances of such materials. For instance, sensor analysis showed that graphene MIPs possessed higher sensitivity along with rapid time response than regular MIPs. This is due to the excellent electronic properties of graphene, which could facilitate electron transfer and enhance signal transduction. Furthermore, the high surface area of graphene has supplied more active sites for the imprinting process and thus brought higher binding capacity and selectivity toward the target molecule [2, 37].

CNTs are another C-based material frequently used in MIPs. The exclusive one-dimensional (1D) structure of CNTs permits the production of MIPs with enhanced mechanical properties and improved thermal stability. CNTs can also promote the formation of a more uniform polymer matrix, leading to better imprinting efficiency and recognition capabilities. Research has demonstrated that CNT-based MIPs exhibit high selectivity and stability in harsh chemical environments, making them appropriate for a range of industrial uses [2, 26]. Activated carbon, with its porous structure and large surface area, is also used in MIPs synthesis. The presence of activated carbon enhances MIPs' adsorption capabilities, allowing for the capture and release of target molecules. This property is particularly beneficial in environmental monitoring and purification processes, where high adsorption capacity and rapid kinetics are essential. Studies have shown that activated C-based MIPs can effectively remove pollutants from water, highlighting their potential in environmental applications [58].

Table 12.2: Carbon-based MIPs, their green synthesis aspect/approaches, and applications.

Formulation	Carbon material	Green synthesis aspect/approach	Benefit of C-material inclusion	Applied niche	References
	Graphene (GR)				
GO@mSiO_2/organic-inorganic polymer (silsesquioxane network)	GO	Simple in situ polymerization/sol-gel technique and reusability	Enhanced adsorption, selectivity, and reusability	Environment (water purification)	[37]
TiO_2/GR-MIP	TiO_2/GR	In situ polymerization using water as a co-solvent	Enhanced photocatalytic efficiency	Environment (water purification)	[38]
N-GQDs/Fe_3O_4 @SiO_2/ IRMOF-1/MIP	N-doped GQDs	Adoption of CA as an N-GQD precursor and the adoption of green methanol for MIP synthesis	Enhanced adsorption, selectivity, and improved reusability	Exclusion of phenylureas from milk, tomato, soybean, cucumber, and radish	[39]
Ethylene glycol dimethyl acrylate (EGDMA)-MIP/IL-GR/glass carbon (GC)	IL-GR	Use of ILs	Enhanced adsorption, selectivity, and improved sensing	Electrochemical sensing of rutin	[40]
Polydopamine (PDA)/GO	GO	The use of green dopamine as a monomer, the use of hexamethylenetetramine (HMTA) as a cross-linker instead of carcinogenic formaldehyde, and the use of NH_3 generated from HMTA as a catalyst in situ	Enhanced adsorption, selectivity, and environmental friendliness	Chlorogenic acid exclusion from water	[41]

(continued)

Table 12.2 (continued)

Formulation	Carbon material	Green synthesis aspect/approach	Benefit of C-material inclusion	Applied niche	References
	C-nanotubes (CNTs)				
CNTs@ILs@MIPs	CNTS	Use of ILs	Fast kinetics and superior adsorbability	Excellent recovery of seven PFCAs (perfluoroalkyl carboxylic acids) from water	[42]
MWCNTs@BSA-MIPILs	MWCNTs	Use of ILs	Enhanced adsorption, selectivity, and reusability	BSA purification from cow-calf serum	[43]
Perfluoroalkyl carboxylic acid (PFCAs)-CNTs-ILs (CNTs@ILs@MIPs)	CNTs	Use of ILs	Enhanced adsorption, selectivity, and reusability	Wastewater remediation	[44]
CNTs@Cu-MIP	Cu@CNTs	In situ polymerization/sol-gel technique using green methanol as a solvent	Enhanced 100% excellent selectivity, adsorption, and reusability	Sensors for chloramphenicol detection	[45]
MWCNT/ polyvinylpyrrolidone (PVP)/ polyacrylic acid (PAA)	MWCNT	Ultrasonication-assisted synthesis using green EtOH solvent, along with biopolymers for electrospinning-MIP and UV-light-induced in situ cross-linking	High selectivity index	Environmental sensor	[46]

SWCNT-MIP	SWCNT	The application of DESs, room-temperature ionic liquids (RTILs), and binary green porogen systems	Good bioavailability and controlled release	Controlled/sustained delivery of fenbufen	[47]
	Activated carbon				
MIP-graphitic carbon nitride (MIP (PMMA)-C_3N_4)	C_3N_4 (C_3N_4–COOH)	The adoption of urea for the synthesis of C-based MIPs	Enhanced photocatalytic efficiency	Environment (water purification)	[23]
	Carbon dots				
Cdots/cellulose nanocrystals (CNCs)-MIP	Cdots	The adoption of wood-derived CNCs and CA-derived Cdots, along with water as a solvent.	Improved adsorption and sensitivity of fluorescence detection	Adsorption and sensitive detection of tetracycline	[6]
Cdots-MIPs	Cdots	The utilization of ultrasonication-based synthesis and water as a solvent, as well as microwave-assisted synthesis	Enhanced fluorescence detection sensitivity and adsorption	Detection of tetracycline in honey	[48]
Cdots@ZIF-8@MIP	Cdots	The utilization of methanol for the synthesis of ZIF-8, the adoption of CA for the synthesis of Cdots, as well as the adoption of ultrasonication-assisted synthesis of the MIPs	Improved adsorption and sensitivity of fluorescence detection	Identification of quercetin in genuine *Ginkgo biloba* extract capsules with extreme precision and focus	[49]
MIP@Si-Cdots	Si-Cdots	Adoption of CA and water for the synthesis of Cdots	Enhanced fluorescence detection sensitivity and adsorption	Adsorption and detection of acetamiprid	[50]

(continued)

Table 12.2 (continued)

Formulation	Carbon material	Green synthesis aspect/approach	Benefit of C-material inclusion	Applied niche	References
Cdots-MIP	Cdots	Water and CA adoption for the production of Cdots, as well as the adoption of methanol as a green solvent for MIP synthesis	Strongly defined mesoporous architectures showed excellent affinity, stability, and scaffolding capacity for coordination sites	Optical detection of tryptamine in meat	[51]
Cdots-MIP	Cdots	Adoption of ammonia solution as a sustainable solvent for the synthesis of MIP, as well as water and CA for Cdot manufacture	Enhanced detection, adsorption, and sensitivity of fluorescence detection	Caffeic acid: optical detection and determination	[52]
Cdots/Fe_3O_4@MIPs	Cdots	The utilization of a methanol-acetic acid green solution for template removal and the adoption of green solvents, such as ammonia and methanol, in the preparation of Cdots/Fe_3O_4@MIPs	Robust fluorescence along with magnetic characteristics for trinitrophenol separation and detection	Detection and recovery of trinitrophenol in spiked tap water and river water	[53]
Cdots-MIPs	Cdots	Adoption of the biopolymer (chitosan) as the monomer	Enhanced selectivity, lower LOD, and naked-eye fluorescence emission	Optical determination of perfluorooctane sulfonate	[54]
Cdots-MIPs	Cdots	Adoption of CA for the synthesis of Cdots and microwave-assisted synthesis of Cdots-MIPs	Improved high sensitivity and selectivity	Detection of tetracycline in milk	[55]
Cdots/SiO_2/MIP	Cdots	Adoption of PEG and IL as precursors for the synthesis of Cdots and Cdots-MIPs	Increased selectivity and sensitivity	Lysozyme detection and cell imaging	[56]

2-(Perfluorooctyl)ethyl acrylate-based MIP-carbon nanodots ($FMIPs@CH_3NH_3PbI_3@CNDs$)	C-nanorods	Utilization of grape juice-derived Cdots and water as the solvent for the synthesis of carbon nanodots	Increased selectivity and sensitivity	Photoelectrochemical detection of cholesterol in human serum	[24]
Fluorinated-MIPs (FMIPs)-$CH_3NH_3PbI_3$@carbon nanodots ($CH_3NH_3PbI_3@CNDs$)	C-nanodots	Using water as the solvent and grape juice as a carbon precursor	Enhanced sensing upon the inclusion of C-nanodots in the synthesized C-based MIPs	Human serum cholesterol photoelectrochemical (PEC) sensor	[57]
N-Cdots	N-Cdots	Utilization of CA as the precursor for N-Cdot synthesis	Enhanced sensing upon the inclusion of C-nanodots in the synthesized C-based MIPs as well as reusability	Optical sensing of pyrraline in fatty foods	[25]

C-based molecularly imprinted polymers represent a resourceful and highly effective class of materials with a broad application. The incorporation of C-based materials like graphene, CNTs, and activated carbon into MIPs enhances their electrical, mechanical, and adsorption properties, leading to improved performance in sensing, separation, and environmental remediation. The advancements in this field hold great promise for the advancement of next-generation materials having customized functionalities and high specificity [26, 58].

12.3.2 Green synthesis methods for C-based materials for MIPs

In the pursuit of sustainable and eco-friendly MIPs, innovative green strategies have emerged to transform their synthesis and preparation [3]. By minimizing environmental impact and promoting resource efficiency, these green approaches offer promising avenues for the development of environmentally benign MIPs. Leveraging techniques such as solvent-free synthesis, microwave or ultrasound-assisted polymerization, bio-based initiators, and green template removal methods, researchers aim to create MIPs with a reduced carbon footprint and enhanced sustainability [3]. This section sets the stage for exploring eco-friendly methodologies reshaping the landscape of MIP synthesis toward a greener future.

12.3.2.1 Solvent-free or solvent-light synthesis

Solvent-free or solvent-light synthesis is a sustainable approach gaining traction for producing green carbon precursors for C-based MIPs [59]. This method minimizes or eliminates using organic solvents, reducing the ecological footprint and improving safety and cost-effectiveness. Solvent-free synthesis typically involves mechanochemical processes, where solid-state reactions are facilitated by grinding or milling the reactants. This approach not only enhances the reaction rate but also ensures uniform mixing of the precursors, leading to high-quality carbon materials. For instance, mechanochemical synthesis has been effectively employed to prepare carbonaceous materials from biomass precursors, leveraging the inherent carbon content of organic wastes such as cellulose, lignin, and chitosan [59, 60]. An excellent instance where a solvent-free technique was adopted for the preparation of C-dots in order to prepare C-based MIPs is presented in the work by Amiri et al. [61]. According to their methodology, thermal pyrolysis was used to create C-dots: 2 g of powdered citric acid were heated to 180 °C in a 50 mL beaker for approximately 15 min, or until the solution turned yellow, in order to create C-dot nanoparticles (NPs) [61]. Once at room temperature, the solution was removed.

Solvent-light synthesis, on the other hand, uses minimal amounts of solvents, often water, to aid in the dispersion and reaction of precursors [59]. This method benefits from the reduced solvent requirement while maintaining the advantages of traditional solution-based methods, such as ease of processing and scalability. An example includes the use of hydrothermal carbonization in an aqueous medium to convert biomass into carbonaceous materials, which are then used as templates for MIPs [59]. Several instances of this approach are discussed in Section 12.1.2.3 [16–18]. In one instance, water was used as a solvent for the preparation of carbon dots from citric acid precursor aimed synthesis of C-based MIPs for selective and high recognition and exclusion of ceftiofur sodium from wastewater systems [62]. Briefly stated, 10 mL of deionized water, 0.9607 g of citric acid, 334 μL of ethylenediamine, 0.005 mol of acrylamide, and 0.005 mol of ethylenediamine were combined, and then ultrasonication was applied for 10 min. The mixture was then moved to a 50 mL autoclave lined with Teflon and heated at 120, 150, 180, 200, and 220 °C for 12 h, respectively. A 0.22 μm membrane was employed to purify the brown aqueous solution that contained the CPDs-NH_2 after it had been chilled to 25 °C. To remove unreacted precursors, dialysis was carried out utilizing a 1,000 Da dialysis bag for approximately 24 h. At last, it was possible to utilize the CPDs-NH_2 solutions. Figure 12.4 displays an illustration of the synthesis of CPDs-NH_2 [62].

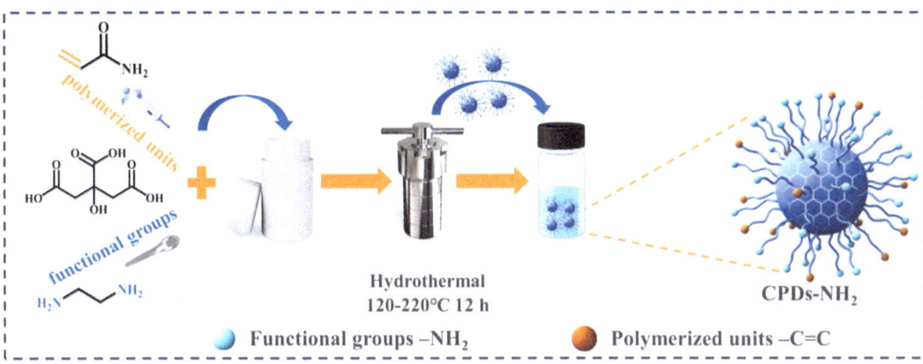

Figure 12.4: Illustration of CPDs-NH_2 synthesis. Reproduced with permission from Wang et al. [62], Copyright 2024, Elsevier Science Ltd.

Both approaches follow the principles of green chemistry, which emphasize waste minimization, energy efficiency, and the utilization of renewable resources. The resulting C-based MIPs from these sustainable processes exhibit excellent selectivity and sensitivity for target molecules, making them suitable for applications in sensing, catalysis, and environmental remediation [1, 5, 8, 10]. As such, solvent-free and solvent-light synthesis methods represent promising strategies for developing environmentally friendly and high-performance carbon materials for MIPs.

12.3.2.2 Microwave-assisted synthesis

Green carbon precursors for C-based MIPs can be produced efficiently and sustainably via microwave-assisted synthesis [1, 10, 59]. This technique leverages microwave radiation to rapidly heat the reactants, promoting uniform and accelerated reaction rates compared to conventional heating methods. The use of microwave energy enhances the synthesis process by reducing reaction times, lowering energy consumption, and often improving the quality and yield of the carbon materials. Such attributes make microwave-assisted synthesis particularly suitable for sustainable and green chemistry applications [1, 10, 59].

In the context of C-based MIPs, microwave-assisted synthesis can effectively convert biomass and other renewable resources into high-quality carbon precursors. Biomass materials such as cellulose, lignin, and chitosan are subjected to microwave irradiation in the presence of activating agents, resulting in the realization of porous carbon structures with high surface areas and tailored functionalities [1, 10, 59]. These carbon materials serve as excellent templates for MIPs, offering enhanced adsorption capacities and selectivity for target molecules. Furthermore, because microwave-assisted synthesis uses less solvent, it promotes the ideas of green chemistry, reducing waste generation, and ensuring energy efficiency. The rapid heating and uniform energy distribution in microwave-assisted processes lead to the formation of carbon materials with superior physicochemical properties, facilitating their application in environmental sensing, catalysis, and pollutant remediation [1, 10, 59]. By integrating this approach, researchers can develop high-performance and environmentally benign C-based MIPs, contributing to sustainable technological advancements. An example of microwave adoption in this regard is presented in the work by Hou et al. [55]. They used 10 mL of ultrapure water and vigorous stirring to dissolve 5 g of citric acid and 1 g of dipotassium hydrogen phosphate (K_2HPO_4). Following a 2-min sonication, the mixture was cooked for 2 min and thirty seconds at 750 W in a standard home microwave oven as illustrated in Figure 12.5. The powders were made pure by repeatedly precipitating ethanol with them and then centrifuging them once they had naturally cooled to room temperature. Using 120 mL of ethanol, pure C-Dots were distributed. Environmentally acceptable and nontoxic raw materials were employed in the manufacture of the C-Dots [55]. This process was immediately followed by the synthesis of C-based MIPs (CDs@MIPs) as illustrated in Figure 12.5. First, a 16 mL C-Dots ethanol solution was mixed with 1.2 mL of APTS. After 30 min of stirring, the mixture was cooked for 4 min at 375 W in a standard home microwave oven. The APTS-modified C-Dots (C-Dots-APTS) were made to cool gradually to ambient temperature before undergoing two rounds of purification using petroleum ether precipitation and then distributed in 40 mL of anhydrous ethyl alcohol for additional testing.

Using a standard sol-gel polymerization, the MIPs were created on the C-Dots coated with silica. While stirring the 12 mL C-Dots-APTS ethanol solution, 100 mg of Tc (template) was added to a 50 mL beaker. Next, injections of 1.5 mL of APTS and 1.2 mL

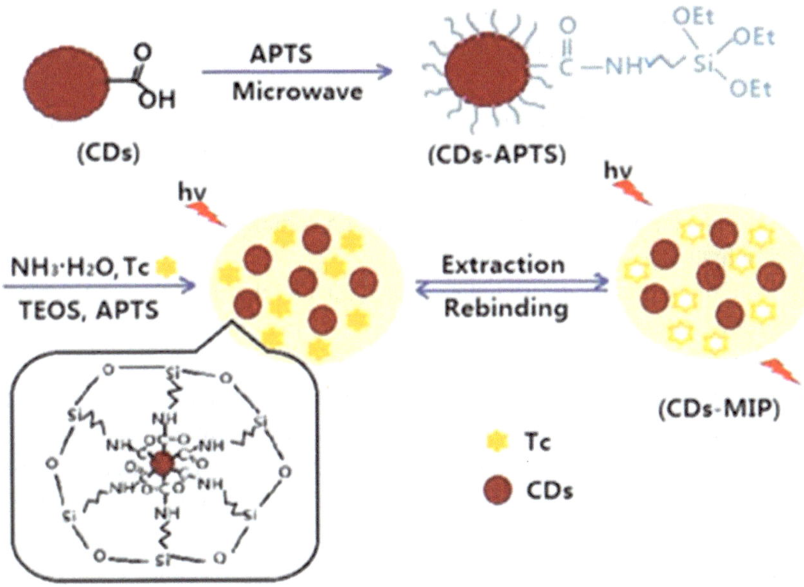

Figure 12.5: Illustration of the preparation of CDs@MIPs. Reproduced with permission from Hou et al. [55], Copyright 2016, Elsevier Science Ltd.

of TEOS were made, respectively. After adding 1.5 mL of 6% $NH_3 \cdot H_2O$ to the reaction beaker, the liquid was swirled for 30 min. The resultant mixture was then processed in a high-pressure microwave oven after being transferred to a microwave digestion tank. The following temperature settings were used for the microwave irradiation: starting at room temperature and increasing to 40 °C in 5 min, 40 °C to 60 °C in 3 min, and finally maintaining at 60 °C for 1 h [55]. After centrifugation, the finished products were collected and carefully cleaned in ultrapure water, methanol, and anhydrous ethanol with ultrasonic agitation, in that order. Until high-performance liquid chromatography confirmed that the washing solvent contained no Tc, methanol and 10% acidified methanol were used to eliminate the original Tc template molecules from the composites. All composites were then dried overnight at 60 °C to ensure complete evaporation of the solvent. Finally, the C-based MIPs were mixed with 40 mL of water before refrigeration at 4 °C prior to use. On the other hand, the synthesis of non-imprinted polymer capped C-Dots, referred to as C-based MIPs in this chapter, did not include any template molecules (Tc).

12.3.2.3 Ultrasound-assisted synthesis

Ultrasound-assisted synthesis is among the modern methods and green chemical approaches for developing carbon precursors applied in the development of C-based

MIPs [1, 10, 59]. With this approach, ultrasonic waves enhance the level of chemical transformation with improved levels of synthesis techniques. The ultrasounds introduce the cavitation effect, including very high localized temperatures and/or pressures, a consequence of collapse cavities showing a fast uniform mixing of all reactants inside without using strict chemicals or intense energy input [1, 10, 59]. This approach was extended for the other types of biomass, namely cellulose and lignin, to yield carbon materials with precursors of a high carbon content suitable for the fabrication of MIP at a high yield [1, 10, 59].

The benefits of the ultrasound-assisted approach include the reduction of reaction time, temperature decrease, and minimizing the usage of solvents in accordance with green chemistry principles. Local energy input by ultrasound reduces the thermal degradation of biomass to a minimum; hence, it preserves the natural structure of biomass and enhances the quality of the final product, carbon material [1, 10, 59]. Moreover, up-scaling would easily be attainable; it therefore holds great potential for further use in industries. Finally, for the MIPs synthesis, sonication methodologies provide one way by which carbon nanostructured materials containing higher surface and porous characteristics – essentially accountable for the choice, adsorption, and determination of analyte – might develop effectively. Better contact between the polymer matrix and the carbon precursor enhances sensitivity and selectivity in MIPs, which is highly effective in environmental monitoring, food safety, and biomedical diagnostics [1, 10, 59].

Thus, ultrasound-assisted synthesis represents a sustainable and cost-effective method for producing high-quality carbon precursors, which promotes the development of high-performance and environmentally friendly C-based MIPs.

12.3.2.4 Photochemical polymerization

Photochemical polymerization provides a green method for the synthesis of C-based MIPs, which employs light-induced polymerization to initiate and propagate chemical reactions [63]. In this process, UV or visible light is used to produce active species such as radicals or cations, which then induce the polymerization of monomers into cross-linked polymers. The most important advantage of photochemical polymerization is that it can occur under mild conditions, without requiring high temperatures or toxic initiators, hence in the scope of the green chemistry principles [63].

Regarding photochemical polymerization, this aspect allows for controlling the kinetics of polymerization processes and the distribution of the formed polymer network with high fidelity to the target molecules' imprint [63]. Furthermore, it introduces photo-initiators, which can decompose upon exposure to light into species that play a role in the rapid generation of polymers, resulting in the formation of well-defined matrices. This strategy has been successfully employed in the production of MIPs using green carbon precursors, including plant-derived phenolic compounds, which act as monomers or cross-linkers [63]. Moreover, photochemical polymerization can be performed either in aqueous systems or in solvent-free systems, making the technique even more environmentally compatible. This technique provides a low environmental impact due to the use of light – a clean and abundant source of energy – from to conventional thermal polymerization methods [63]. Additionally, the ability to control the process by adjusting the intensity and wavelength of light opens up possibilities for extensive variation in the properties of the MIPs obtained. Applications of photochemically synthesized C-based MIPs in selective sensing, catalysis, and environmental remediation have expressed high affinity and specificity toward target analytes [63].

Poma et al. [63] proposed a photochemical polymerization automated synthesizer using an immobilized solid-phase template approach. Operational parameters were specified for the new reactor with the aim of producing nanoMIPs in a controlled, repeatable, and automated manner.

The strategy outlined below, conceptually summarized in Figure 12.6a and including an example of an improved reactor design presented in Figure 12.6b, was selected as a result of the following factors influencing the reactor's design: MIP synthesis is best performed in an organic solvent at a temperature ranging from −30 °C to +10 °C, with the aim of enhancing complexation between the template and monomers. This method is ideal for initiating the polymerization reaction when low temperatures are required. Since this approach seeks to prevent the precipitation of polymers, controlled conditions during the NP production process must be maintained. To avoid contamination caused by a soluble template, the template must be immobilized. Ideally, the template should be recyclable to minimize production costs. Additionally, the reactor design must incorporate a mechanism to separate high-affinity NPs from low-affinity materials and unreacted monomers [63].

It has been possible to manufacture batches of NPs (30–400 nm) with narrow size distributions imprinted with proteins, a peptide ($d = 350$ nm, $K_d = 4.8 \times 10^{-8}$ M), vancomycin ($d = 250$ nm, $K_d = 3.4 \times 10^{-9}$ M), and melamine ($d = 60$ nm). In the instrument, the template was mounted on a column that was filled with glass beads. Because the process parameters were computer-controlled, manual involvement was limited. First, the reliable reuse of molecular templates has been demonstrated with MIP syntheses of ≥30 batches of nanoMIPs without any loss in their performance. While in nanoMIP production, the solid phase served as a template and as the medium for affinity separation, these nanoMIPs are template-free [63].

Thus, photochemical polymerization of green C-based MIPs for various applications seems to become an environment-friendly and low-cost approach.

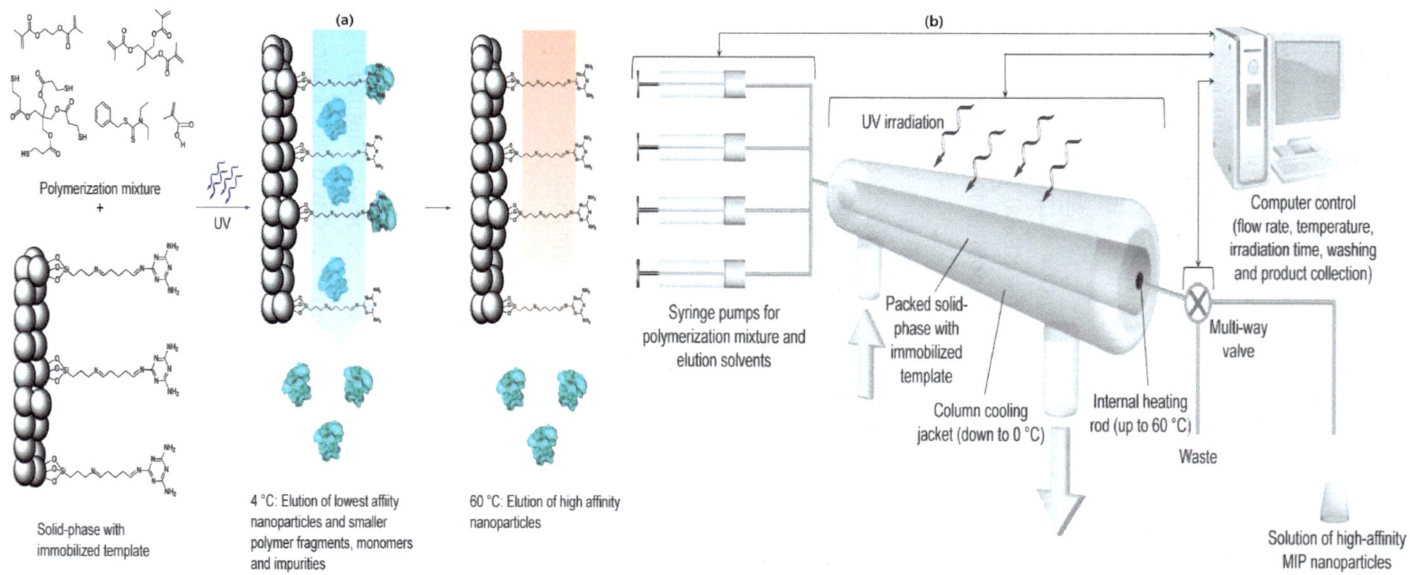

Figure 12.6: (a) Schematic representation of the solid-phase synthesis of MIP nanoparticles. The monomer mixture is injected into the column reactor with an immobilized template, and polymerization is initiated by UV irradiation. The low-affinity particles, as well as unreacted monomers, are eluted at low temperature. The temperature is then increased, and high-affinity particles are eluted from the column for collection. (b) Schematic diagram showing the mode of operation of the automated solid-phase MIP nanoparticle synthesizer. An engineering schematic is provided in the supporting information (b). Typical operational parameters using melamine as the immobilized target are as follows: operation time: 2 h per cycle; yield of high-affinity fraction: 6.6 ± 0.65 mg per cycle; column capacity: 23.5 g of derivatized glass beads (solid phase). Reproduced with permission from Poma et al. [63], Copyright 2013, Wiley.

12.3.2.5 Green template use and its removal methods

Green templates are nontoxic materials that can be used in the development of MIPs and bear less toxicity compared to traditional template molecules. Green templates may be derived from renewable feedstock, including natural polymers such as chitosan and alginate, and biomolecules such as proteins and DNA, which result in lower environmental impact during the synthesis process itself [1, 4, 10, 64]. Additionally, green templates adhere to the principles of green chemistry and contribute toward increasing the biodegradability and biocompatibility of the resulting MIPs [64].

The removal methods for the green template should be environmentally friendly and effective. A widely used method is enzymatic degradation, whereby, with the help of specific enzymes, the material of the template degrades without forming any hazardous by-products. For instance, protease enzymes can degrade protein-based templates, leaving behind the imprinted polymer matrix [1, 4]. Another technique is mild acid or base hydrolysis, which can effectively remove polysaccharide templates such as chitosan or alginate without damaging the polymer structure [1, 4]. There is also thermal decomposition, wherein the template is decomposed at elevated temperatures; this technique must be carefully controlled to avoid degradation of the polymer. Besides, the use of green templates and their harmless removal procedures minimizes the ecological footprint of MIPs. These approaches not only make the synthesis process sustainable but also broaden the applicability of MIPs in environmental monitoring, biological diagnostics, and food safety, owing to the nontoxicity and harmlessness of the materials used for application [1, 3, 4, 15].

12.3.2.6 Greener solvent (such as ILs and DESs)-assisted synthesis

The synthesis of MIPs using greener solvents, such as ILs and DESs, opens up the path to a more sustainable approach to polymerization, thus offering improved selectivity, reduced environmental impact, and better polymer properties [47, 65].

ILs, on the other hand, are salts that consist entirely of ions. Most typically, they have melting points below 100 °C. They exhibit specific properties such as high thermal stability, negligible vapor pressure, and tunable polarity, which makes them interesting as solvents in MIP synthesis. These ILs can serve in MIP synthesis processes as reaction media, template solvents, or porogens. Besides, ILs are able to dissolve both polar and non-polar template molecules, thus providing a homogeneous imprint within the polymer matrix. Due to the high solvation power of ILs, effective template removal after polymerization and accessibility of binding sites increase in the resulting MIPs. Furthermore, ILs can serve as reaction media in MIP polymerization, offering a homogeneous environment for monomer polymerization and template imprinting. These ILs provide benefits such as increased monomer solubility, better control of polymer morphology, and reduced solvent evaporation; this generally increases the yield of the polymerization and the performance of the MIPs [47]. Although ILs are usually considered greener than conventional organic solvents because of their low volatility and potential recyclability, there are concerns regarding the sustainability of their use in MIP synthesis, such as high costs and limited biodegradability. Cations and anions in ILs can be combined in such a way as to provide maximum compatibility between MIPs and target analytes. In addition, recyclability and reusability offer an advantage in the sustainable use of ILs in MIP synthesis.

The following steps were followed in an exceptional example for the production of MIP(PCz-co-PPy)/AuPd/GN-CNTs-IL/GCE: (1) the creation of a three-dimensional (3D) AuPd/GN-CNTs-IL nanocomposite and modification of GCE, (2) Cz and Py copolymerization on AuPd/GN-CNTs-IL/GCE in acetonitrile containing PCM, and (3) the exclusion of PCM template molecules, as illustrated in Figure 12.7a.
On the AuPd/GN-CNTs-IL/GCE, a conductive copolymer (poly(Cz-co-Py)) can be created by electropolymerizing the monomers Cz and Py, which contain pyrrole rings. The cyclic voltammograms corresponding to the electropolymerization of AuPd/GN-CNTs-IL/GCE with Cz (Figure 12.7b**i**), Py (Figure 12.7b**ii**), and Cz-co-Py (Figure 12.7b**iii**) are displayed in Figure 12.7. As the potential cycling was repeated, the current indications progressively increased, signifying the development of the conductive polymer. Since the MIP was generated in an organic solution in this instance, the electrochemical redox of template molecules was effectively inhibited, thereby maintaining the selectivity of the MIP-based sensor. Through H-bond and π-π interactions, PCM molecules combined with both Cz and Py to form a template-monomer complex. PCM molecules were extracted from the MIP film by employing an elution buffer solution to either weaken or replace their non-covalent interactions [65].

Other instances of the utilization of ILs for C-based MIPs synthesis have been also recently investigated by Li et al. [42], Ding et al. [43], and Lu et al. [40].

Conversely, DESs are combinations of two or more substances, usually a metal salt as a hydrogen bond acceptor and a hydrogen bond donor, such as a quaternary ammonium salt, which form a eutectic mixture with a melting point below the melting point of each individual component. DESs exhibit properties similar to ILs, including low vapor pressure, high viscosity, and tunable polarity [47]. DESs offer excellent solvation properties for diverse compounds, allowing for the effective dissolution of template molecules during MIP synthesis [47]. The ability to tailor the composition of DESs enables selective dissolution of target analytes, enhancing the specificity and selectivity of MIPs. Moreover, DESs can serve as reaction media for MIP polymerization, providing a homogeneous environment for monomer polymerization and template imprinting [47]. DES solvents offer advantages such as improved mass transfer, enhanced control over polymer morphology, and reduced environmental impact compared to conventional organic solvents. DESs are considered greener alternatives to ILs and conventional organic solvents because of their lower cost, biodegradability, and ease of synthesis from renewable feedstocks. However, further research is needed to optimize DES formulations and understand their environmental impact

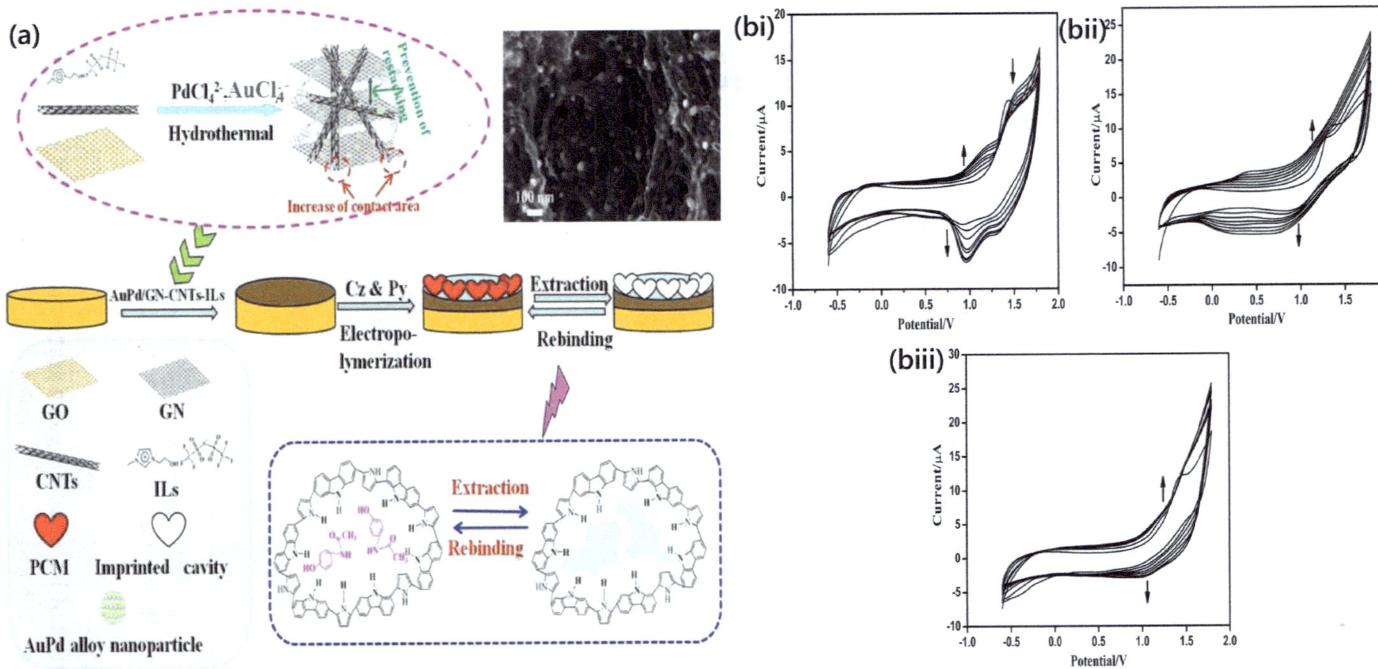

Figure 12.7: (a) Schematic illustration of the preparation of MIP(PCz-co-PPy)/AuPd/GN-CNTs-IL/GCE. Cyclic voltammograms corresponding to the electropolymerization of (bi) 0.50 mM Cz, (bii) 0.50 mM Py, and (biii) 0.07 mM Cz plus 0.43 mM Py in acetonitrile solution containing 25 mM tetrabutylammonium. Scan rate: 75 mV/s; number of cycle scans: 8. Reproduced with permission from Yang et al. [65], Copyright 2021, Elsevier Science Ltd.

throughout their life cycle. A good example in this regard is the use of DESs by Liu et al. [47], as shown in Figure 12.8. Fenbufen (FB)-imprinted monolithic composites were originally created by combining FB, SWCNT, and AIBN (20 mg) in a glass vial to create a pre-polymerization mixture. After that, [BMIM]BF4 and a certain volume of ChCl/EG were introduced. Lastly, the combination above was mixed with EDMA (1359 μL) and 4-VP (192 μL). A stainless-steel column of 100 mm × 4.6 mm I.D. was filled with the mixture following a 15-min ultrasonic session. A 60 °C water bath was used to submerge the column, which had its ends sealed, for a duration of 14 hours. To remove unreacted chemicals, the resulting SWCNT–MIP composite was flushed using acetonitrile (about 50 mL) with an HPLC pump. For the purpose of eliminating the FB molecules, a 9:1 v/v methanol/acetic acid mixture was utilized with a UV spectrophotometer. Even without FB, SWCNT was incorporated into the non-imprinted polymer using the same procedure [47].

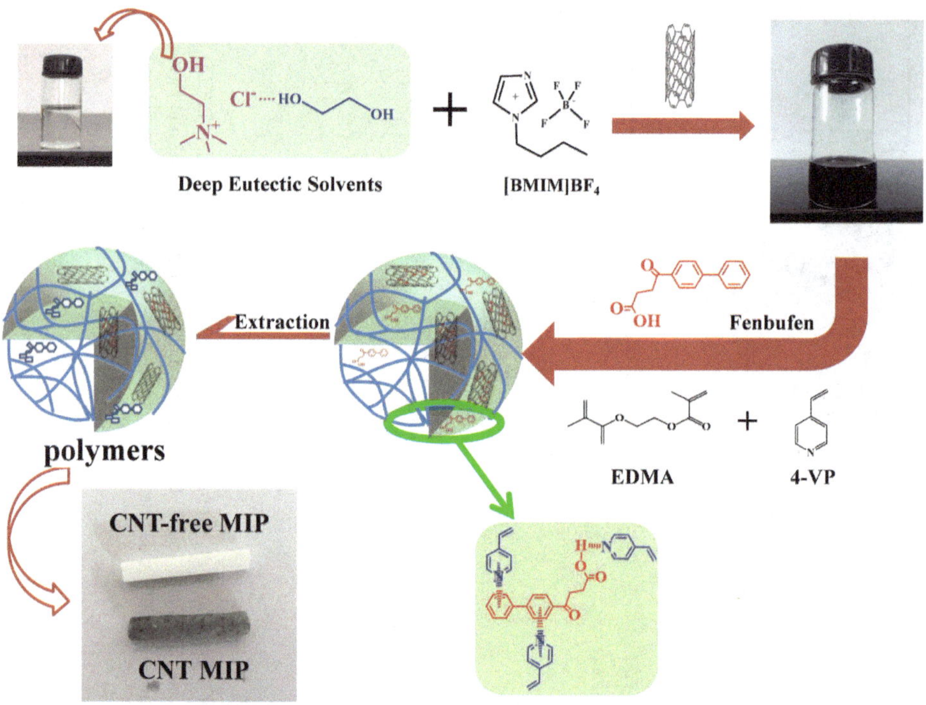

Figure 12.8: The schematic representation of the procedures for synthesizing SWCNT-MIP composites. Reproduced with permission from Liu et al. [47], Copyright 2018, Springer Nature.

Using the unique properties of ILs and DES, some green and sustainable syntheses of MIPs have been developed. The as-generated molecular recognition materials exhibit better performance with a lower environmental impact.

12.3.2.7 Supercritical fluid-assisted synthesis

Supercritical fluid-assisted synthesis of MIPs presents a promising way of manufacturing highly selective and effectual recognition materials used in a wide range of applications [3, 4, 10, 29, 59, 66]. Supercritical fluids like supercritical carbon dioxide (CO_2) ($ScCO_2$) can show inimitable properties at or near the critical point, such as high diffusivity, low viscosity, and tunable solvent power; thus, they are more attractive as alternatives to traditional organic solvents in polymer synthesis [3, 4, 10, 29, 59].

In supercritical fluid-assisted synthesis, CO_2 is pressurized above its critical point (31.1 °C, 73.8 atm) to achieve a state in which it exhibits both gas-like and liquid-like properties, as pictorially shown in Figure 12.9a [66].

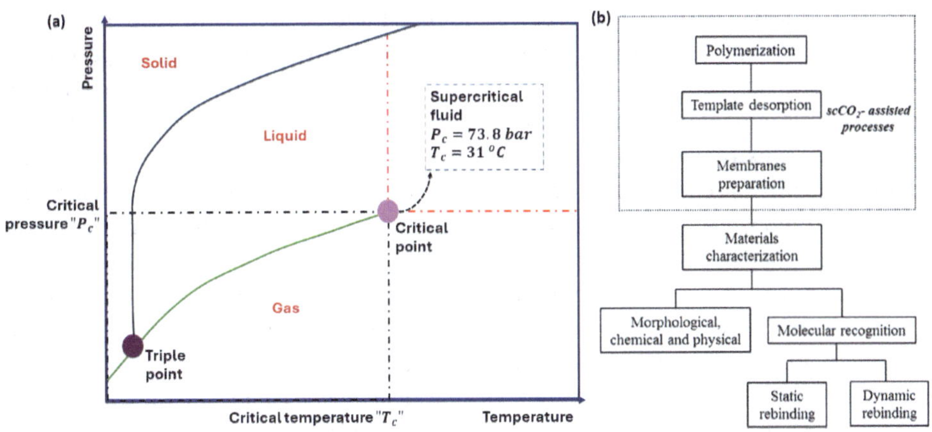

Figure 12.9: (a) Schematic illustration of CO_2 phase diagram. (b) Schematic of the experimental procedure used in this work. Reproduced with permission from da Silva et al. [67], Copyright 2011, Elsevier Science Ltd.

CO_2 in its supercritical state serves as a green and tunable solvent for polymerization reactions, enabling the dissolution and diffusion of monomers, template molecules, and cross-linkers within the polymerization medium, as illustrated in Figure 12.9b [66, 67]. By changing the pressure and temperature, the solvation power of supercritical CO_2 can be adjusted, thereby allowing the kinetics and morphology of polymerization to be controlled, yielding MIPs with tailored properties. A great example is shown by Soares da Silva et al. [67]. To create MIP membranes, a Teflon cap was used to hold the casting solution, which contained 30 wt% of a polymer blend made up of 70:30 PMMA and MIPs or NIPs, in 5 mL of DMF. The cap was then placed inside the high-pressure cell. To distribute CO_2 on top of the casting solution, the cell had a porous structure supporting a Raschig ring bed. The cell was submerged in a water bath with a thermostat, heated by a controller that maintained the temperature within ±0.01 °C, to produce membranes at 45 °C. Using a piston pump from Gilson 305, CO_2 was added until a 20 MPa operable pressure was obtained. A back-pressure regulator was utilized to establish a pressure of 20 MPa, thereby isolating the CO_2 from the DMF contained in the casting solution. In all the tests, a flow rate of 9.8 g/min of CO_2 was fed into the system for three hours. The internal pressure of the system was measured by a pressure transducer with an accuracy of ±0.100 kPa. After slow depressurization of the system for 20 min, PMMA and MIP (PMMA MIP) or NIP (PMMA NIP) were combined to create a thin, homogeneous membrane. With the exception of not adding pre-synthesized polymer to the casting solution, the identical process was used to manufacture the pure PMMA membrane [67]. These authors first synthesized the MIPs in $scCO_2$ followed by the fabrication of the hybrid material.

Supercritical CO_2 is nontoxic, noninflammable, and can easily be sourced as a by-product of industrial processes; thus, it is said to be environmentally benign for the synthesis of MIPs. Besides, it can be re-circulated and re-used with ease, producing minimal waste and reducing its environmental impact to a minimum. Supercritical fluids exhibit superior mass transport properties compared with other conventional solvents; thus, monomers and template molecules can diffuse into the polymerization medium at an extremely fast and efficient rate. With improved mass transfer in templating, the MIPs prepared could be expected to exhibit high capacity and selectivity due to the homogeneous distribution of functional groups and template molecules in the polymer matrix. The synthesis using supercritical fluid involves relatively benign temperatures and pressures without resorting to elevated temperatures or other harsh reaction conditions that most standard polymerization techniques require. Additionally, under such mild conditions, the stability of sensitive biomolecules and functional groups remains unaffected. This, in turn, opens routes for the synthesis of MIPs with an increasing degree of both biocompatibility and specificity. The variable solvent power in supercritical CO_2 allows modulation in morphology and structure, including particle size, porosity, and surface area. By altering process variables such as pressure, temperature, and CO_2 density, the physical and chemical characteristics of MIPs can be tailored to meet particular application needs, such as increased adsorption capacity or faster kinetics [66].

More importantly, it has been suggested that to synthesize approximately 1 g of MIPs using $scCO_2$ (99.998% pure) and a co-solvent like acetonitrile as a solvent, you will need the following materials [66]:

- A high-pressure, 33 mL cell with steel or sapphire panes is used. The choice between sapphire and steel is based on whether one needs to visually observe the reaction, as sapphire windows are transparent. Appropriate O-rings that can support high pressure and are compatible with supercritical CO_2 must also be used. Suitable materials generally include Viton or Teflon, both of which offer chemical resistance in addition to providing durability.
- Application of high-pressure valves and fittings suitable for CO_2 applications, typically made of stainless steel due to its strength and resistance to corrosion.

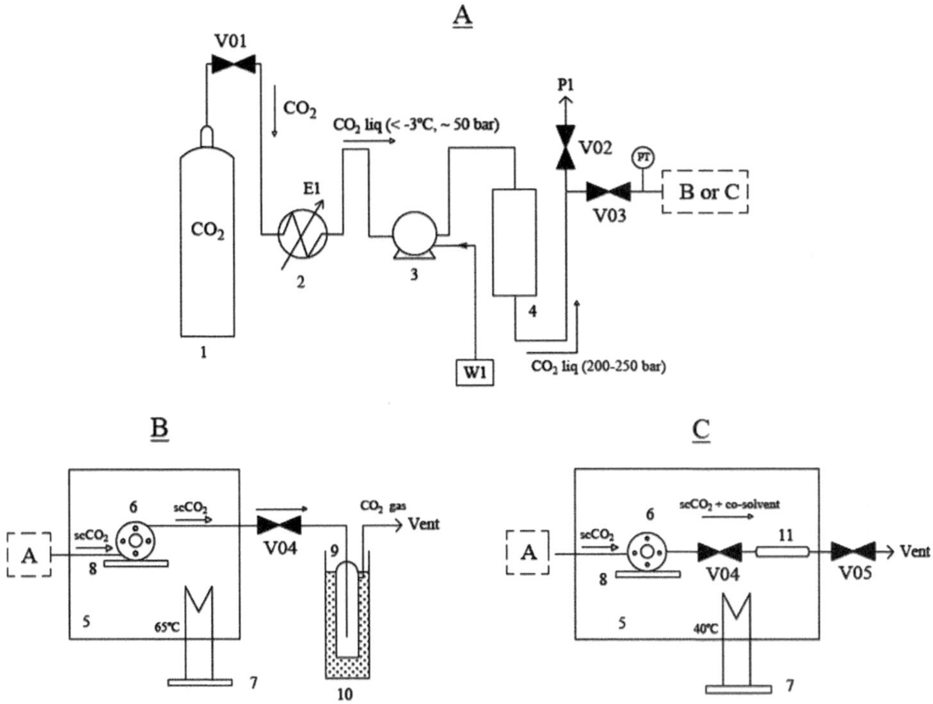

Figure 12.10: Schematic of the experimental apparatus for MIP synthesis in $scCO_2$ (**a**, **b**) and for $scCO_2$-assisted template desorption (**a–c**). (1) CO_2 supply; (2) cryostat; (3) CO_2 liquid pump; (4) column packed with molecular sieve; (5) water bath; (6) high-pressure cell; (7) heating bath circulator; (8) magnetic stirrer; (9) expansion vessel; (10) freezing mixture; (11) MIP-packed column; V01–V05: high-pressure valves; W1: compressed air supply; P1: outlet chain; PT: pressure transducer. Reproduced with permission from Furtado et al. [66], Copyright 2021, Springer Nature.

- Functional monomer: Normally, the use of MAA, ITA, bio-based monomers, and/or MAM are used because these monomers show the ability of hydrogen bond interaction with template molecules.
- Cross-linker: One of the most common cross-linkers is ethylene glycol dimethacrylate (EGDMA), which confers rigidity and stability to the polymer matrix.
- Template molecule: The molecule that you want MIP can recognize and bind. Commonly, pharmaceutical compounds, pesticides, or any target analyte like metronidazole, acetamide, and bisphenol A are used.
- Initiator: Azobisisobutyronitrile (AIBN) is often used as the thermal initiator for free radical polymerization.
- Other materials needed: Teflon-coated magnetic stirring bar and magnetic stirrer; water bath with a thermostat for temperature control; pressure transducer; heating bath circulator to maintain a stable bath temperature; compressed air and

CO_2 liquid pump; cryostat, a column packed with molecular sieve (to remove remaining water molecules from CO_2 source); Schlenk tube; and freezing mixture (ice). The assembly is shown in Figure 12.10.

The general procedure often involves the dissolution of the functional monomer, cross-linker, template molecule, and initiator in $scCO_2$ [66]. This may involve initially dissolving them in a small amount of an organic co-solvent to enhance solubility. This phase is immediately followed by the polymerization reaction under supercritical conditions, typically at temperatures around 60–80 °C and pressures of 100–300 bar. After polymerization, the template molecule is extracted, often using $scCO_2$ or an appropriate solvent, to leave behind the imprinted cavities [66]. For further reading about this process, we refer readers to the work by Furtado et al. [66]. The safety of researchers or industrialists working with these systems is very important. For instance, during depressurization and recovery, one should carefully depressurize the reactor and recover the synthesized MIP. Wash and dry the polymer to remove residual solvent and unreacted monomers. Further, it is important to note that procedures must be undertaken in a safety-minded manner since both high-pressure systems and supercritical CO_2 can be hazardous. Wear suitable personal protective equipment (PPE) and ensure that the reactor is rated for the conditions of operation.

Therefore, synthesis with the assistance of supercritical fluids offers a greener and more sustainable route toward MIPs with designed properties for various applications in molecular recognition, drug delivery (DD), and separation technologies. Utilizing the unique properties of supercritical CO_2 can enable researchers to take another step toward low-cost, environmentally friendly MIP development for several industrial and scientific uses [66]. Such eco-friendly approaches in the synthesis and preparation of MIPs greatly enable researchers to develop a class of benign, sustainable molecular recognition materials that will have applications ranging from environmental remediation to DD, sensing, and separation technologies.

12.4 C-based materials in MIPs: environmental benefits

So far, C-based materials have received tremendous interest within the area of MIPs for their outstanding properties and environmental advantages [3, 4, 10, 15, 68]. Adding carbonaceous materials into the polymer matrix could further enhance the adsorption capacity, mechanical stability, and selectivity of MIPs with such elements as graphene, CNTs, carbon/graphene dots, or activated carbon as important sorbents in the cleanup of the environment. Utilizing C-based resources in MIPs not only increases the effectiveness in pollutant removal from air, water, and soil but also involves sustainability by engendering resource efficiency and waste reduction. The goal of this chapter is to outline the environmental benefits concerned with the use of C-based materials in MIPs by considering synthesis, properties, and their applications in mitigation against critical environmental challenges.

C-based materials, including activated carbon, CNTs, graphene, and carbonaceous NPs, are characterized by large surface area, porosity, and surface reactivity. They are perfect prospects because of all these qualities to enhance the adsorption capability of MIPs toward environmental pollutants [62]. Besides, the molecular recognition sites, which are tailor-made in C-based MIPs, can effectively remove target pollutants from complex matrices with minimal excessive purification steps and waste generation [62]. Further, some green synthesis routes for the C-based MIPs-microwave-assisted synthesis, solvent-free-ensure at low energy consumption and usage of the organic solvents to minimize produced waste, improving the environmental sustainability [3, 4, 10, 15, 69]. On the other side, there are C-based MIPs able to realize value recovery in complex waste streams relevant to the resource efficiency/circular economy theme since they make full use of wastes and mitigate the environmental pollutants [3, 4, 10, 15, 68].

A sustainable and efficient solution could be developed through the environmental benefits provided by C-based material in MIPs for the remediation, sensing, and recovery of environmental resources toward natural ecosystem protection.

12.5 Applications of green C-based MIPs

Green C-based MIPs hold immense potential across various environmental applications, owing to their sustainable synthesis routes, eco-friendly properties, and tailored molecular recognition capabilities [3, 4, 10, 15, 68].

12.5.1 Water/soil remediation

Because of their exceptional qualities and environmental friendliness, green-synthesized C-based MIPs have become very practical materials for a variety of applications. These MIPs are particularly valuable in sensing and detection technologies: For example, they have been utilized to develop sensitive fluorimetric sensors for detecting pesticides like thiabendazole and/or pyrraline in food products, offering rapid and selective measurement capabilities [25]. They also provide great specificity and reusability for the selective identification of contaminants and biomarkers in the biological and environmental monitoring domains [2, 17, 25, 67]. Their green synthesis ensures minimal environmental impact while maintaining high efficiency and performance. Instances as per literature illustrating the application of C-based MIPs and their application in water/soil remediation are discussed in this section. An innovative green method for producing a water-compatible superhydrophilic MIP resin on graphene oxide (SMIR/PGO) was investigated in this specific case [41]. Dopamine (DA) played two roles in the design of SMIR/PGO: (1) It self-polymerized on the GO surface to create polydopamine, which served as an eco-friendly bridge for SMIR covalent grafting; and (2) it acted as green hydrophilic monomers, substituting for conventional phenolic monomers to obtain SMIR. A straightforward one-pot synthesis is made possible by the dual function of DA, which removes the need for laborious processes like the separation, washing, and drying of intermediate components. Furthermore, formaldehyde, the carcinogen, is avoided by substituting hexamethylenetetramine (HMTA) for the conventional cross-linker. In the synthesis of SMIR/PGO, the NH_3 generated by HMTA breakdown functions as a catalyst, eliminating the need for further catalysts. In accordance with the principles of green chemistry, this one-pot synthesis takes place in an aqueous matrix and completely eschews the use of organic solvents, guaranteeing compatibility with aqueous solutions. Therefore, SMIR/PGO combines the large specific surface area of GO, the multifunctionality of polydopamine, the outstanding water compatibility, and the great selectivity of SMIR. Chlorogenic acid can be selectively extracted from the aqueous matrix by a variety of adsorption interactions made possible by this combination. Superhydrophilic MIP materials with remarkable recognition and adsorption characteristics in watery environments were claimed to be effectively created using this ecologically friendly synthetic approach [41].

It was shown in a different study that it is necessary to identify indoxacarb (IXC), a frequent chemical pollutant, in environmental and biological samples quickly and effectively [70]. A modern optical sensor was developed for IXC based on eco-friendly MIP coated on silica-CQDs (SiCQDs). Highly fluorescent SiCQDs were prepared using a hydrothermal process, and then MIP was created on the surface (MIP@SiCQDs) by a sol-gel technique. Under ideal circumstances, Figure 12.11a and b shows a linear relationship with a 1-nM detection limit between the fluorescence quenching effect and rising IXC concentration for the range of 4–102 nM. As demonstrated in Figure 17.11c and d, MIP@SiCQDs exhibit a linear response to IXC in the range of 4–102 nM, with a 1-nM limit of detection (LOD). For five duplicate detections at 21 and 60 nM IXC, respectively, the precisions were 2.3% and 4.5%. Recoveries in the range of 95–106%, with a relative standard deviation of <6.0%, confirmed the sensor's applicability for IXC quantification in biological and environmental samples [70].

Once more, Li et al. [42] emphasized how difficult it is to precisely identify low concentrations of perfluoroalkyl carboxylic acid molecules (PFCAs) in environmental water, despite their significant impact. Their research was motivated by this challenge, and they used surface imprinting and dummy-template imprinting techniques to selectively fabricate CNTs@ILs@MIPs using perfluorotetradecanoic acid as the dummy template. In conjunction with gas chromatography-mass spectrometry measurements, the produced CNTs@ILs@MIPs were employed as a selective extraction adsorbent to extract PFCAs from environmental water. Extensive research was con-

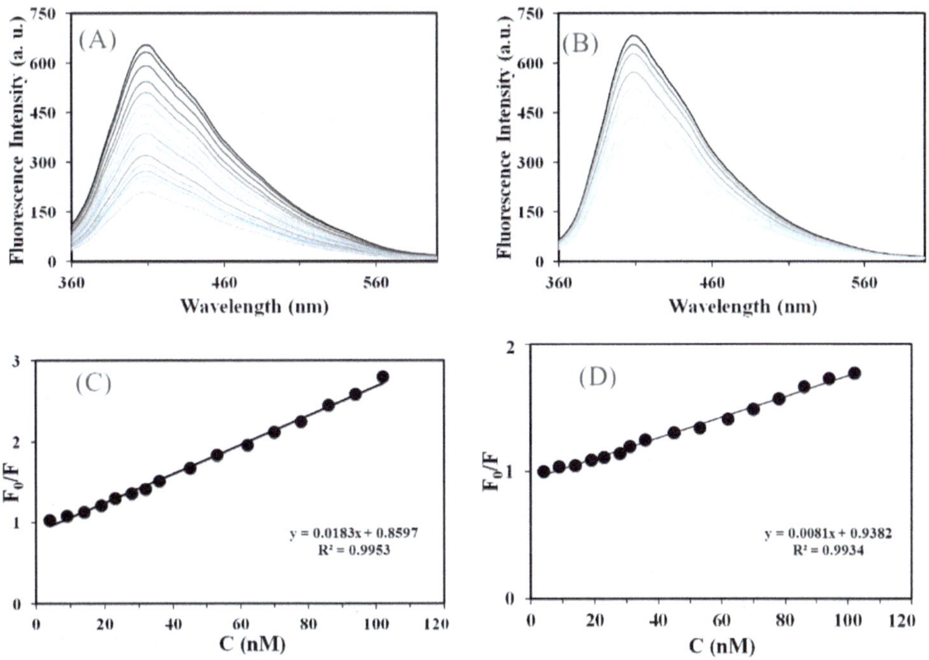

Figure 12.11: Fluorescence spectra of (A) MIP@SiCQDs and (B) NIP@SiCQDs with the addition of different concentrations of IXC (4.0–102). The Stern-Volmer plots for (C) MIP@SiCQDs and (D) NIP@SiCQDs with different concentrations of IXC under optimum conditions. Reproduced with permission from Shirani et al. [70], Copyright 2021, Elsevier Science Ltd.

ducted on the primary parameters of preparation and extraction settings. The established approach demonstrated low LODs ranging from 0.60 to 1.64 ng/L, wide linearity with R^2 values exceeding 0.9994, and satisfactory adsorption recoveries (80.5–112.5%) for seven PFCAs. The CNTs@ILs@MIPs exhibited excellent adsorptivity. By leveraging the strong affinity and recognition capacity of functional materials modified with ILs to create molecularly imprinted CNTs, this proposed method offers a novel and adaptable strategy for the detection of targeted contaminants in environmental water [42].

Green C-based MIPs are adept at detecting and consequently removing contaminants from water sources such as heavy metals, organic contaminants, and/or emerging micropollutants. Their high selectivity enables the effective removal of specific pollutants, contributing to the purification of drinking water, wastewater treatment, and remediation of polluted aquatic environments. Also, Green C-based MIPs offer novel solutions for soil remediation, particularly in the removal of organic contaminants, pesticides, and heavy metals from soil matrices. Their tailored molecular recognition sites enable selective adsorption of target pollutants, facilitating the restoration of contaminated soils and the preservation of agricultural land.

12.5.2 Sensing and monitoring

C-based MIPs serve as robust platforms for environmental sensing and monitoring applications, enabling the detection and quantification of target analytes with high sensitivity and selectivity [40, 59, 65]. By incorporating functionalized carbon nanomaterials into the MIP matrix, these sensor devices offer real-time monitoring of environmental parameters, aiding in pollution detection, environmental surveillance, and risk assessment. By electropolymerization imprinting on 3D AuPd NPs –IL functionalized graphene-CNTs nanocomposite (AuPd/GN/CNTs/IL) modified glassy carbon electrode, for example, a novel electrochemical sensor for paracetamol (PCM) determination was created [65]. The one-pot hydrothermal method was used to prepare the GN-CNTs supported AuPd alloy NPs in the presence of IL, or 1-hydroxyethyl-3-methyl imidazolium bis[(trifluoromethyl) sulfonyl] imide. This IL not only encouraged the formation of small AuPd alloy NPs but also served as a "spacer" to stop graphene sheets and CNTs from aggregating and stacking π-π. Large surface area and strong electrocatalysis were present in the resultant composite. High stability and good recognition of PCM were displayed by the PCM imprinted poly(carbazole-co-pyrrole). Utilizing the combined effects of 3D AuPd/GN/CNTs/IL nanocomposite and PCM imprinted copolymer, an extremely sensitive and selective electrochemical sensor was developed. From 0.10 to 10 μM, it showed a strong linear relationship with a low detection limit of 50 nM ($S/N = 3$) as shown in Figure 12.12. The sensor's acceptable recoveries (84.5–102%) make it suitable for use in biological sample PCM detection. Furthermore, it was effectively employed to track the level of PCM in a patient's urine who had a fever and cold [65].

In a different study, IL [HO_2CMMIm]Cl was synthesized using a one-step ultrasound process to create an IL-GR composite, which served as an electrode-sensitive element [40]. Subsequently, through free radical polymerization, MIPIL was developed using ILs as a cross-linker and functional monomer. The IL-GR/GCE demonstrated a large surface area and superior electrical conductivity. As illustrated in Figure 12.12b, the MIPIL sensor was utilized under optimal conditions to detect rutin in BR buffer (0.04 M, pH = 2). The sample was incubated for 10 min. Figure 12.12c shows the linear relationship observed between the rutin concentration at the MIPIL sensor and the DPV signal, ranging from 0.03 to 1 μM. For rutin, the LOD at the MIPIL sensor was 0.01 μM ($S/N = 3$). At the MIPIL sensor, the linear regression equation for rutin was $I(\mu A) = 0.183C(\mu M) - 0.0026$ ($R^2 = 0.9946$). The linearities of the NIPIL, MBA-MIP, and EGDMA-MIP sensors for rutin were examined to demonstrate the superiority of the MIPIL sensor, as shown in Figure 12.12. Linear correlations were observed between the NIPIL sensor, MBA-MIP sensor, and EGDMA-MIP sensor and rutin concentrations ranging from 0.05 to 1 μM, 0.3 to 1 μM, and 0.3 to 1 μM, respectively. The corresponding LODs for $S/N = 3$ were 0.02, 0.17, and 0.12 μM, respectively. Due to its strong conductivity and identification ability, the MIPIL sensor demonstrated the best

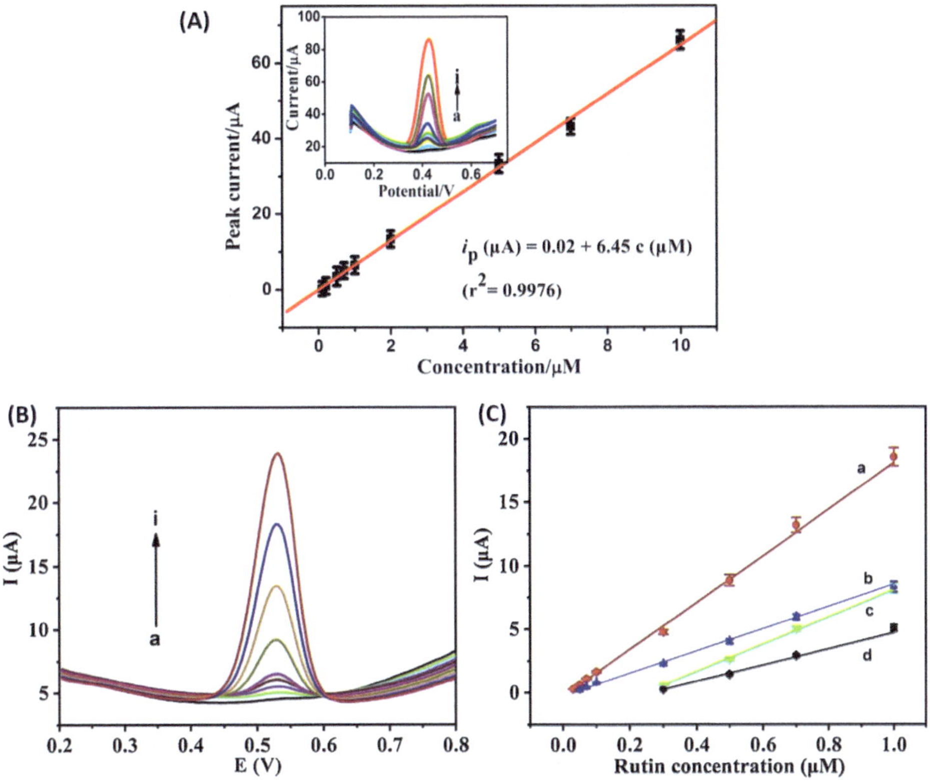

Figure 12.12: (A) The calibration curve of PCM. Inset: voltammetric response of MIP(PCz-co-PPy)/AuPd/GN-CNTs-IL/GCE to PCM in 0.10 M PBS (pH = 6.5). PCM concentrations: 0.10, 0.20, 0.50, 0.70, 1.0, 2.0, 5.0, 7.0, and 10 μM (from bottom to top). Reproduced with permission from Yang et al. [65], Copyright 2021, Elsevier Science Ltd. (B) DPV curves of the MIPIL sensor for rutin detection. Rutin concentrations: 0, 0.03, 0.05, 0.07, 0.1, 0.3, 0.5, 0.7, and 1 μM rutin (a–i). (C) Calibration curves for rutin analysis with MIPIL (a), NIPIL (b), MBA-MIP (c), and EGDMA-MIP sensors (d). Reproduced with permission from Lu et al. [40], Copyright 2021, Elsevier Science Ltd.

performance in rutin detection. Because of its high rutin selectivity, the MIPIL sensor is proposed as a promising option for rutin detection in tablets [40].

12.5.2.1 Environmental sensors

Plant health and the environmental conditions influencing it can be studied in great detail by detecting VOCs released by various plant species and their organs. One biomarker for tracking plant health and identifying stress is the emission of limonene. New approaches have been proposed because traditional methods for VOC detection face challenges. To create limonene sensors, a team of researchers suggested combining conductive nanofibers, molecular imprinting, and electrospinning in one study [46]. In particular, multiwalled CNTs (MWCNTs) improved conductivity, while polyvinylpyrrolidone (PVP) and polyacrylic acid (PAA) functioned as fiber and cavity formers, respectively. Using an electrospinning process, the authors created one-step monolithic molecularly imprinted fibers where the target molecule was *S*(−)-limonene. The target molecule was washed after the functional cavities were fixed using a UV curing procedure. This process enabled the creation of limonene recognition sites within the nanofiber matrix, improving sensor functionality and simplifying manufacturing. For the sensor to function, humidity was essential, with 50% relative humidity being ideal. *S*(−)-limonene caused the sensors to react quickly, reaching a plateau within 200s. An LOD of 137 ppb was achieved by increasing fiber density, which also enhanced sensor performance. However, high fiber density made it harder to access active sites, which diminished sensitivity. Surprisingly, when compared to other VOCs such as ethyl alcohol (EtOH), aromatic compounds like toluene, and two other monoterpenes with similar structures (α-pinene and linalool), the thinnest mat on the fibrous sensors produced the highest selectivity for limonene (selectivity index: 72%) [46]. These results highlighted how the proposed integrated strategy could be applied for selective VOC detection in environmental monitoring and precision agriculture applications.

For example, among the nitrophenol species found in wastewater, *p*-nitrophenol (4-NP) is recognized as the most persistent and hazardous. With the goal of detecting 4-NP in water, Wang et al. [71] combined carbon dots (CQDs) with superior optical qualities and MIP with advantageous selectivity to create a new fluorescent probe for 4-NP detection in aqueous environments. Using *o*-phenylenediamine as a surface modification agent and citric acid (CA) hydrate as a carbon source, the hydrothermal approach was employed to create the CQDs. Then, the sol-gel imprinting technique was used to create the MIP coating on the CQDs, referred to as MIP@CQDs. To evaluate the selectivity by comparing the values of F0/F, several aromatic compounds with comparable chemical structures, such as 2-NP, 3-NP, NB, p-PD, AN, 4-CP, PG, and HQ, were chosen as possible interfering substances. As seen in Figure 12.13a, MIP@CQDs exhibited the highest fluorescence quenching for 4-NP, followed by 2-NP and 3-NP. 4-CP showed higher fluorescence quenching than NB, p-PD, AN, PG, and HQ, while the quenching of NB, *p*-PD, AN, PG, and HQ was very low and closely spaced from one another. With the exception of 4-NP, 2-NP, and 3-NP, the quenching of NIP@CQDs for the remaining organics was modest and resembled that of MIP@CQDs. The outcomes demonstrated that the proposed strategy had good selectivity for 4-NP. More importantly, the probe could distinguish 4-NP from 2-NP and 3-NP with precision. The selectivity of common inorganic ions typically found in aqueous samples, such as Na^+, Ca^{2+}, Mg^{2+}, Cu^{2+}, Al^{3+}, Fe^{3+}, Co^{2+}, Ni^{2+}, CO_3^{2-}, SO_4^{2-}, and NO_3^-, was also examined. As shown in Figure 12.13b, these common inorganic ions had no discernible effect on the detection of 4-NP. This is because the synthesized MIP@CQDs possess numerous recognition sites for 4-NP, allowing 4-NP to bind tightly to the MIP@CQDs, resulting in a reduction in luminescence intensity. Figure 12.13c illustrates how the internal filtering effect and dynamic quenching upon encountering 4-NP are thought to be the causes of the fluorescence quenching of MIP@CQDs. The probe was suitable for the quantitative detection of trace 4-NP in actual aqueous samples, including tap water, wastewater, and seawater, due to its high selectivity, low detection limit of 0.41 µmol/L, wide detection linear range between 0 and 144 µmol/L, and satisfying recoveries ranging from 95.1% to 107.8%. The detection outcomes align with the findings obtained by the HPLC technique. This work offers an easy-to-use, quick, and efficient fluorescence technique for detecting 4-NP in an aqueous environment [71].

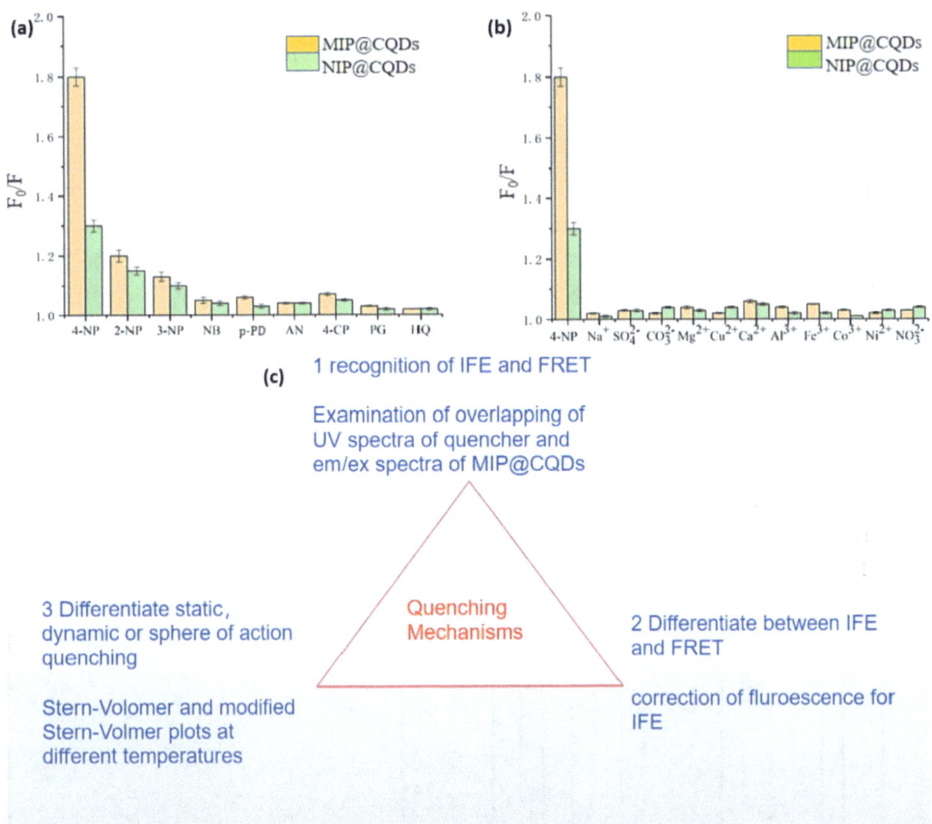

Figure 12.13: The fluorescence quenching efficiency of 4-NP (36 µmol/L) and analog (36 µmol/L) on MIP@CQDs and NIP@CQDs (a), as well as the fluorescence quenching efficiency of 4-NP (36 µmol/L) and common inorganic ions (36 µmol/L) on MIP@CQDs and NIP@CQDs (b). All experiments were performed in a pH 8.5 buffer. (c) Infographic illustration of the main steps to elucidate the mechanisms of fluorescence quenching of MIP@CQDs by 4-NP. Reproduced with permission from Wang et al. [71], Copyright 2023, Elsevier Science Ltd.

12.5.2.2 Biosensors

Because of their green synthesis, eco-friendliness, and specific recognition ability, C-based MIPs have gained great application as biosensors. These types of MIPs were tailored to recognize some biological entities via binding with them, thus proving to be one of the superior platforms for detection purposes related to pathogens, toxins, and biomarkers. As an illustration, carbon dot-integrated MIPs were also developed for highly sensitive glucose and pesticide detection with high sensitivity and selectivity. Besides, they are also used in food safety, environmental monitoring, and medical diagnostics, bringing about an environment-friendly and effective way for real-time analysis on-site [24, 72]. Using, for instance, fluorinated MIPs (FMIPs) functionalized $CH_3NH_3PbI_3$@carbon nanodots ($CH_3NH_3PbI_3$@CNDs) as a platform, a novel ultrasensitive cholesterol photoelectrochemical sensor has been supposedly constructed [24]. Because 2-(perfluorooctyl) ethyl acrylate (PFEA) contains a long hydrophobic chain of fluorine, it was chosen as a functional monomer in the study to enhance the aqueous stability of $CH_3NH_3PbI_3$ and to confer high selectivity. Additionally, by reducing defect density through the polymer passivation of perovskite grain boundaries, the FMIPs can lengthen the carrier lifetime and suppress the recombination of electron holes, which will significantly improve charge transfer efficiency. As demonstrated in Figure 12.14a, under optimal conditions, the photocurrent response of the sensor decreased as the concentration of cholesterol increased. The logarithm of cholesterol concentration (log C) and the current change (ΔI) displayed a linear relationship within the range of 1.0×10^{-13}–1.0×10^{-9} mol/L. According to Figure 12.14b (curve a), the PEC sensor's linear equation was ΔI (μA) = 3.1443 + 0.2253 log C ($R^2 = 0.9949$). The slope (LOD = 3Sa/b) and the standard deviation of the response led to the estimation of the LOD, which was 8.91×10^{-14} mol/L [24].

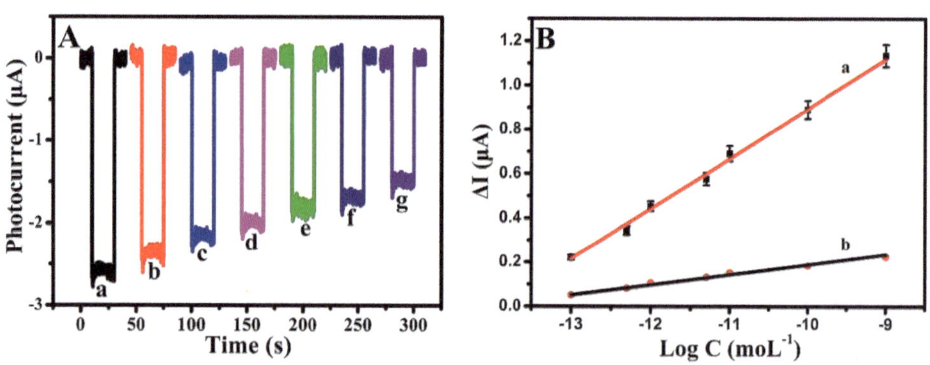

Figure 12.14: (A) Photoelectric response of FMIPs@$CH_3NH_3PbI_3$@CNDs/ITO PEC sensor to different concentrations of cholesterol. (B) Linear relationship curve of FMIPs@$CH_3NH_3PbI_3$@CNDs/ITO. Reproduced with permission from Yu et al. [24], Copyright 2024, Elsevier Science Ltd.

For the highly sensitive detection of tiny amounts of SM2 in food, a simple one-step electro-copolymerization process was successfully employed to produce the MIP/GQDs-PtNPs/GCE sensing system in a different study [72]. The DFT-D approach was utilized to integrate the solvation effect and molecular docking in the screening of bifunctional monomers for the logical design of molecularly imprinted materials. This was achieved using a comprehensive computer simulation strategy. Along with analyzing the stoichiometry, π-π interaction, molecular docking, and thermodynamic features such as binding energy, the DFT-D approach was employed to investigate the interactions between monomers and templates. Two advancements were incorporated into the developed sensor. First, it was the initial material to be molecularly imprinted using EDOT and MAA as dual-purpose monomers. Second, this was the first report on the design of the SM2 electrochemical sensor using computer simulation. Additionally, the sensor demonstrated superior performance in terms of cost, ease of use, stability, specificity, and sensitivity. A robust integration of the MIP film with a transducer was achieved through a single-step electro-copolymerization procedure. Consequently, this approach reduced costs and simplified the preparation of MIP materials. Using differential pulse voltammetry (DPV), it was possible to quantitatively detect SM2 in solutions under optimal experimental conditions. The DPV responses of the MIP/GQDs-PtNPs/GCE sensor for SM2 are shown in Figure 12.15a. The peak current (i) exhibited a linear relationship with the SM2 concentration within the 0.1 nM to 0.1 mM range, and the oxidation potential was 0.22 V. As illustrated in Figure 12.15b, the first equation was $i\ (\mu A) = 0.6796c\ (nM) + 7.9362$ in the concentration range of 0.1 nM to 100 nM. In Figure 12.15c, the second equation was expressed as follows: $i\ (\mu A) = 0.997c\ (\mu M) + 82.469$ in the concentration range of 0.5 μM to 100 μM. Moreover, the detection limit was determined to be 0.023 nM ($S/N = 3$), which was significantly lower than the Chinese food quality standard value and earlier electrochemical sensor results. Additionally, based on the electrode surface area and the slope of the calibration curve, the sensitivity of the MIP/GQDs-PtNPs/GCE sensor was determined to be 161.8 $\mu A \cdot \mu/M \cdot cm^2$. The SM2 sensor's performance was also compared with that of the MIP/GQDs-PtNPs/GCE sensor. Overall, the MIP/GQDs-PtNPs/GCE sensor outperformed others in terms of selectivity, lower detection limit, stability, cost, and operational simplicity [72].

12.5.3 Resource recovery

Green C-based MIPs have significant potential in resource recovery applications due to their high selectivity, environmental friendliness, and competent recognition capabilities [73, 74]. By utilizing green solvents and biomass-derived carbon sources, these MIPs align with sustainability principles while providing effective solutions for resource recovery from various waste streams.

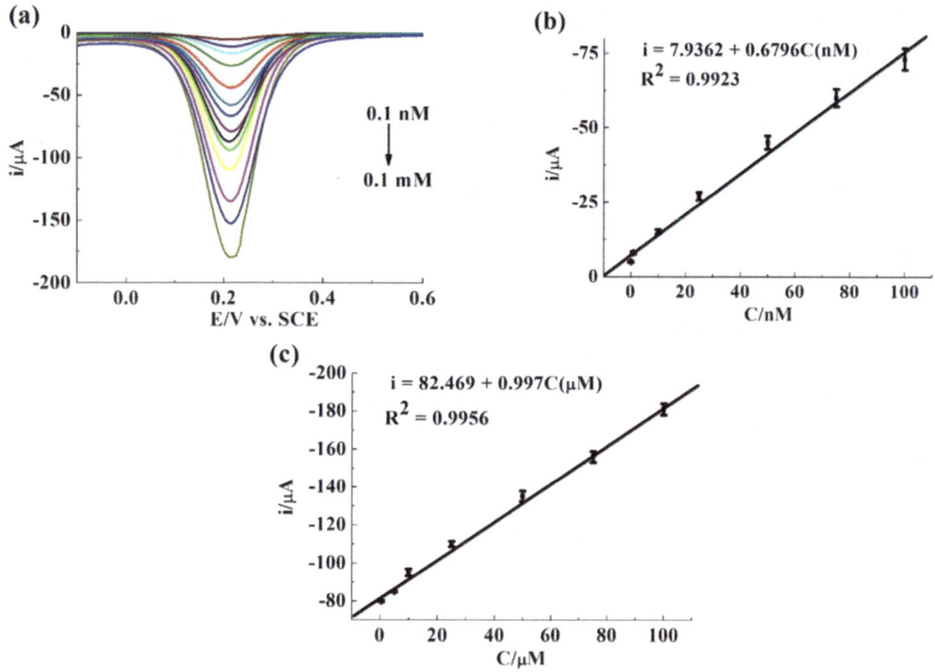

Figure 12.15: (a) DPV responses of the MIP/GQDs-PtNPs/GCE sensor for different concentrations of SM2. From bottom to top, the concentrations of SM2 are 0.1 nM, 1 nM, 10 nM, 25 nM, 50 nM, 75 nM, 100 nM, 500 nM, 5 μM, 10 μM, 25 μM, 50 μM, 75 μM, and 100 μM. (b) The calibration curve corresponding to SM2 in the concentration range from 0.1 nM to 100 nM. (c) The calibration curve corresponding to SM2 in the concentration range from 0.1 μM to 0.1 mM. Reproduced with permission from Zhang et al. [72], Copyright 2020, Elsevier Science Ltd.

Green C-based MIPs can selectively bind and recover heavy metal ions like lead, mercury, and cadmium from industrial wastewater. The high surface area and tailored functional groups of these MIPs enhance their binding affinity for specific metal ions, enabling proper separation and recovery. This process not only prevents environmental contamination but also allows the reclamation of valuable metals for reuse in industrial processes [73]. Moreover, the recovery of rare earth elements (REEs) from electronic waste and industrial effluents is critical due to their essential role in innovative technologies. C-based MIPs synthesized using green methods can be designed to target specific REEs, facilitating their selective extraction and concentration. This approach contributes to the sustainable management of REE resources, reducing the need for primary mining and its associated environmental impacts [73]. Again, green C-based MIPs are effective in capturing and recovering organic pollutants, such as dyes, pharmaceuticals, and pesticides, from wastewater. The high selectivity and adsorption capacity of MIPs enables the effective removal of these contaminants, which can then be recovered and potentially reused or safely disposed of. This application is particularly relevant in wastewater treatment and environmental remediation efforts [74].

Recovering nutrients such as nitrogen and phosphorus from agricultural runoff and wastewater streams is crucial for sustainable agriculture and water management [75]. Green C-based MIPs can selectively adsorb these nutrients, allowing their recovery and reuse as fertilizers. This not only mitigates water pollution but also promotes the circular economy in agricultural practices [75].

In one case, phenylureas were extracted and then enriched using a highly selective composite nanosorbent that was devised and fabricated [39]. The nanosorbent consisted of magnetite NPs embedded in an MIP, N-doped GQDs (N-GQDs), and a zinc metal-organic framework (IRMOF-1). Using a dispersive magnetic solid-phase extraction (d-MSPE) method, the synthesized nanosorbent was studied and utilized to extract and enrich three phenylureas. The target phenylureas were detected in trace amounts by the highly sensitive devised approach. The detection limit was set at 1.0 μg/L, whereas the linear range extended from 1.0 to 150 μg/L. Highly effective adsorption was achieved due to the nanosorbent's large specific surface area. Recoveries ranged from 80.2% to 95.8% when used to extract phenylureas from vegetable samples, with relative standard deviations $\leq$8.0%. The fabricated nanosorbent demonstrated excellent stability, allowing for four reuses, as well as high selectivity and efficiency in extraction. It was also very simple to produce. Furthermore, the developed technique can be modified to extract and determine other organic components in a variety of complex samples [39].

Zhang et al. [75] claimed that by creating carbon nanodots (CNDs) from leftover traditional Chinese medicine (TCM), they could fully utilize the variety of chemical compositions found in TCM and reduce the amounts needed for processing and disposal. This idea was influenced by green chemistry. Urea, thiourea, and boric acid were used as carbon, nitrogen, sulfur, and boron sources, respectively, in the hydrothermal synthesis of N-, S-, and B-doped CNDs using waste Clematis Chinensis Osbeck. A ratiometric fluorescence probe, or CNDs@MIPs, was created by encasing red CNDs, N-, S-, and B-doped CNDs in molecular imprinted polymers (MIPs). From 0.39% to 18.39%, the relative quantum yield of N-, S-, and B-doped CNDs could be achieved by fine-tuning the preparation parameters (reaction temperature, reaction time, and mass ratio). Analgesic and anti-inflammatory active components may be present in the doped CNDs, as shown by the time-of-flight secondary ion mass spectrometry, which also validated the elemental distribution of N, S, and B. Furazolidone (FZD) could be sensed by the N, S, and B doped CNDs in a linear range of 5–700 mg/L with a detection limit (LOD) of 4.82 mg/L, thanks to the inner filter effect and static quenching mechanism. The FZD could be selectively detected with a LOD of 1.98 mg/L by the CNDs@MIPs probe in the 5–400 mg/L range. Fluorescence sensors are very applicable for FZD sensing, as evidenced by the 90%–110% recoveries of N, S, and B-doped CNDs and CNDs@MIPs in the various dietary samples [75].

The application of green C-based molecularly imprinted polymers in resource recovery offers a sustainable and economical approach to managing waste and recovering valuable materials. By leveraging green chemistry principles and sustainable materials, these MIPs can play a critical role in addressing environmental challenges and supporting resource efficiency in various industries.

12.5.4 Green catalysis

Green C-based MIPs have emerged as promising materials for green catalysis, offering environmentally friendly solutions for various catalytic processes [69, 76, 77]. These polymers, synthesized using biomass-derived carbon sources and green solvents, align with sustainable chemistry principles by minimizing the environmental footprint of catalyst production and usage. They can mimic the activity of natural enzymes, offering a cost-effective and robust alternative. By imprinting the polymer with specific substrates, MIPs can achieve high catalytic efficiency and selectivity in biochemical reactions. This application is crucial for sustainable industrial processes, reducing reliance on expensive and fragile natural enzymes [76]. These MIPs can be designed to have specific active sites tailored for particular catalytic reactions. The high surface area and porous structure of these MIPs provide numerous active sites, enhancing the overall catalytic efficiency. For example, MIPs can be used in oxidation, reduction, and photocatalytic processes, facilitating the conversion of substrates under mild conditions and with high selectivity [69, 77].

The robust nature of C-based MIPs ensures excellent thermal and chemical stability, making them suitable for repeated use in catalytic cycles without significant loss of activity. This reusability aligns with the goals of green chemistry by waste reduction and the need for frequent catalyst replacement [69, 77]. Moreover, the use of green solvents and biomass-derived carbon sources in the synthesis of MIPs reduces the reliance on toxic chemicals and non-renewable resources. This sustainable approach not only lowers the environmental impact of catalyst production but also promotes the utilization of renewable materials [1, 15, 69, 77]. Additionally, green C-based MIPs can mimic the activity of natural enzymes, offering a cost-effective and robust alternative. By imprinting the polymer with specific substrates, MIPs can achieve high catalytic efficiency and selectivity in biochemical reactions [78, 79]. This application is crucial for sustainable industrial processes, reducing reliance on expensive and fragile natural enzymes [78, 79].

Incorporating biomass-derived carbon materials into MIPs enhances their photocatalytic properties. These MIPs can efficiently harness solar energy to drive chemical reactions, such as the degradation of pollutants or the synthesis of fine chemicals. The use of green MIPs in photocatalysis promotes cleaner production processes and reduces the environmental footprint of chemical manufacturing [35]. Also, green C-based MIPs can be functionalized with metal NPs, serving as effective catalysts in various organic transformations, including hydrogenation, oxidation, and coupling reactions. The high surface area and tailored pore structures of MIPs facilitate the dispersion of metal catalysts, enhancing their reactivity and stability. This strategy adheres to the principles of green chemistry via waste minimization and energy consumption.

Materials that are both practical and effective are considered crucial for reducing environmental pollution [76]. Thus, using surface molecular imprinting in conjunction with photocatalytic degradation and magnetic separation, a novel magnetic MIP ($CoFe_2O_4$@TiO_2-MMIP) was created. For $CoFe_2O_4$@TiO_2-MMIP, the highest adsorption capacity is 14.26 mg/g, and the adsorption rate constant is 0.21 g/mg/min. The adsorption characteristics of the magnetic molecularly imprinted polymer were examined in relation to the impacts of experimental conditions. Fluoroquinolones were the only molecules that $CoFe_2O_4$@TiO_2-MMIP could selectively adsorb. Norfloxacin's removal performance was clearly impacted by another drug in a binary adsorption system, and the adsorption efficiency was closely correlated with the target contaminant's molecular structure, molecular weight, polarity, and functional groups. Good reusability was indicated by the adsorption-photocatalytic recycling experiment, which confirmed that $CoFe_2O_4$@TiO_2-MMIP could complete in-situ regeneration and pollutant degradation simultaneously. This substance would be highly appealing for use in the elimination of fluoroquinolones due to its selective adsorption and photocatalytic regeneration properties [76].

Thus, green C-based MIPs offer a sustainable and inexpensive platform for green catalysis, combining high catalytic performance with environmental benefits.

12.5.5 Drug delivery

Green C-based MIPs offer a promising approach for DDSs, providing high specificity, controlled release, and sustainability [80]. These MIPs, synthesized using renewable biomass-derived carbon sources and green solvents, can be engineered to possess specific binding sites that recognize and bind target drugs with high affinity.

Green C-based MIPs can be tailored to target specific tissues or cells, enhancing the therapeutic efficacy and minimizing side effects. The imprinted cavities within the MIPs ensure that the drug is selectively released at the desired site, improving the precision of treatment [80]. Additionally, the structure of these MIPs allows for the controlled release of drugs over an extended period. This is especially advantageous for improving patient compliance, lowering the frequency of administration, and preserving the body's ideal medication concentrations. The biodegradable nature of biomass-derived carbon materials also ensures that the DDS is environmentally friendly [80]. Moreover, green C-based MIPs are designed to be biocompatible and stable under physiological conditions. This stability ensures that the drug remains protected until it reaches the target site, while the biocompatibility minimizes adverse reactions, making these MIPs suitable for various biomedical applications [80]. One study created a MIP system for controlled release and selective drug adsorption using chitosan, a biocompatible and biodegradable substitute for acrylate derivatives and other non-ecofriendly monomers [81]. In order to increase the drug loading capacity, GQDs were introduced to the CS matrix. Thus, the molecular imprinting approach was utilized to successfully manufacture CS-GQD MIP using CS-GQD as the matrix and oxaliplatin as the template. Moreover, CS MIP and NIPs (without templates) were made and employed as controls in all of the trials. In comparison to CS MIP and NIPs, the adsorbability of CS-GQD MIP was noticeably better at pH 6.5, the ideal pH. The results of the adsorption kinetics and isotherm investigations showed that the adsorption behavior was in good agreement with the Langmuir model (maximum adsorption capacity of 17.15 mg/g) and Lagergren first order (rate constant of 0.038 min^{-1}). When compared to other anticancer medications, OXAL had a relatively high imprinting factor (α 2.88) and specificity factor, which verified the selectivity of the produced MIPs. The in vitro drug release data demonstrated that, as shown in Figure 12.16, CS-GQD-Drug had a regulated release profile of OXAL at pH 1.2, with an 80% release within 8 h and a mechanism consistent with Fickian diffusion. Targeting using imprinted polymers is expected to be a successful medication carrier for controlled-release anticancer delivery [81].

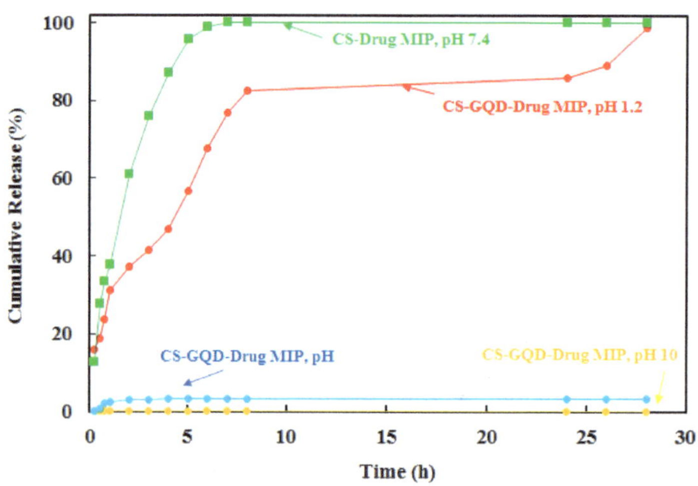

Figure 12.16: The in vitro release profile of OXAL from MIPs at different pH levels. Reproduced with permission from Farshi Azhar et al. [81], Copyright 2014, Taylor & Francis.

Liu et al. [47] also created a DDS based on the simple synthesis of a binary green porogen system doped with single-walled CNTs (SWCNTs) and room-temperature ILs (RTILs) at room temperature, as well as DESs. When SWCNTs were present, 1-butyl-3-methylimidazolium tetrafluoroborate, choline chloride/ethylene glycol, and 4-vinylpyridine (4-VP) were employed as functional and cross-linking monomers, respectively, with fenbufen (FB) serving as the template molecule. Controlling the number of SWCNTs, the ratio of RTILs to DES, and the composition of DES improved the imprinting impact of the SWCNT-MIP composites. UV band blue shifts provided compelling evidence that SWCNT doping during the self-assembly process can enhance the interaction between 4-VP and FB. This reinforced imprinting effect of CNT-doped MIP allows it to achieve superior controlled-release properties. In particular, the imprinting factor of the SWCNT-MIP composites doubled compared to the control MIP that was prepared without SWCNTs. In the examination of the FB release kinetics from all samples, the lowest drug diffusivity value was observed in the SWCNT-reinforced MIP. While the control MIP and SWCNT-non-MIP composites showed relative bioavailability of only 48.3% and 44.4%, respectively, the SWCNT-MIP composites exhibited the highest value of 143.3% when compared with the commercial FB tablet. According to the findings, the SWCNT-MIP nanocomposites created here may be used as a controlled-release device [47].

Applications of green C-based MIPs in DDSs have represented an outstanding step toward green healthcare. Thus, they possess the properties to provide DD with targeted, controlled, and stable manners according to the basic approach of green chemistry, possessing tremendous potential in providing better therapeutic output with the least environmental impact.

12.5.6 Material purification selectively

Green C-based MIPs are revolutionizing material purification through their ability to selectively bind and remove specific contaminants [1, 59]. These MIPs, synthesized using sustainable biomass-derived carbon sources and green solvents, offer environmentally friendly solutions for high-precision purification processes. Moreover, green C-based MIPs can be engineered to target and remove specific pollutants from complex mixtures, such as heavy metals, organic dyes, and pharmaceutical residues in water. Their high selectivity is achieved through tailored binding sites that match the size, shape, plus functional groups of the target molecules, enabling resourceful purification even in trace amounts. On the other hand, MIPs are applicable in the natural product purification field by extracting bioactive compounds from plant extracts. Therefore, MIP-selective binding properties could provide the high-purity required compound, ensuring the exclusion of impurities by its nature of binding, to afford a much more effective end product [1]. In some industrial applications, using green C-based MIPs, raw material, and intermediate purification is done to remove contaminants that would interfere in the subsequent stages of processing. This selective purification improves product quality by reducing waste and increasing efficiency in manufacturing processes generally [1].

Using bovine serum albumin (BSA) as a template, acrylamide-modified MWCNTs (MWCNTs-AAm) as substrates, and allyl-functionalized ILs as monomers, a variety of new protein MIPs were produced through a surface molecular imprinting process in another study [43]. Using Fourier transform infrared spectroscopy (FTIR), scanning electron microscopy, and Brunauer-Emmett-Teller analyses, the MWCNTs@BSA-MIPILs synthesized with varying monomer/template/cross-linker ratios were characterized. The adsorption kinetics and isotherm, imprinting effect, competitiveness and selectivity, reusability, and practical applicability were all thoroughly evaluated. Studies were conducted to examine the imprinting effect of MWCNTs@BSA-MIPILs, as well as the influence of ILs' anion types and MWCNTs' outer diameters. Lower MWCNT diameters (<8 nm and 10–20 nm) were found to be more advantageous in the synthesis of surface-imprinted polymers. Accordingly, MIPs prepared using IL monomers with anions of low nucleophilicity and hydrogen bond basicity (PF_6^-) and high steric effect ($CF_3SO_3^-$) exhibited a better imprinting effect compared to those prepared using IL and conventional acrylamide monomers with Cl^- and BF_4^- as counter anions. Additionally, these MIPs demonstrated better selectivity for BSA compared to trypsin, lysozyme, human serum albumin, and bovine hemoglobin. A binary protein solution containing BSA and bovine hemoglobin significantly enhanced the imprinting and selectivity parameters. Furthermore, BSA from bovine calf serum was successfully purified using the synthesized MWCNTs@BSA-MIPILs. Due to their significant role in stabilizing biomacromolecules, these findings suggest that PF_6^- and $CF_3SO_3^-$-based ILs are suitable functional monomers for creating protein molecule MIPs that are biocompatible [43].

The application of green C-based MIPs in material purification showcases their potential for achieving high selectivity and efficiency in removing specific contaminants. By embracing green chemistry principles, the MIPs are offering answers to effective purification in a sustainable manner that contributes to environmental management and industrial processes.

According to the literature reviewed, applications of green C-based MIPs cover such environmental science niches as sustainable solutions for pollution control, resource management, and environmental monitoring. With their tailored molecular recognition properties coupled with eco-friendly properties, these materials are very promising in developing green technologies and moving toward a sustainable future.

12.6 Characterization techniques for green C-based MIPs in a green manner

The definition of green characterization techniques, within the realm of material science, involves a group of analytical methods used to investigate the properties of a material by prioritizing environmental friendliness with minimal environmental impacts [82–85]. This may be accomplished by the elimination or reduction of hazardous solvents, as green techniques often operate in water or other less toxic or flammable solvents than those employed in traditional organic analyses. The generation of waste is minimized because techniques are designed to reduce the amount of waste arising from the characterization procedure, and energy consumption is lower as green techniques are designed to use less energy compared to typical methods. Green methodologies should characterize C-based MIPs in order for them to be effective and viable ecologically. The depth of their structure, functionality, and performance requires green characterization techniques. This section discusses some of the innovative techniques that are being developed for the complete analysis of these materials with minimal environmental impact, based on our understanding and also presented in Table 12.3. Green characterization techniques such as spectroscopy, chromatography, and microscopy will enable researchers to get information about the structural, morphological, and functional properties of C-based MIPs. This analysis has traditionally been carried out using techniques such as FTIR, X-ray photoelectron spectroscopy, Raman spectroscopy, and nitrogen gas adsorption (for porosity analysis), the latter particularly for porosity analyses.

Table 12.3: Some common characterization techniques for C-based MIPs in a green manner.

Characterization technique	Description	References
Scanning electron microscopy (SEM)	It provides high-resolution images of the MIP surface morphology and thus allows for the visualization of particle size, shape, and porosity.	[43, 81]
Transmission electron microscopy (TEM)	It offers nanoscale-resolution imaging of internal structures in MIPs and provides information on morphology and dispersion for carbonaceous materials.	[56, 76]
Fourier transform infrared spectroscopy (FTIR)	This technique elucidates the functional groups of MIPs with minimal sample preparation and without the use of solvents. It helps in identifying whether the desired functional groups are present and in monitoring the processes of polymerization. It typically identifies the functional groups and chemical bonds present within MIPs, thereby facilitating the characterization of polymer composition and the interaction between carbon materials and functional monomers.	[43, 84]
X-ray photoelectron spectroscopy (XPS)	This method provides information about the elemental composition and chemical state of elements on the MIP surface. Green variants of XPS are emerging that use lower X-ray power, thereby reducing energy consumption. It analyzes surface chemistry and elemental composition of MIPs, elucidating the distribution and oxidation states of carbon species and other elements within the polymer matrix.	[85]

Table 12.3 (continued)

Characterization technique	Description	References
Raman spectroscopy	This non-destructive approach yields information about the vibrational modes of molecules in the MIP. It can reveal details about the bonding between the polymer and the template molecule. This technique probes the vibrational modes of C-based materials in MIPs, providing information on structural defects, graphitic content, and functionalization of carbon surfaces.	[86]
Brunauer-Emmett-Teller (BET) analysis	It determines the pore size distribution, specific surface area, and pore volume of carbon materials in MIPs, thereby quantifying their adsorption capacity and porosity.	[37]
Thermogravimetric analysis (TGA)	Thermal stability and decomposition behaviors of MIPs, the degradation temperature of the carbonaceous components, and the associated weight loss are investigated.	[81]
Differential scanning calorimetry (DSC)	Measures heat flow changes in MIPs upon heating or cooling, providing information on glass transition temperature, crystallinity, and thermal transitions of the polymer matrix and carbon materials. Mostly, low-temperature differential scanning calorimetry (LT-DSC) is utilized.	[87]
Atomic force microscopy (AFM)	Offers three-dimensional imaging of MIP surfaces with nanometer resolution, revealing surface roughness, topography, and interactions between C-based materials and template molecules.	[84]
Low-energy ion scattering (LEIS)	Surface composition and structure analysis constitute one of the applications of C-based MIP study using LEIS. From the LEIS analysis, highly sensitive detection and elemental depth profiling become important factors in understanding not only surface interaction but also surface performance optimization regarding their applications toward sensors and catalysis.	[84]
X-ray spectroscopy	For that reason, in the study of C-based MIPs, X-ray spectroscopy such as XPS and XAS plays an important role. These allow for the procurement of detailed information on the elemental composition, chemical bonding, and surface structure in MIPs, thus assisting in their characterization and the study of the mechanisms of interaction with target molecules.	[84]
UV-vis spectroscopy	UV-vis spectroscopy characterizes C-based MIPs by means of measuring absorbance at certain wavelengths. It assesses MIP formation, template removal efficiency, and binding site characteristics. This technique will help in optimizing MIP synthesis for applications such as sensors, separations, and DDSs.	[55, 76]

These characterization techniques are thus crucial for explaining the structural, morphological, chemical, and thermal properties of C-based MIPs, thereby enabling their optimization and application in various niches: environmental remediation, sensing, and separation technologies.

It is worth noting that green analytical chemistry principles focus on reducing the environmental impact of chemical analysis, while also ensuring sustainability in laboratory practices [82, 83, 85]. These principles aim to minimize the use of hazardous substances, reduce waste, and improve the efficiency and safety of the analytical process. Some of the major principles of green analytical chemistry include the prevention of waste, the use of safer solvents and reagents, energy efficiency, renewable feedstocks, reduction of derivatization, performing in-situ and on-site analysis, designing for degradation, atom economy, the use of greener chemical syntheses, renewable analytical processes, recycling and re-use, and education and training, as illustrated in Figure 12.17 [82–85].

Prevention of waste: The analytical methodology shall be designed not to generate waste and by-products. It can also mean, under certain circumstances, the use of smaller samples and reagents, as well as the application of micro-sampling and microextraction techniques.

Safer solvents and reagents: This means the use of nontoxic, biodegradable solvents and reagents wherever possible, as well as the substitution of nonhazardous chemicals for dangerous ones to reduce risks to human health and the environment.

Energy efficiency: This means developing and using analytical methods that consume less energy.

Implementing techniques such as ambient-temperature reactions and room-temperature ILs.

Renewable feedstocks: This means utilizing renewable resources for the development of analytical reagents and materials, promoting the use of bio-based solvents and reagents derived from renewable sources.

Reduction of derivatization: This involves avoiding unnecessary derivatization steps that increase the consumption of reagents and generate additional waste by employing direct analytical methods that eliminate or reduce the need for chemical modification of analytes.

In situ and on-site analysis: This involves developing portable and miniaturized analytical instruments for on-site analysis to reduce the need for sample transportation and preservation. It also includes implementing real-time monitoring and in-situ analysis to minimize sample handling and associated environmental impacts.

Design for degradation: This means creating analytical methods and materials that break down into nontoxic components after use.

Ensuring that any waste generated is easily degradable and does not persist in the environment.

Atom economy: Maximizing the incorporation of all materials used in the analytical process into the final measurement, thereby reducing the generation of waste.

Developing efficient synthetic pathways for reagents and standards with minimal by-products.

Safer chemical syntheses: These refer to designing synthesis methods for analytical reagents that minimize the use and generation of hazardous substances. It also entails using safer, more environmentally friendly synthesis methods, such as solvent-free processes and microwave-assisted synthesis.

Renewable analytical processes: This means using renewable analytical methods that rely on sustainable resources and have a lower environmental footprint. It also refers to employing techniques such as green chromatography and eco-friendly spectroscopy.

Recycling and reuse: Encourage recycling and reusing solvents, reagents, and materials in analytical procedures. Closed-loop systems and recovery processes should be implemented to reduce the consumption of new resources.

Education, training, and use of computational chemistry: Education, training, and use of computational chemistry are important factors in enhancing green characterization and analytical chemistry of C-based MIPs. These computational methods can now predict virtually and optimize the synthesis and performance of MIPs through the use of techniques like molecular dynamics simulation and density functional theory calculations. This minimizes experimental trial-and-error testing and reduces chemical waste. Such training of scientists in these techniques would contribute to sustainability by reducing resource use and, consequently, environmental impact while speeding up the development of highly effective and selective MIPs tailored for applications in green chemistry, including environmental remediation and biomedical diagnostics.

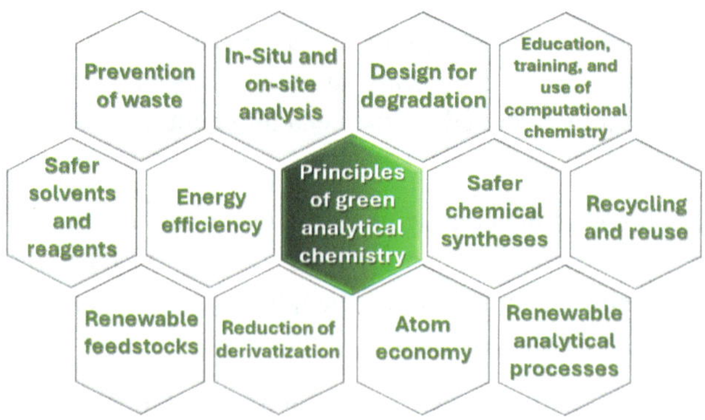

Figure 12.17: Principles of green analytical chemistry.

For a better understanding of this context, assessment tools such as the Analytical Eco-Scale, Green Analytical Procedure Index, and/or Analytical Greenness Metric must be applied, as discussed in the work by Sajid and Płotka-Wasylka [85], Sedehi et al. [88], and de la Guardia and Garrigues (book) [84]. The tendency in this direction is to develop, within green analytical chemistry, methods that are sustainable, safe, and efficient, with as little environmental impact as possible, while at the same time maintaining a high level of scientific accuracy and reliability. These techniques of characterization may not cover all aspects of green analytical chemistry, but they are, to a great extent, favorable for most of the discussed principles.

12.7 Challenges and future directions

This chapter has focused on the green aspects of C-based MIPs and in this regard one can appreciate the achievements to date as well as consider the areas that require further improvement. In this section, I discuss the last obstacles to filling the gap and propose further steps to develop the work in a sustainable way. A few among them are scalability, cost, and regulatory concerns requiring improvisation [69]. Possible future directions can be optimization of synthesis protocols, investigation of new green materials, and coupling with new emerging technologies. This discussion is intended to map a way forward toward greener and more effective applications of C-based MIPs.

12.7.1 Challenges

Scalability and cost-effectiveness: The production of C-based MIPs at industrial levels faces major obstacles because of unmet requirements for scalability and cost-effectiveness during green synthesis [3, 4, 15]. Adopting manufacturing tactics for process efficiency as well as raw material cost reduction and workflow simplification will make C-based MIPs more accessible to industries.

Template leakage and imprint stability: Molecular imprint stability in C-based MIPs remains a challenge when operated in conditions that change over time [10]. The implementation of important methods to protect imprints must be combined with measures to prevent template loss and enhance MIP resilience under diverse environmental conditions because this supports practical application needs.

Tailored molecular recognition: Controlling how MIPs recognize molecules by selectivity, affinity, and binding speed remains the main obstacle for C-based MIP development. Thus, multiple optimization strategies combined with computer models and automated testing systems must be developed to improve this process toward precise sensor fabrication [69].

Multifunctionality and integration: Extending the scope of C-based MIPs toward multifunctionality like catalysis, sensing, or stimuli responsiveness offers new possibilities along with challenges. While the critical importance of structural performance within the MIP matrix still has to be addressed to achieve a mixed-functional MIP, it remains a challenge to achieve [69].

Low quantum yield: The practical and large-scale applications of CDs may be limited due to their low quantum yields [59]. As a result, creative synthetic methods should be carefully devised for the effective synthesis of CDs with high purity and quantum yields. Additionally, some C-MIPs based on CDs have a long-term impact on the environment since they are not biodegradable or take time to degrade. Furthermore, non-biodegradable materials lead to permanent pollution, as such materials discarded after use are too dangerous. New techniques for synthesis and surface modification are required to enhance the biocompatibility and homogeneity of CDs. However, the complex polymerization processes involved in the preparation of MIPs require highly demanding optimization. This can be simplified by integrating molecular simulation technology into experimental research. Research should focus on creating sensitive and resourceful sensor platforms through the development and synthesis of new functional monomers, cross-linkers, and competent polymerization techniques [59].

Recyclability and reusability: While MIPs are designed to be reusable, the efficiency of these materials often decreases over multiple uses, necessitating frequent replacements. This issue raises concerns about the sustainability of their lifecycle.
Biodegradability: Most C-based MIPs are non-biodegradable or their degradation takes many years; thus, the short-term environmental impact is visible. Non-biodegradable nature leads to long-lasting pollution when these substances are disposed of after usage.

Environmental impact of synthesis: The MIPs based on C contain many aggressive chemicals and require high energy consumption during their synthesis, thus impacting the environment negatively [3, 10, 29, 89]. For example, organic solvents are widely used in the polymerization process, which poses a major environmental threat.

12.7.2 Future directions

Advanced characterization techniques: The characterization techniques, including atomic force microscopy, electron paramagnetic resonance spectroscopy, and in situ spectroscopy, aid in understanding the molecular structure of C-based MIPs. Thus, the deeper study of these techniques contributes to the study and optimization of MIPs for specific environmental applications.

Bioinspired design and biomimetic strategies: Inspired by biological systems and governed by biomimetic principles, one can think of new approaches for the design and synthesis of C-based MIPs with better performance and functionality [68]. Mimicry of natural patterns in regard to recognition motifs, hierarchical structures, and self-assembly will open up chances for MIPs with better molecular recognition and environmental adaptability.

Smart and responsive materials: The incorporation of stimuli-responsive components like pH, temperature, and light-sensitive polymers or molecules (e.g., based on photo-, thermo-, or photo-responsive (P) polymers) into C-based MIPs could also lead to SMART materials with adaptive properties [4, 10, 59]. The future of stimulus-based MIPs as environmental sensors, controlled release agents, and dynamic pollutant scavengers for responsive and self-adjusting environmental microdevices is an area that merits exploration [4, 10, 59].

Multiscale modeling and simulation: Using multiscale modeling and simulation methods, including density functional theory calculations, molecular dynamics simulations, and computational fluid dynamics simulations, is helpful for the predictive design and optimization of C-based MIPs [10]. By combining computational tools with experimental techniques, the screening for MIPs with designed properties and performance is accelerated.

Green synthesis methods: It is important to develop eco-friendly synthesis methods. In this regard, the use of water as a solvent and the application of green chemistry principles to reduce waste and energy consumption will have much lesser environmental impacts than C-based MIPs [1, 4, 10, 15].

Biodegradable MIPs: The investigation of biodegradable C-based MIPs is critical. The use of biodegradable linkers or bio-based monomers can give rise to materials that can decompose naturally, minimizing environmental impact [1, 3, 4, 10, 15, 58].

Enhanced lifecycle management: Strategies to design MIPs that are more recyclable, sustainable, or reusable, such as developing new binding sites based on the type of polymer used, which allow the MIPs to perform well even after repeated use by being engineered to be more efficient, can be implemented [1, 3, 4, 10, 15, 58].

Application in green technologies: Wider application of C-based MIPs in green technologies such as water purification, CO_2 capture, and sustainable agriculture would allow the full exploitation of their potential with minimal further environmental impact [10, 89].

With that in mind, the green features of C-based MIPs are key factors in their continuing sustainability. By designing new synthesis approaches together with biodegradation possibilities and disposal of polymer products' stocks today's problems may be solved opening the path to more effective use of these materials. Thus, focusing on these aspects of the field can provide a lot to the overall mission of sustainable chemistry and environmental protection.

12.8 Conclusion

In conclusion, the emergence of the green aspects related to the theoretical analysis of C-based MIPs indicates that the utilization of eco-friendly and sustainable materials can solve many currently existing environmental problems. It is, therefore, possible to have molecularly imprinted polymers with tailored functionalities for application in environmental remediation, resource recovery, and monitoring of pollution with the incorporation of carbonaceous substances like graphene, CNTs, or activated carbon into the polymer matrix. Green synthesis routes also known as clean synthesis techniques, which include solventless process, microwave-assisted synthesis, and supercritical fluid technologies help in the optimization of resource utilization, minimization of waste generation, and practicing environment-friendly synthesis methods. In addition, the selectivity and flow-through nature of C-based MIP can be applied to cleaning the air, water, and ground from pollutant substances and contribute to the protection of ecosystems and people's health. It is possible to conclude that the utilization of green chemistry and sustainable material design strategies will help chemists further enhance the opportunities of C-based MIPs in the context of environmental responsibility and stability. Bound to the bewildering synergy between human actions and the environment, green aspects in MIPs point toward a brighter and environmentally friendly future.

References

[1] Del Sole R., et al. (2021). Green aspects in molecularly imprinted polymers by biomass waste utilization, *Polymers*. 13, doi: 10.3390/polym13152430.

[2] Li F., et al. (2024). Application of molecularly imprinted polymers in the water environmental field: A review on the detection and efficient removal of emerging contaminants, *Materials Today Sustainability*. 27, 100904.

[3] Yıldız Ü. Y., et al. (2024). Chapter 3 – Green approaches for the preparation of molecularly imprinted polymers, Hussain C. M., & Keçili R. (eds.) *Green Imprinted Materials*. Elsevier, 69–94.

[4] Donato L., et al. (2022). Green chemistry and molecularly imprinted membranes, *Membranes*. 12, doi: 10.3390/membranes12050472.

[5] Anastas P., & Eghbali N. (2010). Green chemistry: Principles and practice, *Chemical Society Reviews*. 39(1), 301–312.

[6] Luo Q., et al. (2022). Non-toxic fluorescent molecularly imprinted hydrogel based on wood-derived cellulose nanocrystals and carbon dots for efficient sorption and sensitive detection of tetracycline, *Industrial Crops and Products*. 177, 114528.

[7] Espinosa-García B. M., et al. (2007). Molecularly imprinted chitosan– Genipin hydrogels with recognition capacity toward o-xylene, *Biomacromolecules*. 8(11), 3355–3364.

[8] Keçili R., Hussain C. G., & Hussain C. M. (2024). Molecularly imprinted polymer (MIP)-coated green carbon dots-based sensing platforms, *Green Imprinted Materials*. Elsevier, 435–452.

[9] Li Y., et al. (2024). Smartphone-integrated molecularly imprinted ratiometric fluorescent sensor for selective and visual detection of doxycycline in lake water and foodstuff, *ACS Sustainable Chemistry & Engineering*. 12(2), 1062–1071.

[10] Arabi M., et al. (2021). Molecular imprinting: Green perspectives and strategies, *Advanced Materials*. 33(30), 2100543.

[11] Le Goff N., et al. (2020). Renewable plant oil-based molecularly imprinted polymers as biopesticide delivery systems, *ACS Sustainable Chemistry & Engineering*. 8(42), 15927–15935.

[12] Deng H., Wei Z., & Wang X. (2017). Enhanced adsorption of active brilliant red X-3B dye on chitosan molecularly imprinted polymer functionalized with Ti (IV) as Lewis acid, *Carbohydrate Polymers*. 157, 1190–1197.

[13] Yoon Y. H., et al. (2023). Effects of hydrothermal treatment of cellulose nanocrystal templated TiO2 films on their photodegradation activity of methylene blue, methyl orange, and rhodamine B, *Ceramics International*. 49(2), 2911–2922.

[14] Chattopadhyay D., et al. (2024). *Stimuli-responsive Hydrogels for Ophthalmic Drug Delivery*, Woodhead Publishing.

[15] Keçili R., Hussain C. G., & Hussain C. M. (2024). Chapter 14 – Molecularly imprinted polymer (MIP)-coated green carbon dots-based sensing platforms, Hussain C. M., & Keçili R. (eds.) *Green Imprinted Materials*. Elsevier, 435–452.

[16] Liu H., et al. (2019). A facile, green synthesis of biomass carbon dots coupled with molecularly imprinted polymers for highly selective detection of oxytetracycline, *Journal of Industrial and Engineering Chemistry*. 69, 455–463.

[17] Shariati R., et al. (2019). Application of coated green source carbon dots with silica molecularly imprinted polymers as a fluorescence probe for selective and sensitive determination of phenobarbital, *Talanta*. 194, 143–149.

[18] Fang M., et al. (2019). A molecularly imprinted polymers/carbon dots-grafted paper sensor for 3-monochloropropane-1,2-diol determination, *Food Chemistry*. 274, 156–161.

[19] Jalili R., & Amjadi M. (2018). Bio-inspired molecularly imprinted polymer–green emitting carbon dot composite for selective and sensitive detection of 3-nitrotyrosine as a biomarker. *Sensors and Actuators B: Chemical*, 255, 1072–1078.

[20] Demir B., et al. (2018). Tracking hyaluronan: Molecularly imprinted polymer coated carbon dots for cancer cell targeting and imaging, *ACS Applied Materials & Interfaces*. 10(4), 3305–3313.

[21] Sun X., et al. (2019). Synthesis of molecularly imprinted fluorescent probe based on biomass-derived carbon quantum dots for detection of mesotrione, *Analytical and Bioanalytical Chemistry*. 411(21), 5519–5530.

[22] Dai J., Dong X., & Fidalgo de Cortalezzi M. (2017). Molecularly imprinted polymers labeled with amino-functionalized carbon dots for fluorescent determination of 2,4-dinitrotoluene, *Microchimica Acta*. 184(5), 1369–1377.

[23] Yuan Q., et al. (2020). Selective adsorption and photocatalytic degradation of extracellular antibiotic resistance genes by molecularly-imprinted graphitic carbon nitride, *Environmental Science & Technology*. 54(7), 4621–4630.

[24] Yu L., et al. (2024). 2-(Perfluorooctyl) ethyl acrylate functionalized pervoskite: A selective and stable photoelectrochemical platform for ultrasensitive detection of cholesterol, *Microchemical Journal*. 202, 110792.
[25] Xie C., et al. (2023). Red emissive N-doped carbon dots encapsulated within molecularly imprinted polymers for optosensing of pyrraline in fatty foods, *Microchimica Acta*. 190(3), 88.
[26] Haupt K. (2003). Peer reviewed: Molecularly imprinted polymers: The next generation, *Analytical Chemistry*. 75(17), 376A–383A.
[27] Jablonský M., et al. (2020). Involvement of deep eutectic solvents in extraction by molecularly imprinted polymers – A minireview, *Crystals*. 10(3), 217.
[28] Díaz-Álvarez A. E., et al. (2014). Recent advances in the use of glycerol as green solvent for synthetic organic chemistry, *Current Green Chemistry*. 1(1), 51–65.
[29] Wood C. D., Cooper A. I., & DeSimone J. M. (2004). Green synthesis of polymers using supercritical carbon dioxide, *Current Opinion in Solid State and Materials Science*. 8(5), 325–331.
[30] Smith E. L., Abbott A. P., & Ryder K. S. (2014). Deep eutectic solvents (DESs) and their applications, *Chemical Reviews*. 114(21), 11060–11082.
[31] O'Neal K. L., et al. (2010). Fluorous media for extraction and transport, *Journal of Chromatography A*. 1217(16), 2287–2295.
[32] Mateti S., et al. (2021). Mechanochemistry: A force in disguise and conditional effects towards chemical reactions, *Chemical Communications*. 57(9), 1080–1092.
[33] Krajewski M. (2017). Magnetic-field-induced synthesis of magnetic wire-like micro-and nanostructures, *Nanoscale*. 9(43), 16511–16545.
[34] Dąbrowska S., et al. (2018). Current trends in the development of microwave reactors for the synthesis of nanomaterials in laboratories and industries: A review, *Crystals*. 8(10), 379.
[35] Ding Q., et al. (2021). Biomass activated carbon–derived imprinted polymer with multi-boronic acid sites for selective capture of glycoprotein, *Journal of Colloid and Interface Science*. 596, 225–232.
[36] Unterlass M. M. (2016). Green synthesis of inorganic–organic hybrid materials: State of the art and future perspectives, *European Journal of Inorganic Chemistry*. 2016(8), 1135–1156.
[37] Guo L., et al. (2019). Preparation of dual-dummy-template molecularly imprinted polymers coated magnetic graphene oxide for separation and enrichment of phthalate esters in water, *Chemical Engineering Journal*. 361, 245–255.
[38] Lai C., et al. (2016). Synthesis of surface molecular imprinted TiO2/graphene photocatalyst and its highly efficient photocatalytic degradation of target pollutant under visible light irradiation, *Applied Surface Science*. 390, 368–376.
[39] Sa-nguanprang S., Phuruangrat A., & Bunkoed O. (2024). A magnetic adsorbent of nitrogen-doped graphene quantum dots, zinc metal-organic framework and molecularly imprinted polymer to extract phenylureas, *Journal of Food Composition and Analysis*. 126, 105911.
[40] Lu Y., et al. (2020). Electrochemical determination of rutin based on molecularly imprinted poly (ionic liquid) with ionic liquid-graphene as a sensitive element, *Sensors and Actuators B: Chemical*. 311, 127911.
[41] Yuan Y., et al. (2024). Green synthesis of water-compatible molecularly imprinted resin on graphene oxide for highly selective extraction of chlorogenic acid in aqueous systems††Electronic supplementary information (ESI) available, *Green Chemistry*. 26(11), 6787–6798, *See* doi: https://doi.org/10.1039/d4gc00626g.
[42] Li G., et al. (2024). Fabrication of molecularly imprinted carbon nanotubes integrating ionic liquids for efficient detection of perfluoroalkyl carboxylic acid in environmental water. *Talanta*, 275, 126017.
[43] Ding H., et al. (2016). Preparation and characterization of biocompatible molecularly imprinted poly (ionic liquid) films on the surface of multi-walled carbon nanotubes, *RSC Advances*. 6(49), 43526–43538.

[44] Liu H., et al. (2019). A fluorescent nanoprobe for 4-ethylguaiacol based on the use of a molecularly imprinted polymer doped with a covalent organic framework grafted onto carbon nanodots, *Microchimica Acta*. 186, 1–10.

[45] Munawar A., et al. (2018). Investigating nanohybrid material based on 3D CNTs@Cu nanoparticle composite and imprinted polymer for highly selective detection of chloramphenicol, *Journal of Hazardous Materials*. 342, 96–106.

[46] Macagnano A., et al. (2024). Nanofibrous conductive sensor for limonene: One-step synthesis via electrospinning and molecular imprinting, *Nanomaterials*. 14, doi: 10.3390/nano14131123.

[47] Liu X.-L., et al. (2018). Green synthesis of carbon nanotubes-reinforced molecularly imprinted polymer composites for drug delivery of fenbufen, *AAPS PharmSciTech*. 19(8), 3895–3906.

[48] Li H., et al. (2018). Single-hole hollow molecularly imprinted polymer embedded carbon dot for fast detection of tetracycline in honey, *Talanta*. 185, 542–549.

[49] Xu L., et al. (2019). Carbon dots embedded metal-organic framework@ molecularly imprinted nanoparticles for highly sensitive and selective detection of quercetin, *Sensors and Actuators B: Chemical*. 286, 321–327.

[50] Shirani M. P., Rezaei B., & Ensafi A. A. (2019). A novel optical sensor based on carbon dots embedded molecularly imprinted silica for selective acetamiprid detection. *Spectrochimica Acta Part A, Molecular and Biomolecular Spectroscopy*. 210, 36–43.

[51] Zhang D., et al. (2019). Rapid detection of tryptamine by optosensor with molecularly imprinted polymers based on carbon dots-embedded covalent-organic frameworks, *Sensors and Actuators B: Chemical*. 285, 546–552.

[52] Xu X., et al. (2018). Carbon dots coated with molecularly imprinted polymers: A facile bioprobe for fluorescent determination of caffeic acid, *Journal of Colloid and Interface Science*. 529, 568–574.

[53] Wang M., et al. (2019). A magnetic and carbon dot based molecularly imprinted composite for fluorometric detection of 2,4,6-trinitrophenol, *Microchimica Acta*. 186(2), 86.

[54] Jiao Z., et al. (2018). A molecularly imprinted chitosan doped with carbon quantum dots for fluorometric determination of perfluorooctane sulfonate, *Microchimica Acta*. 185, 1–9.

[55] Hou J., et al. (2016). Rapid microwave-assisted synthesis of molecularly imprinted polymers on carbon quantum dots for fluorescent sensing of tetracycline in milk, *Talanta*. 146, 34–40.

[56] Fang M., et al. (2019). Fluorescent probe based on carbon dots/silica/molecularly imprinted polymer for lysozyme detection and cell imaging, *Analytical and Bioanalytical Chemistry*. 411, 5799–5807.

[57] Lv P., Xie D., & Zhang Z. (2018). Magnetic carbon dots based molecularly imprinted polymers for fluorescent detection of bovine hemoglobin. *Talanta*, 188, 145–151.

[58] Villarreal-Lucio D. S., et al. (2022). Molecularly imprinted polymers for environmental adsorption applications, *Environmental Science and Pollution Research*. 29(60), 89923–89942.

[59] Keçili R., Hussain C. G., & Hussain C. M. (2023). Fluorescent nanosensors based on green carbon dots (CDs) and molecularly imprinted polymers (MIPs) for environmental pollutants: Emerging trends and future prospects, *Trends in Environmental Analytical Chemistry*. 40, e00213.

[60] Shen F., et al. (2020). Recent advances in mechanochemical production of chemicals and carbon materials from sustainable biomass resources, *Renewable and Sustainable Energy Reviews*, 130, 109944.

[61] Amiri A., Faridbod F., & Zoughi S. (2024). Selective and rapid optical detection of citalopram using a fluorescent probe based on carbon quantum dots embedded in silica molecularly imprinted polymer, *Journal of Fluorescence*. 34(3), 1171–1181.

[62] Wang L., et al. (2024). Carbonized polymer dots-based molecular imprinting: An adsorbent with enhanced selectivity for highly efficient recognition and removal of ceftiofur sodium from complex samples, *Journal of Hazardous Materials*. 473, 134637.

[63] Poma A., et al. (2013). Solid-phase synthesis of molecularly imprinted polymer nanoparticles with a reusable template–"plastic antibodies", *Advanced Functional Materials*. 23(22), 2821–2827.

[64] Ansari S., & Karimi M. (2017). Recent configurations and progressive uses of magnetic molecularly imprinted polymers for drug analysis, *Talanta*. 167, 470–485.
[65] Yang L., et al. (2021). Ionic liquid functionalized 3D graphene-carbon nanotubes–AuPd nanoparticles–molecularly imprinted copolymer based paracetamol electrochemical sensor: Preparation, characterization and application, *Talanta*. 224, 121845.
[66] Furtado A. I., Viveiros R., & Casimiro T. (2021). MIPMolecularly imprinted polymers (MIP)synthesis and processing using supercritical fluids, Martín-Esteban A. (ed.) *Molecularly Imprinted Polymers: Methods and Protocols*. Springer US, New York, NY, 19–42.
[67] Soares da Silva M., et al. (2012). Supercritical CO2-assisted preparation of a PMMA composite membrane for bisphenol A recognition in aqueous environment, *Chemical Engineering Science*. 68(1), 94–100.
[68] Cui J., et al. (2016). Bio-inspired synthesis of molecularly imprinted nanocomposite membrane for selective recognition and separation of artemisinin, *Journal of Applied Polymer Science*. 133(19), 43405.
[69] Aoulad El Hadj Ali Y., et al. (2024). Progress and prospects in the green synthesis of molecularly imprinted polymers for sorptive extraction and sensing applications toward emerging contaminants in various sample matrices, *TrAC Trends in Analytical Chemistry*. 170, 117466.
[70] Shirani M. P., et al. (2021). Development of an eco-friendly fluorescence nanosensor based on molecularly imprinted polymer on silica-carbon quantum dot for the rapid indoxacarb detection, *Food Chemistry*. 339, 127920.
[71] Wang K., et al. (2023). A molecular imprinted fluorescence sensor based on carbon quantum dots for selective detection of 4-nitrophenol in aqueous environments, *Marine Pollution Bulletin*. 187, 114587.
[72] Zhang L., et al. (2020). Theoretical and experimental studies of a novel electrochemical sensor based on molecularly imprinted polymer and GQDs-PtNPs nanocomposite, *Microchemical Journal*. 158, 105196.
[73] Huang Y., et al. (2024). A perspective on molecular recognition technology for recovering critical metals from minerals and processing wastes, *Separation and Purification Technology*. 347, 127734.
[74] Musarurwa H., & Tavengwa N. T. (2021). Sustainable extraction of pesticides in food and environmental samples using emerging green adsorbents, *Sustainable Chemistry and Pharmacy*. 24, 100545.
[75] Zhang S., et al. (2023). Ratiometric fluorescence probe molecularly imprinted polymer encapsulating N, S, B doped carbon nanodots from waste Clematis Chinensis Osbeck for sensing furazolidone, *Carbon*. 213, 118213.
[76] Fang L., et al. (2021). Efficient removal of norfloxacin in water using magnetic molecularly imprinted polymer, *Chemosphere*. 262, 128032.
[77] Li X., et al. (2021). Targeted degradation of refractory organic compounds in wastewaters based on molecular imprinting catalysts, *Water Research*. 203, 117541.
[78] Shen X., et al. (2009). Photocatalytic removal of pentachlorophenol by means of an enzyme-like molecular imprinted photocatalyst and inhibition of the generation of highly toxic intermediates, *New Journal of Chemistry*. 33(11), 2278–2285.
[79] Okhokhonin A., et al. (2021). Enzymeless electrochemical glucose sensor based on carboxylated multiwalled carbon nanotubes decorated with nickel(II) electrocatalyst and self-assembled molecularly imprinted polyethylenimine, *Electroanalysis*. 33(1), 111–119.
[80] Rai V. K., Nishchay K., & Gupta G. D. (2023). Chapter 9 – Molecularly imprinted polymers for drug delivery and biomedical applications, Vyas S. P., Agrawal U., & Sharma R. (eds.) *Smart Polymeric Nano-Constructs in Drug Delivery*. Academic Press (Elsevier), 249–287.
[81] Farshi Azhar F., Ahmadi M., & Khoshmaram L. (2024). Chitosan-graphene quantum dot-based molecular imprinted polymer for oxaliplatin release. *Journal of Biomaterials Science*, Polymer Edition, 35(14), 2115–2136.

[82] Keith L. H., Gron L. U., & Young J. L. (2007). Green analytical methodologies, *Chemical Reviews*. 107(6), 2695–2708.

[83] Kaya S. I., Cetinkaya A., & Ozkan S. A. (2022). Green analytical chemistry approaches on environmental analysis, *Trends in Environmental Analytical Chemistry*. 33, e00157.

[84] de la Guardia M., & Garrigues S. (2012). *Handbook of Green Analytical Chemistry*. John Wiley & Sons, 794.

[85] Sajid M., & Płotka-Wasylka J. (2022). Green analytical chemistry metrics: A review, *Talanta*. 238, 123046.

[86] Yang L., et al. (2024). Extraction and electrochemical sensing of anthocyanins in berry fruits by use of carbon nanotube-based electrode, *Journal of Food Measurement and Characterization*. 18, 7019–7028.

[87] Ma X., et al. (2024). Exploring the influence of sodium nitrite on the early-age freeze resistance of low-carbon sulphoaluminate cement (SAC), *Journal of Building Engineering*. 84, 108489.

[88] Sedehi S., et al. (2024). A comparative evaluation of analytical green metrics for microextraction techniques based on polymeric and gel membranes as solid support, *TrAC Trends in Analytical Chemistry*. 178, 117840.

[89] Preda D., et al. (2022). Recent trends in the development of carbon-based electrodes modified with molecularly imprinted polymers for antibiotic electroanalysis, *Chemosensors*. 10, doi: 10.3390/chemosensors10070243.

Smrita Singh*, Tanzeel Ahmed, and Ashutosh Singh Chauhan

Chapter 13
Challenges and future prospects of carbon-based molecularly imprinted polymers

Abstract: Molecularly imprinted polymers (MIPs) are a hot topic in science because they excel at grabbing and holding onto specific molecules. These man-made receptors are like tiny, custom-designed gloves, perfect for applications ranging from delivering medicine to speeding up chemical reactions. Scientists are particularly excited about MIPs made with carbon. These "carbon-based MIPs" offer advantages over traditional MIPs, such as being tougher in extreme temperatures and resistant to harsh chemicals. This resilience makes them ideal for use in challenging environments. Carbon's versatility is another win. Whether it's using carbon nanotubes, graphene, or activated carbon, scientists can tweak the physical and chemical properties of carbon-based MIPs. This fine-tuning creates a better "mold" for target molecules, leading to more effective binding. However, even carbon-based MIPs have hurdles to overcome. The limited attachment points on carbon's surface can make it tricky to create strong bonds with the target molecules, potentially reducing the number of available binding sites. Additionally, carbon's rigid structure can sometimes make it difficult for the target molecules to reach those binding sites. Researchers are actively searching for solutions to these challenges. One approach is to modify the carbon surface to increase its reactivity, while another focuses on refining the production process to improve binding efficiency. Scientists are also exploring hybrid materials, combining carbon with other polymers or inorganic materials to enhance performance. The future of carbon-based MIPs is bright. As research progresses, we can anticipate exciting discoveries of new materials and techniques that will further enhance their selectivity, sensitivity, and versatility. These advancements will pave the way for their widespread use in various industries, from purifying chemicals and detecting pollutants to delivering drugs with pinpoint accuracy.

Acknowledgement: I sincerely thank Prof. M. P. Pandey, Dr. Sanjeev Agrawal, and Prof. Vaibhav Trivedi for their steadfast support in the writing of this book chapter. I also gratefully acknowledge the contributions of the authors whose work greatly informed this study, including Vasapollo et al. (2011), Cabaleiro-Lago et al. (2023), and Preda et al. (2022). and team MDPI.

***Corresponding author: Smrita Singh,** School of Biotechnology, IFTM University, Lodhipur Rajput, Moradabad, Uttar Pradesh, India, email id: smrita1@gmail.com
Tanzeel Ahmed, Professor and Director, School of Biotechnology, IFTM University, Moradabad, Uttar Pradesh, India
Ashutosh Singh Chauhan, Air Pollution and Stone Conservation Laboratory, Archaeological Survey of India, Agra,Uttar Pradesh, India, e-mail: ashu.chauhan1979@gmail.com

https://doi.org/10.1515/9783111503325-013

Keywords: Molecularly imprinted polymers, graphene, nanotubes, functional monomer, hybrid system, synthetic receptors

13.1 Introduction of molecularly imprinted polymers and their history

Molecularly imprinted polymers (MIPs) are synthetic materials engineered to mimic the recognition abilities of biological macromolecules like antibodies and enzymes [1]. The concept of molecular imprinting was first introduced in the 1930s, but gained significant attention in the 1970s. The key principle underlying MIPs is the creation of specific binding sites within a polymeric matrix that are complementary in size, shape, and functionality to a target molecule of interest.

The process of producing MIPs typically involves three main steps: forming a complex between a template molecule and functional monomers, polymerizing this complex within a cross-linking matrix, and subsequently removing the template to leave behind binding sites specific to the target molecule. These binding sites can then selectively recognize and bind the target molecule, even in complex mixtures.

MIPs have been explored for a wide range of applications, including drug delivery, sensing, separation, and catalysis [2–4]. In the biomedical field, they have shown promise for controlled drug release, diagnostic assays, and tissue engineering. One innovative application is their use in contact lenses, where researchers have developed silicone hydrogel materials with MIP binding sites to control the release of therapeutic agents that may improve contact lens comfort. Another area of interest is the use of MIPs for the localized and on-demand inhibition of matrix metalloproteinases, which are involved in extracellular matrix degradation and various diseases [5].

The development of MIPs has been influenced by the growing interest in biomimetic and bioinspired materials, with researchers drawing inspiration from natural systems like the adhesive properties of mussels to create enhanced polymeric materials.

MIPs have garnered significant attention in the scientific community due to their remarkable selectivity and affinity towards target analytes [6]. These synthetic receptors have found applications in diverse fields, from drug delivery to catalysis [7]. One particular research focus is the development of carbon-based MIPs, which hold promise for addressing various challenges associated with traditional imprinted polymers.

The enhanced thermal and chemical stability of carbon-based MIPs makes them suitable for applications in harsh environments [8]. Furthermore, the versatility of carbon materials, such as carbon nanotubes (CNTs), graphene, and activated carbon, allows for the tailoring of their physical and chemical properties to optimize the imprinting process and improve the recognition capabilities of the resulting polymers [9, 10].

However, the development of carbon-based MIPs is not without its challenges. The limited number of functional groups on the carbon surface can hinder the formation of strong interactions between the template molecule and the functional monomers, leading to a decreased number of recognition sites. Additionally, the accessibility of the imprinted cavities to the target analyte can be compromised due to the inherent rigidity and structural complexity of carbon materials.

To address these challenges, researchers have explored strategies such as the incorporation of additional functional groups or the optimization of the imprinting process to enhance the recognition capabilities of carbon-based MIPs. Furthermore, the integration of carbon-based materials with other polymeric or inorganic materials has been investigated to create hybrid systems with improved performance.

Looking to the future, the continued development of carbon-based MIPs holds great promise for advancements in areas such as selective separation, sensing, and drug delivery. As research in this field progresses, we can expect the emergence of novel materials and strategies that will further enhance the selectivity, sensitivity, and versatility of these synthetic receptors, ultimately leading to their widespread application in various industries and scientific domains.

13.2 Synthesis and characterization of molecularly imprinted polymers

MIPs have emerged as a promising class of materials with significant applications in various fields, including analytical chemistry, drug delivery, and biosensing. The synthesis of these polymers involves the creation of specific molecular cavities within a polymeric matrix, which can selectively bind to a target analyte [1]. The scheme for molecular imprinting is shown in Figure 13.1.

13.2.1 Templates and monomers

The process of synthesizing MIPs typically starts with the selection of a template molecule, which is the target analyte that the polymer is designed to recognize. The template molecule is then combined with functional monomers, which can interact with the template through covalent or non-covalent interactions. Some most common structures of monomers are shown in Figure 13.2, which is used for molecular imprinting.

13.2.2 Polymerization

The functional monomers are then cross-linked using a cross-linking agent, forming a rigid polymeric structure that retains the shape and binding sites of the template molecule. Some cross-linking agents (structures) are displayed in Figure 13.3.

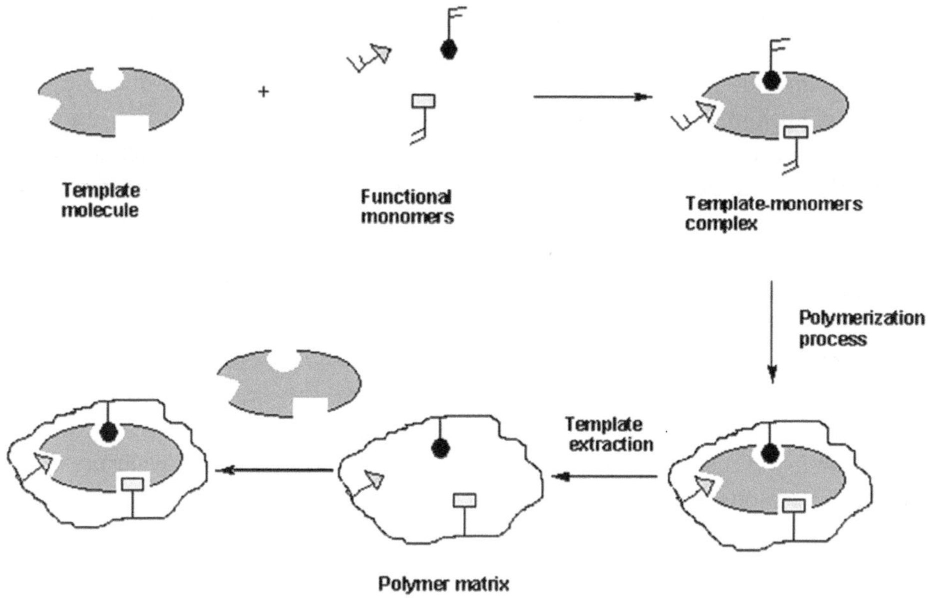

Figure 13.1: Schematic representation of molecular imprinting. Adapted from [9].

13.2.3 Template removal

After the polymerization is complete, the template molecule is removed, leaving behind the imprinted cavities that can selectively bind to the target analyte.

The design and optimization of the molecular imprinting process is crucial for the successful synthesis of these polymers. Factors such as the choice of template molecule, functional monomers, cross-linking agent, and polymerization conditions can all impact the performance of the final polymer. Additionally, the controlled fabrication of hierarchical morphologies, such as dendrites, can enhance the surface area and improve the applicability of these polymers in advanced industrial applications [10].

The unique properties of MIPs, such as their optical and photothermal characteristics, have also been explored for various applications. For instance, the functionalization of light-responsive polymer gels with photo-responsive chromophores can enable shape reconfiguration and motion, making them promising candidates for a range of applications. Furthermore, the development of biomimetic surface modifications, such as mussel-inspired polydopamine coatings, can enhance the water dispersibility, low cytotoxicity, and controlled drug release properties of these polymers, making them highly desirable for biomedical applications.

Additionally, the sequence-controlled synthesis of polymers, such as polyurethane block copolymers, has shown the ability to influence the adsorption of specific pro-

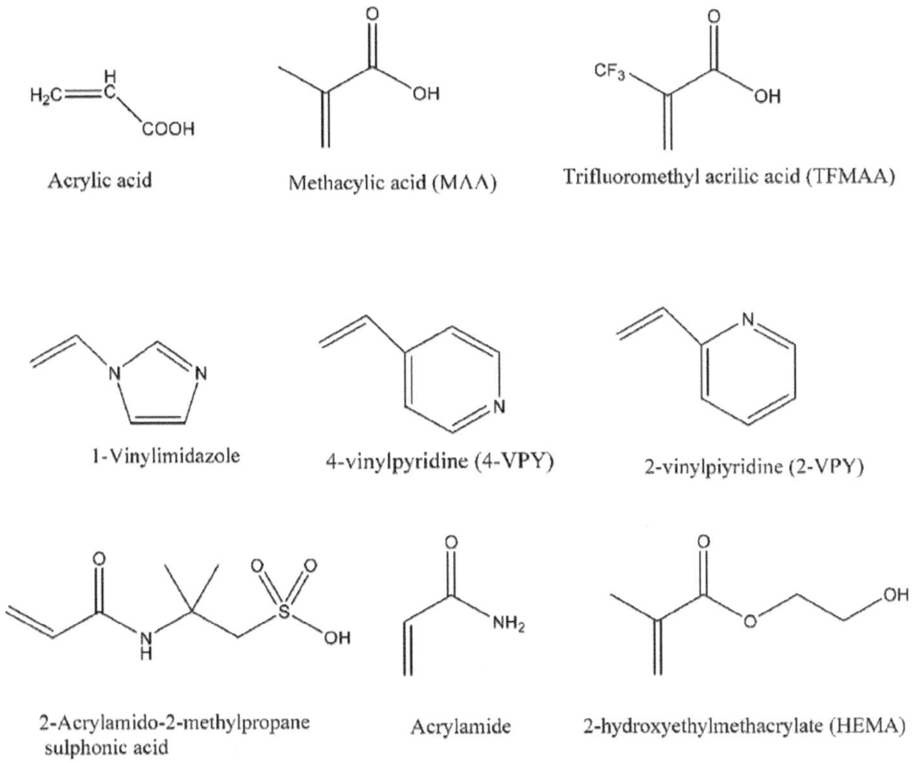

Figure 13.2: Common monomer structures used for molecular imprinting. Adapted from [9].

teins, which is important for the design of immunomodulatory biomaterials [11]. This highlights the importance of understanding the role of monomer sequence.

13.3 Synthesis of carbon-based molecularly imprinted polymers

The synthesis of carbon-based MIPs typically involves three main phases: template selection, monomer selection, and polymerization [12]. The choice of template molecule is critical, as it dictates the specific binding sites and recognition capabilities of the imprinted polymer. Catechol derivatives, both non-phytogenic and phytogenic, have emerged as promising templates due to their versatile adhesive properties and structural features [13].

In the present study, a catechol-based compound was selected as the template molecule. The choice of monomer is equally important, as it must have the ability to interact with the template and form a stable complex during the polymerization pro-

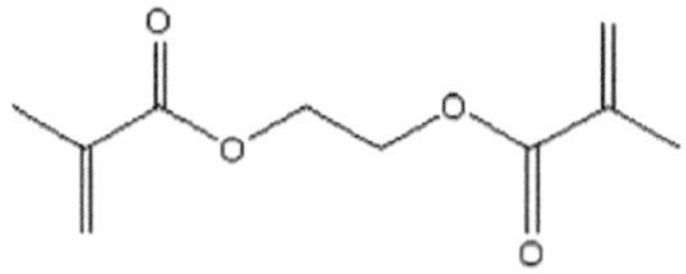

Ethyleneglycoldimethacrylate (EGDMA)

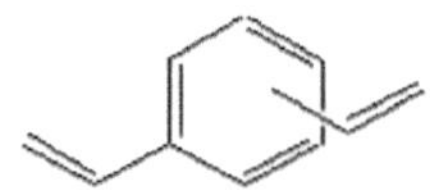

Divinylbenzene (DVB)

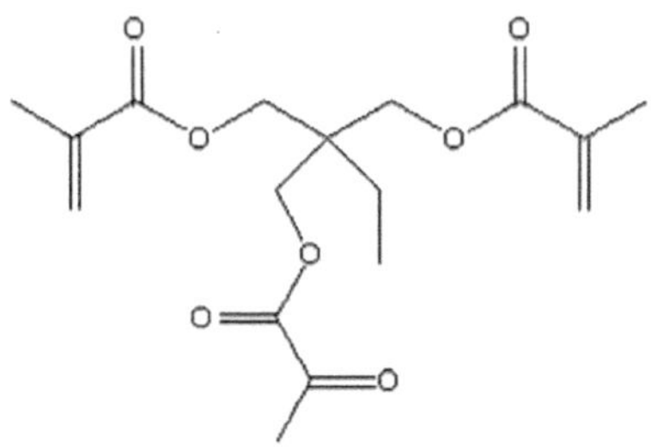

Trimethylolpropane trimethacrylate (TRIM)

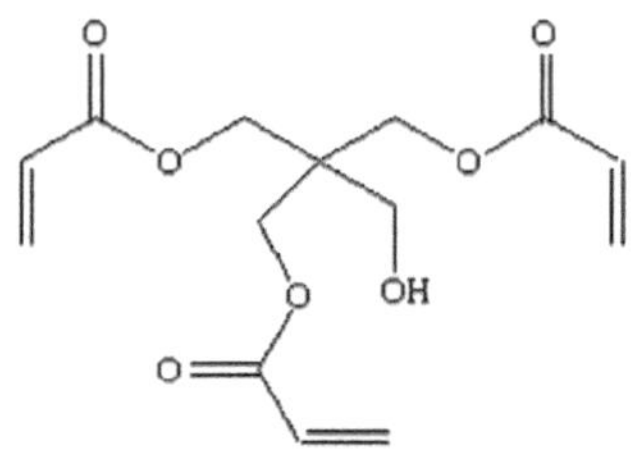

Pentaerythritol triacrylate (PETRA)

Figure 13.3: Cross-linking structures that are commonly used in molecular imprinting. Adapted from [9].

cess. Various monomers, such as acrylic acid, methacrylic acid, and vinylpyrrolidone, have been explored in the synthesis of carbon-based MIPs [11, 15].

The polymerization step involves the formation of a cross-linked polymer network around the template molecule. This is typically achieved through a radical polymerization process, which can be facilitated by initiators or organocatalysts [14]. After polymerization, the template molecule must be removed from the polymer matrix to create the desired imprinted binding sites. This can be accomplished through various extraction methods, such as solvent extraction or thermal treatment.

The synthesized carbon-based MIPs are characterized using a range of techniques, including Fourier-transform infrared spectroscopy, scanning electron microscopy, and nitrogen adsorption-desorption analysis. These characterizations provide insights into the chemical composition, morphology, and porosity of the imprinted polymers [15].

The developed carbon-based MIPs can be applied in various fields, such as sensor development, adsorbent materials, and catalysis [16]. Their high specificity and selectivity towards target analyses make them attractive candidates for numerous industrial and environmental applications.

13.3.1 Synthesis of carbon-imprinted polymers: a comprehensive approach

Carbon-imprinted polymers are a versatile class of materials with diverse applications, ranging from water purification and gas separation to catalysis and sensor development. The synthesis of these polymers follows a systematic approach that integrates the principles of molecular imprinting and carbon materials engineering [10, 14, 17, 18].

The first step in the synthesis process is the selection of a suitable carbon precursor, such as phenolic resins, furfuryl alcohol, or sucrose, and a cross-linking agent like epoxy resins, divinylbenzene, or ethylene glycol dimethacrylate. The choice of these components depends on the desired properties of the final polymer, including pore size, surface area, and mechanical stability.

Next, the imprinting template is prepared by incorporating a target molecule, which can be a small organic molecule, a metal ion, or even a macromolecule, into the polymer matrix during the synthesis.

The polymer is then subjected to a carbonization step, typically at high temperatures under an inert atmosphere, resulting in the formation of a highly porous carbon structure with well-defined pore size and distribution.

The final step involves the removal of the imprinting template, which can be achieved through various methods such as solvent extraction, thermal treatment, or chemical etching, depending on the nature of the template [19].

The resulting carbon-imprinted polymer exhibits a high degree of specificity towards the target molecule, making it a valuable tool for a variety of applications, including selective adsorption, catalysis, and sensing. The versatility of this synthesis approach allows for the tailoring of the polymer's properties to meet the specific requirements of the desired application.

The synthesis process begins with the selection of a suitable carbon precursor, such as sucrose, glucose, or phenolic resins, which is then subjected to carbonization, typically through pyrolysis, to produce a carbon-based material. The next stage involves the imprinting process, where the carbon material is exposed to a target molecule or template, which serves as the imprinting agent. During this step, the template interacts with the carbon surface, creating specific binding sites that can selectively recognize and bind the target molecule.

13.3.2 Use of carbon in molecularly imprinted polymers

MIPs have garnered substantial attention in the scientific community due to their remarkable ability to selectively recognize and bind target molecules [20, 21]. A critical component in the design and development of these polymers is the incorporation of diverse carbon-based materials, such as CNTs, graphene, and activated carbon [22–24].

The utilization of carbon in MIPs serves several vital purposes. Carbon materials exhibit high surface areas, excellent adsorption capabilities, and exceptional mechanical and thermal stability, rendering them ideal for enhancing the performance of these polymers. CNTs, in particular, possess unique structural and electronic properties that can significantly improve the sensitivity, selectivity, and response time of MIP-based sensor. Similarly, graphene's high surface area and exceptional electrical conductivity have been employed to enhance the charge transfer and electrochemical performance of MIPs. Activated carbon, on the other hand, is a versatile form of carbon that can be tailored to have specific pore sizes and surface functionalities, allowing for the optimization of MIP adsorption and recognition properties.

The incorporation of these diverse carbon materials into MIPs has led to the development of highly efficient and selective sensing devices for a wide range of applications, including environmental monitoring, biomedical diagnostics, and pharmaceutical analysis.

13.3.3 Carbon nanotubes

CNTs are cylindrical structures made of carbon atoms arranged in a hexagonal lattice, much like a rolled-up sheet of graphene. They are incredibly small, with diameters typically measuring in nanometers (one billionth of a meter) [25].

13.3.3.1 Types of carbon nanotubes

There are two main types of CNTs:

1. **Single-walled CNTs:** These consist of a single layer of graphene rolled into a tube.
2. **Multi-walled CNTs:** These are made up of multiple concentric layers of graphene, resembling a stack of rolled-up tubes [25].

13.3.3.2 Remarkable properties and applications

CNTs possess a unique combination of properties that make them highly desirable for various applications [32]:

- **Exceptional strength and stiffness:** CNTs are significantly stronger and stiffer than steel, making them ideal for reinforcing composite materials [34].
- **High electrical conductivity:** Depending on their structure, CNTs can exhibit metallic or semiconducting properties, making them suitable for electronics and sensors [32].
- **Excellent thermal conductivity:** CNTs can efficiently conduct heat, making them useful for thermal management applications.
- **Large surface area:** Their high surface area-to-volume ratio makes them suitable for applications like catalysis and energy storage [26].

13.3.3.3 Potential applications

The remarkable properties of CNTs have sparked interest in a wide range of fields, including:

- **Electronics:** CNTs can be used to create transistors, sensors, and other electronic components.
- **Materials science:** They can reinforce composites, enhance the strength of polymers, and create lightweight, high-performance materials [33].
- **Energy storage:** CNTs show promise for use in batteries, supercapacitors, and fuel cells [26].
- **Biomedical applications:** Their unique properties make them attractive for drug delivery, biosensing, and tissue engineering.

13.3.3.4 Challenges and future directions

Despite their potential, challenges remain in realizing the full potential of CNTs, including:

- **Controlled synthesis:** Producing CNTs with specific properties and chirality remains a challenge.
- **Cost-effective production:** Scaling up production to meet commercial demands while maintaining affordability is crucial.
- **Toxicity concerns:** Research is ongoing to understand the potential environmental and health impacts of CNTs.

Despite these challenges, research on CNTs continues to advance, driven by their extraordinary properties and the promise of revolutionary applications [27]. CNTs, in particular, have unique structural and electronic properties that can significantly improve the sensitivity, selectivity, and response time of MIP-based sensors [23].

13.3.4 Graphene

Graphene is a two-dimensional material consisting of a single layer of carbon atoms arranged in a hexagonal lattice [40]. This unique structure gives graphene an array of remarkable properties that have captivated scientists and engineers alike.

13.3.4.1 Exceptional properties

- **Exceptional strength:** Graphene is considered the strongest material ever measured, surpassing even diamond in terms of intrinsic strength [40].
- **High electron mobility:** Electrons can move through graphene with remarkable ease, exhibiting exceptionally high electron mobility at room temperature [40].

- **Excellent thermal conductivity:** Graphene conducts heat more efficiently than any other known material, surpassing even diamond and copper [40].
- **Optical transparency:** Despite its atomic thinness, graphene absorbs a surprisingly small amount of light, making it nearly transparent [40].
- **Impermeability:** Graphene acts as a perfect barrier to gases, even the smallest atoms like helium cannot penetrate its tightly packed structure [40].

13.3.4.2 Potential applications

Graphene's extraordinary properties have sparked immense interest in various fields, including:

- **Electronics:** Graphene's high electron mobility makes it a promising material for high-speed transistors, flexible electronics, and advanced sensors [28].
- **Energy storage:** Its large surface area and excellent electrical conductivity make it suitable for applications in batteries, supercapacitors, and solar cells [29].
- **Materials science:** Graphene can be used to reinforce composites, create ultra-strong and lightweight materials, and develop advanced coatings [35].
- **Biomedical applications:** Graphene's unique properties make it attractive for drug delivery, biosensing, and tissue engineering [30].

13.3.4.3 Challenges and future outlook

Despite its immense potential, challenges remain in realizing the full-scale commercialization of graphene:

- **Cost-effective production:** Producing high-quality graphene on a large scale at an affordable cost remains a challenge [38].
- **Integration into devices:** Integrating graphene into existing manufacturing processes and devices presents technical hurdles.
- **Environmental concerns:** Research is ongoing to understand the potential environmental impact of graphene production and disposal.

Despite these challenges, graphene remains a material of intense research and development. Its exceptional properties continue to inspire scientists and engineers to explore its full potential and drive innovation across various fields [38].

Similarly, graphene, with its high surface area and exceptional electrical conductivity, has been employed to improve the charge transfer and electrochemical performance of MIPs [24].

13.3.5 Activated carbon

Activated carbon is a form of carbon that has been processed to create a highly porous structure with an incredibly large surface area [39]. This unique characteristic makes it exceptionally effective at adsorbing a wide range of substances, both in liquid and gaseous phases.

13.3.5.1 Production process

Activated carbon can be produced from various carbonaceous materials, including:
- **Coconut shells**
- **Nutshells**
- **Coal**
- **Peat**
- **Wood**

The production process typically involves two main steps:
1. **Carbonization:** The raw material is heated to high temperatures in the absence of oxygen, removing volatile components and leaving behind a carbon-rich char.
2. **Activation:** The char is treated with oxidizing agents, such as steam or carbon dioxide, at elevated temperatures. This process creates a network of pores within the carbon structure, significantly increasing its surface area.

13.3.5.2 Adsorption mechanism

Activated carbon's remarkable adsorption capacity stems from its porous structure and the presence of various surface functional groups. These features create a large surface area for interactions with other molecules.

The adsorption process occurs through two main mechanisms:
- **Physical adsorption:** Weak van der Waals forces attract and hold molecules to the surface of the activated carbon.
- **Chemical adsorption:** Stronger chemical bonds form between the adsorbate molecules and the surface functional groups of the activated carbon.

13.3.5.3 Applications

The unique properties of activated carbon make it a versatile material with a wide range of applications, including:
- **Water treatment:** Removing impurities from drinking water and wastewater, such as chlorine, pesticides, and organic compounds.
- **Air purification:** Filtering out volatile organic compounds, odors, and pollutants from air streams.

- **Gas separation and purification:** Separating and purifying gases in industrial processes, such as natural gas processing and hydrogen production.
- **Medical applications:** Treating drug overdoses and poisoning by adsorbing toxins in the digestive system.
- **Food and beverage industry:** Decolorizing sugar solutions, removing unwanted flavors and odors from beverages, and purifying edible oils.

13.3.5.4 Types and characteristics

Activated carbon is available in various forms, including:
- **Powdered activated carbon**
- **Granular activated carbon**
- **Extruded activated carbon**
- **Impregnated activated carbon**

The choice of activated carbon type depends on the specific application and the properties required, such as pore size distribution, surface area, and surface chemistry.

Activated carbon, on the other hand, is a versatile form of carbon that can be tailored to have specific pore sizes and surface functionalities, allowing for the optimization of MIP adsorption and recognition properties [20, 21].

The incorporation of these various carbon materials into MIPs has led to the development of highly efficient and selective sensing devices for a wide range of applications, including environmental monitoring, biomedical diagnostics, and pharmaceutical analysis.

13.4 Carbon in MIPs: enhancing performance for tailored recognition

Because of their special qualities, carbon-based materials such as CNTs, graphene, and activated carbon can be included into MIPs with considerable advantages.

The reasons are described further.

13.4.1 Increased surface area for enhanced binding

- **The problem:** The quantity of binding sites that are available for the target molecule has a significant impact on MIP performance. More surface area means that there are more possible binding sites.
- **Carbon's solution**: CNTs [25], graphene [40], and activated carbon [39] all have extremely high surface area-to-volume ratios.

- **The impact**: Integrating these materials into the MIP matrix significantly boosts the available surface area for imprinting, resulting in higher binding capacity and sensitivity.

13.4.2 Improved selectivity through tailored interactions

- **The problem**: MIPs must be highly selective for their target molecule, even when similar compounds are present.
- **Carbon's solution**: The surface chemistry of carbon materials can be easily altered, allowing for the addition of specific functional groups that interact more selectively with the target molecule.
- **The impact**: This precise tuning of interactions improves the selectivity of the MIP, making it more effective at distinguishing between the target and other molecules.

13.4.3 Enhanced electrical properties for sensor applications

- **The problem**: Many MIP applications, especially in sensing, depend on detecting changes in electrical signals upon target binding.
- **Carbon's solution**: CNTs and graphene are excellent electrical conductors [32].
- **The impact**: Incorporating these materials into MIPs can enhance conductivity and signal transduction, leading to more sensitive and responsive sensors.

13.4.4 Mechanical and thermal stability for durability

- **The problem**: MIPs need to endure various conditions, particularly in real-world applications.
- **Carbon's solution**: Carbon materials are renowned for their excellent mechanical strength and thermal stability [34].
- **The impact**: Adding carbon to MIPs can improve their durability, making them suitable for harsh environments and extending their operational lifespan.

Incorporating carbon-based materials into MIPs is akin to adding secret ingredients to a recipe. They enhance the sensitivity, selectivity, responsiveness, and durability of MIPs. This opens up exciting possibilities for applications in various fields, such as catalysis, separation, and sensing.

13.5 MIP applications: tailored recognition for various fields

MIPs selectively bind to specific target molecules, similar to synthetic antibodies. This unique capability makes them highly beneficial across various fields. Applications of MIPs in various fields are shown in Figure 13.4.

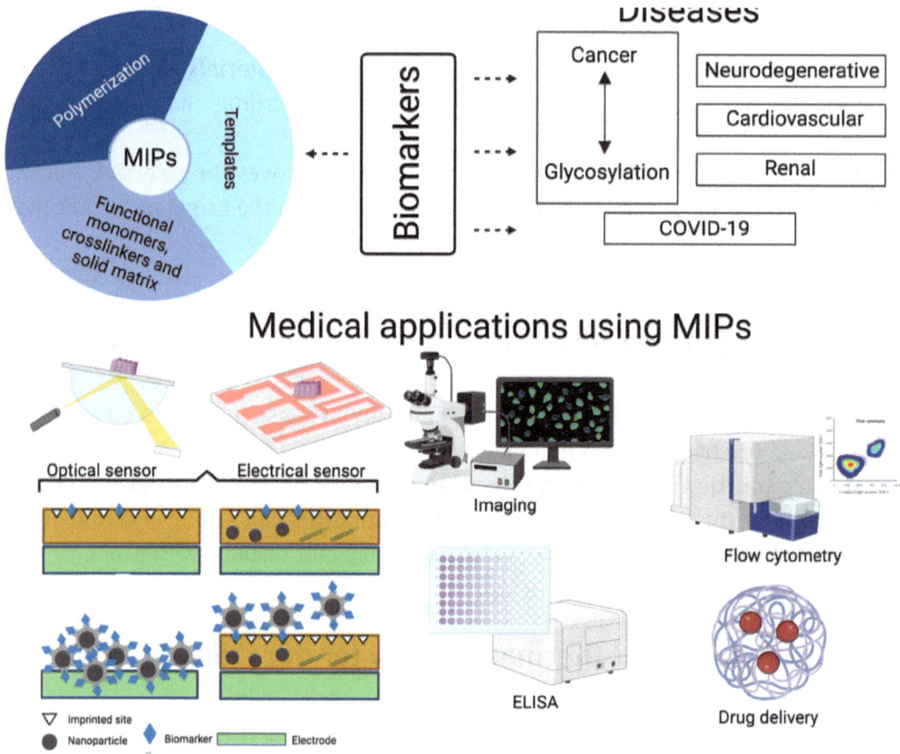

Figure 13.4: Applications of MIPs in various fields. Adapted from Cabaleiro-Lago [41].

13.5.1 Sensors: detecting with specificity

- Environmental monitoring: MIP-based sensors can detect pollutants like pesticides [21], heavy metals, and pharmaceuticals in water and soil samples with high sensitivity and selectivity.
- Food safety: These sensors are capable of identifying contaminants, toxins, and allergens in food products, ensuring consumer safety.
- Medical diagnostics: MIPs have the potential to create quick and affordable diagnostic tools for diseases by detecting biomarkers in bodily fluids.

- Security applications: MIP-based sensors can identify explosives, drugs, and other hazardous substances.

Main constituents of sensing device are shown in Figure 13.5.

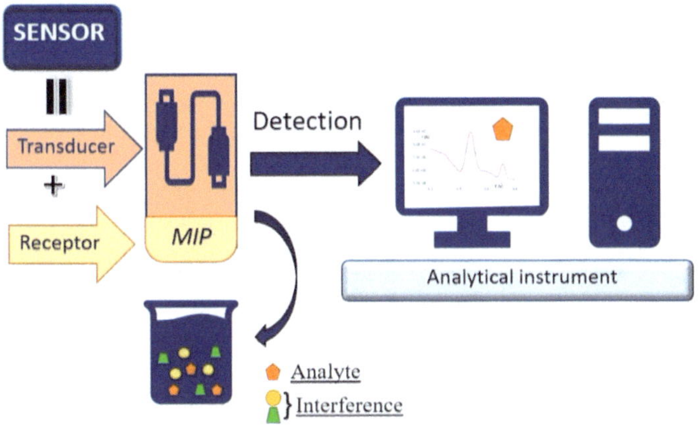

Figure 13.5: Main constituent of sensing device. Adapted from Preda [42].

13.5.2 Separation science: precision purification

- Solid-phase extraction: MIPs can selectively extract target compounds from complex mixtures, such as isolating pharmaceuticals from plant extracts or removing pollutants from water samples [31].
- Chromatographic separations: Using MIPs in chromatographic columns allow highly specific separations of complex mixtures, such as chiral compounds or biomolecules.

There are four steps in molecular imprinting separation phase extraction shown in Figure 13.6.

13.5.3 Drug delivery: targeted therapeutics

- Controlled release: MIPs can be designed to deliver drugs in a controlled manner at specific target locations, improving drug effectiveness and reducing side effects [36].
- Stimuli-responsive delivery: MIPs can be engineered to release drugs at the target site in response to specific stimuli, such as changes in pH or temperature.

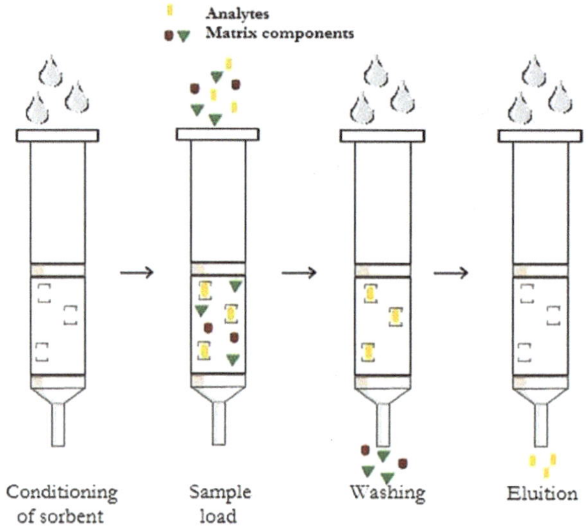

Figure 13.6: Steps of molecular imprinting separation phase extraction. Adapted from Vasapollo et al. [9].

13.5.4 Catalysis: mimicking nature's enzymes

- **Catalytic MIPs:** MIPs can be engineered with catalytic sites that mimic natural enzyme activity, allowing them to catalyze specific chemical reactions [37].
- **Biomimetic catalysis:** MIPs provide a durable and cost-effective alternative to natural enzymes for various industrial and biotechnological applications.

13.5.5 Other emerging applications

- **Anti-counterfeiting:** MIPs can be used to create unique tags and markers for authentication and anti-counterfeiting purposes.
- **Membrane technology:** Incorporating MIPs into membranes can enhance their selectivity for specific molecules, enabling efficient separations and purifications.

Future directions: The field of MIPs is rapidly evolving, with ongoing research uncovering new applications and expanding their potential. As the technology advances and production costs decrease, MIPs are expected to become increasingly integral to various aspects of our lives. Despite their great promise, MIPs also present certain challenges that researchers are actively working to overcome.

13.6 MIP challenges: hurdles on the path to wider adoption

Despite their remarkable capabilities, MIPs face several challenges that hinder their widespread application.

13.6.1 Template-related challenges

- Getting the right template: Finding enough of the pure target molecule to act as a mold for the MIP can be costly and time-consuming, especially for large, complex molecules found in living things.
- Removing the template: After the MIP is formed, getting rid of every trace of the original template molecule can be tricky. Leftover template can mess up the binding sites and make the sensor less effective.
- Template durability: Some target molecules, especially those from living organisms, are delicate. The harsh conditions needed to create MIPs can damage or destroy these sensitive molecules.

13.6.2 Synthesis and optimization hurdles

- Creating consistent binding sites: It's difficult to make all the binding sites within the MIP exactly alike. Variations in their ability to attract and hold onto the target molecule can impact how well the MIP performs.
- Finding the perfect recipe: Figuring out the best combination of ingredients and conditions (like the ratio of starting materials, the liquid used, and the temperature) for a specific target molecule is a complex puzzle that takes time to solve.
- Scaling up production: Taking a process from small-scale lab experiments to large-scale industrial production while ensuring each batch is identical is a major challenge.

13.6.3 Performance obstacles

- Unwanted binding: Sometimes MIPs can bind to molecules other than the intended target, which reduces their accuracy and selectivity.
- Limited capacity: MIPs may not be able to hold as much of the target molecule as natural receptors or antibodies, especially if the target has a strong attraction to the MIP.

- Shape-shifting: Changes in the surrounding liquid or its acidity can cause MIPs to swell or shrink, altering their binding ability and stability.

13.6.4 Proving MIP's work as intended

- **Understanding the binding sites:** Figuring out exactly how the binding sites within a MIP work – how strongly they attract the target, how picky they are, and how many there are – is still tricky.
- **Real-world testing:** To gain approval for widespread use, scientists need to develop reliable ways to prove that MIPs perform well in real-world situations, not just in the lab.

Researchers are actively searching for ways to overcome these obstacles, including:
- **Thinking outside the template box:** Instead of using the full target molecule, scientists are exploring smaller, sturdier molecules that resemble parts of the target as templates.
- **Improving polymerization:** New and improved methods for creating MIPs are being developed to make more consistent binding sites and control the overall shape and structure of the MIP.
- **Customizing MIP design:** By tweaking the chemical makeup and physical structure of MIPs, researchers aim to boost their capacity to hold the target, improve their selectivity, and make them more stable.

As scientists continue to tackle these challenges, MIPs are moving closer to fulfilling their promise of revolutionizing various fields by mimicking nature's remarkable ability to recognize and interact with specific molecules.

13.7 MIPs: emerging trends and a bright future

The field of MIPs is dynamic and rapidly evolving. Here are some exciting emerging trends and future prospects:

13.7.1 New frontiers in material design

13.7.1.1 The next generation of MIPs

- **Smart MIPs:** Imagine MIPs that can change their behavior on demand! These "stimuli-responsive" MIPs can adjust how strongly they bind or how easily mole-

cules pass through them when triggered by things like temperature changes, acidity levels, light, or magnetic fields. This technology opens up possibilities for precisely controlling drug delivery, creating super-sensitive sensors, and developing adaptable systems for separating molecules.
- **MIPs Go Nano:** Shrinking MIPs down to the nanoscale, creating things like tiny particles or ultra-thin fibers, dramatically increases their surface area [36]. This boost translates to a greater capacity to capture target molecules, faster response times, and enhanced sensitivity.
- **Hybrid power:** Combining the unique abilities of MIPs with other materials like metal nanoparticles, quantum dots, or CNTs creates powerful hybrid materials. These combinations unlock synergistic properties, leading to superior performance in sensing, speeding up chemical reactions (catalysis), and targeted drug delivery.

13.7.2 Expanding applications in biomedicine

Targeted drug delivery: MIPs are being investigated for their potential to deliver drugs directly to cancer cells or other disease sites, enhancing therapeutic effectiveness and reducing side effects.
Biosensing and diagnostics: MIP-based sensors are promising for early disease detection, personalized medicine, and point-of-care diagnostics due to their high sensitivity, selectivity, and affordability.
Bioimaging and theranostics: MIPs can be tailored to target specific tissues or cells for imaging purposes or for combined imaging and therapeutic applications (theranostics).

13.7.3 MIPs for a sustainable future

Environmental remediation: MIPs can help clean up the environment by removing pollutants like heavy metals, pesticides, and pharmaceuticals from water and soil.
Resource recovery: MIPs can selectively extract valuable substances, such as precious metals or rare earth elements, from industrial waste streams, supporting resource recovery and circular economy practices.
Food safety and security: MIP-based sensors can ensure food safety by detecting contaminants and toxins, and they can also help maintain food security by monitoring food quality and freshness.

13.7.4 Advancements in characterization and modeling

Computational modeling: Computational methods are increasingly used to design MIPs with better binding properties and to predict their performance, speeding up the development process.
Advanced characterization techniques: New techniques are emerging to gain a deeper understanding of the structure and properties of MIPs at the molecular level, allowing for more rational design and optimization.

13.7.5 Commercialization and industrial adoption

- Cost-effective production: With advancements in research and production techniques, the cost of manufacturing MIPs is anticipated to drop, enhancing their commercial viability.
- Regulatory approval: Initiatives are in progress to create standardized protocols for characterizing and validating MIP performance, facilitating regulatory approval and broader adoption across various industries.

13.7.6 A bright future ahead

The outlook for MIPs is promising, fueled by their adaptability, precision, and customizable nature. As scientists keep advancing the design and applications of MIPs, we can anticipate the emergence of these extraordinary materials.

13.8 Conclusion: carbon-based MIPs – navigating challenges toward a brighter future

Carbon-based MIPs are at an exciting juncture, with the potential to transform various fields despite facing significant obstacles to widespread adoption. The integration of CNTs, graphene, and activated carbon into MIPs has led to notable improvements in sensitivity, selectivity, and mechanical properties. However, challenges such as template synthesis, binding site heterogeneity, and performance optimization remain.

Nevertheless, the future of carbon-based MIPs is promising. Innovations in material design, including stimuli-responsive MIPs and nanostructured composites, are expected to overcome current limitations and unlock new capabilities. As research advances, addressing issues related to cost-effective production, standardization, and regulatory approval will be essential to fully realize the potential of these materials. With ongoing innovation and a focus on applying laboratory successes to real-world scenarios, carbon-based MIPs are poised to address critical challenges in areas such as environmental remediation, biosensing, drug delivery, and beyond.

References

[1] Omidian H., Wilson R., & Babanejad N. (2023, August 1). *Bioinspired polymers: transformative applications in biomedicine and regenerative medicine*. Multidisciplinary Digital Publishing Institute, 13(8), 1673–1673, Basel, Switzerland, https://doi.org/10.3390/life13081673.

[2] Jones L., Byrne M. E., Ciolino J. B., Legerton J., Markoulli M., Papas E., & Subbaraman L. N. (2016, April 1). Revolutionary future uses of contact lenses, *Lippincott Williams & Wilkins*. 93(4), 325–327, https://doi.org/10.1097/opx.0000000000000847.

[3] Wales D. J., Cao Q., Kastner K., Karjalainen E., Newton G. N., & Sans V. (2018, April 30). 3D-printable photochromic molecular materials for reversible information storage. *Advanced Materials*, 30(26), https://doi.org/10.1002/adma.201800159.

[4] Samanta S. K., & Isaacs L. (2020, May 1). *Biomedical applications of metal organic polygons and polyhedra (Mops)*. Elsevier BV, vol. 410, 213181–213181, https://doi.org/10.1016/j.ccr.2020.213181.

[5] Purcell B. P., Lobb D. C., Charati M. B., Dorsey S. M., Wade R. J., Zellars K. N., Doviak H., Pettaway S., Logdon C. B., Shuman J., Freels P. D., Gorman J. H., Gorman R. C., Spinale F. G., & Burdick J. A. (2014, March 30). Injectable and bioresponsive hydrogels for on-demand matrix metalloproteinase inhibition, *Nature Portfolio*. 13(6), 653–661, https://doi.org/10.1038/nmat3922.

[6] Xiao J. S., et al. (2013). *Research progress of molecular imprinting technology*. Trans Tech Publications, 750–752, 1678–1681, Bäch, Switzerland, Available at https://doi.org/10.4028/www.scientific.net/amr.750-752.1678.

[7] Sellergren B., & Allender J. C. (2005). *Molecularly imprinted polymers: a bridge to advanced drug delivery*. Elsevier BV, 57(12), 1733–1741, Amsterdam, Netherlands, Available at https://doi.org/10.1016/j.addr.2005.07.010.

[8] Trotta F., Biasizzo M., & Caldera F. (2012). *Molecularly imprinted membranes*. Multidisciplinary Digital Publishing Institute, vol. 2(3), 440–477, Basel, Switzerland, Available at https://doi.org/10.3390/membranes2030440.

[9] Vasapollo G., Sole R. D., Mergola L., Lazzoi M. R., Scardino A., Scorrano S., & Mele G. (2011). Molecularly imprinted polymers: present and future prospective, *International Journal of Molecular Sciences*. 12, 5908–5945, https://doi.org/10.3390/ijms12095908.

[10] Zhang H. (2023, March 2). *Functional polymeric systems for advanced industrial applications*. Multidisciplinary Digital Publishing Institute, 15(5), 1277–1277, Basel, Switzerland, https://doi.org/10.3390/polym15051277.

[11] Zhao S., Battiston K. G., & Santerre J. P. (2020, July 8). Sequence-controlled polyurethane block copolymer displays differentiated immunoglobulin-G adsorption that influences human monocyte adhesion and activity, *American Chemical Society*. 6(8), 4433–4445, https://doi.org/10.1021/acsbiomaterials.0c00496.

[12] Kumar S., Jain S., Nehra M., Dilbaghi N., Marrazza G., & Kim K. (2020, October 1). *Green synthesis of metal–organic frameworks: a state-of-the-art review of potential environmental and medical applications*. Elsevier BV, vol. 420, 213407–213407, Amsterdam, Netherlands, https://doi.org/10.1016/j.ccr.2020.213407.

[13] Wang H., Wang L., Zhang S., Zhang W., Li J., & Han Y. (2021, April 23). *Mussel-inspired Polymer Materials Derived from Nonphytogenic and Phytogenic Catechol Derivatives and Their Applications*, Wiley-Blackwell, vol. 70(9), 1209–1224, Hoboken, New Jersey, United States, https://doi.org/10.1002/pi.6230.

[14] Kiesewetter M. K., Shin E. J., Hedrick J. L., & Waymouth R. M. (2010, February 16). Organocatalysis: Opportunities and challenges for polymer synthesis, *American Chemical Society*. 43(5), 2093–2107, https://doi.org/10.1021/ma9025948.

[15] Biswas A., Kim S., Selling G. W., & Cheng H. N. (2014, September 1). *Conversion of agricultural residues to carboxymethylcellulose and carboxymethylcellulose acetate*. Elsevier BV, vol. 60, 259–265, Amsterdam, Netherlands, https://doi.org/10.1016/j.indcrop.2014.06.004.

[16] Kröger S., Turner A. P., Mosbach K., & Haupt K. (1999, August 3). Imprinted polymer-based sensor system for herbicides using differential-pulse voltammetry on screen-printed electrodes, *American Chemical Society*. 71(17), 3698–3702, https://doi.org/10.1021/ac9811827.

[17] Khosla A., Shah S., Shiblee M. N. I., Mir S. H., Nagahara L. A., Thundat T., Shekar P. K., Kawakami M., & Furukawa H. (2018, August 17). Carbon fiber doped thermosetting elastomer for flexible sensors: Physical properties and microfabrication, *Nature Portfolio*. 8(1), https://doi.org/10.1038/s41598-018-30846-3.

[18] Yue Z., & Economy J. (2017, January 1). *Carbonization and activation for production of activated carbon fibers*. Elsevier BV, 61–139, Amsterdam, Netherlands, https://doi.org/10.1016/b978-0-08-100660-3.00004-3.

[19] Pourebrahimi S. (2022, December 1). *Upcycling face mask wastes generated during COVID-19 into value-added engineering materials: a review*, Elsevier BV, 851, 158396–158396, Amsterdam, Netherlands, https://doi.org/10.1016/j.scitotenv.2022.158396

[20] Piletsky S. A., Sappati S., & Turner A. (2001, December 1). *application of molecularly imprinted polymers in sensors for the environment and biotechnology*. Emerald Publishing Limited, vol. 21(4), 292–296, Bingley, United Kingdom, https://doi.org/10.1108/eum0000000005998.

[21] Azizi A., & Bottaro C. S. (2020, March 1). *A critical review of molecularly imprinted polymers for the analysis of organic pollutants in environmental water samples*. Elsevier BV, 1614, 460603–460603, Amsterdam, Netherlands, https://doi.org/10.1016/j.chroma.2019.460603.

[22] Shah S. S., & Aziz M. A. (2020, December 11). Agricultural product-derived carbon for energy, sensing, and environmental applications: A mini-review, *Bangladesh Association of Plant Taxonomists*. 27(2), 467–478, https://doi.org/10.3329/bjpt.v27i2.50686.

[23] Mugadza K., Stark A., Ndungu P., & Nyamori V. O. (2020, September 7). *Synthesis of carbon nanomaterials from biomass utilizing ionic liquids for potential application in solar energy conversion and storage*, Multidisciplinary Digital Publishing Institute, 13(18), 3945–3945, Basel, Switzerland, https://doi.org/10.3390/ma13183945.

[24] Wu Z., Xu S., Yan Q., Chen Z., Ding Y., Li C., Liang H., & Yu S. (2018, July 6). Transition metal–assisted carbonization of small organic molecules toward functional carbon materials, *American Association for the Advancement of Science*. 4(7), https://doi.org/10.1126/sciadv.aat0788.

[25] Mohd Nurazzi N., Asyraf M. R. M., Khalina A., Abdullah N., Sabaruddin F. A., Kamarudin S. H., Ahmad S., Mahat A. M., Lee C. L., Aisyah H. A., et al. (2021). Fabrication, functionalization, and application of carbon nanotube-reinforced polymer composite: An overview, *Polymers*. 13, 1047, https://doi.org/10.3390/polym13071047.

[26] Raffaelle R. P., Landi B. J., Harris J. D., Bailey S. G., & Hepp A. F. (2005). Carbon nanotubes for power applications, *Materials Science and Engineering: B*. 116(3), 233–243, ISSN 0921-5107 https://doi.org/10.1016/j.mseb.2004.09.034.

[27] Iijima S. (2002). Carbon nanotubes: Past, present, and future, *Physica B: Condensed Matter*. 323(1–4), 1–5.

[28] Westervelt R. M. (18th April 2008). Graphene nanoelectronics, *Science*. 320, 324–325.

[29] Urade A. R., Lahiri I., & Suresh K. S. (14 October 2022). Graphene properties, synthesis and applications: A review, *Journal of Materials Research*. 37, 1–28.

[30] Dutta V., Singh P., Shandilya P., Sharma S., Raizada P., Saini A. K., Gupta V. K., Hosseini-Bandegharaei A., Agarwal S., & Rahmani-Sani A. (August 2019). Review on advances in photocatalytic water disinfection utilizing graphene and graphene derivatives-based nanocomposites, *Journal of Environmental Chemical Engineering*. 7, 103081.

[31] Pichon V., & Chapuis-Hugon F. (August 2008). Role of molecularly imprinted polymers for selective determination of environmental pollutants – A review, *Analytica Chimica Acta*. 622(1–2), 48–61, https://doi.org/10.1016/j.aca.2008.05.086.

[32] Baughman R. H., Zakhidov A. A., & De Heer W. A. (2002). Carbon nanotubes-the route toward applications, *Science*. 297(Issue 5582), 787–793.

[33] De Volder M. F. L., Tawfick S. H., Baughman R. H., & Hart A. J. (2013). Carbon nanotubes: Present and future commercial applications, *Science*. 339(6119), 535, doi: 10.1126/science.1222453.

[34] Zhu Y. T. (October 2004). Ultralong single-wall carbon nanotubes, *Zeitschrift Für Metallkunde*. 95, 988–994, https://doi.org/10.3139/146.101792.

[35] Novoselov K. S., Fal'ko V. I., Colombo L., Gellert P. R., Schwab M. G., & Kim K. (11 October 2012). A roadmap for graphene, *Nature*. 490, 192–200.

[36] Wackerlig J., & Schirhagl R. (5 November 2015). Applications of molecularly imprinted polymer nanoparticles and their advances toward industrial use: A review, *Analytical Chemistry*. 88, 250–261, https://doi.org/10.1021/acs.analchem.5b03434.

[37] Zhao M., & Shen G. (July 2015). The application of molecularly imprinted polymers, *Separation and Purification Reviews*. 44, 1–15, https://doi.org/10.1080/15422119.2013.856612.

[38] Reiss T., Hjelt K., & Ferrari A. C. (Oct 2019). Graphene is on track to deliver on its promises, *Nature Nanotechnology*. 14, 10–12.

[39] Henning K. D., & Von Kienle H. Carbon, 5. Activated carbon, *Ullmann's Encyclopedia of Industrial Chemistry*. Wiley-VCH Verlag GmbH & Co. KGaA.

[40] Geim A. K. (19 Jun 2009). Graphene: Status and prospects, *Science*. 324, 1530–1534, https://doi.org/10.1126/science.1158877.

[41] Cabaleiro-Lago, C., Hasterok, S., Gjörloff Wingren, A., & Tassidis, H. (2023). Recent Advances in Molecularly Imprinted Polymers and Their Disease-Related Applications, *Polymers*. 15(21), 4199. https://doi.org/10.3390/polym15214199.

[42] Preda, D., David, I. G., Popa, D.-E., Buleandra, M., & Radu, G. L. (2022). Recent Trends in the Development of Carbon-Based Electrodes Modified with Molecularly Imprinted Polymers for Antibiotic Electroanalysis, Chemosensors. 10(7), 243. https://doi.org/10.3390/chemosensors10070243.

About the editors

Chaudhery Mustansar Hussain, PhD, is an adjunct professor and director of laboratories in the Department of Chemistry and Environmental Sciences at the New Jersey Institute of Technology (NJIT), Newark, New Jersey, USA. His research focused on the applications of nanotechnology and advanced materials, environmental management, sustainability, analytical chemistry, and other various industries. Dr. Hussain is the author of numerous papers in peer-reviewed journals as well as a prolific author and editor of around 200 books, including scientific monographs and handbooks in his research areas. He has published with Elsevier, American Chemical Society, Royal Society of Chemistry, John Wiley & Sons, CRC Press, and Springer.

Rüstem Keçili received his MSc in analytical chemistry from the Anadolu University, Turkey, and he earned his PhD in chemistry under Marie Curie Actions from the Technical University of Dortmund, Germany. He worked as an early-stage researcher at MIP Technologies AB (a subsidiary of Biotage AB) in Lund, Sweden. He was a visiting researcher at the University of Manchester, UK. He is currently an associate professor and vice director at the Department of Medical Services and Techniques, Yunus Emre Vocational School of Health Services, Anadolu University, Turkey. His multidisciplinary professional background covers green and sustainable chemistry, nanotechnology, advanced materials, molecularly imprinted polymers (MIPs), molecular recognition, MIP-based sensor systems, affinity chromatography, and solid-phase extraction. Dr. Keçili is the author of numerous research and review papers in prestigious peer-reviewed journals as well as the editor of one scientific book and author of two scientific books and several book chapters published by the American Chemical Society, Royal Society of Chemistry, Elsevier, John Wiley & Sons, and Springer.

Rüstem Keçili
Anadolu University Yunus Emre Vocational School of Health Services
Yunus Emre Campus Tepebaşı, 26470 Eskişehir, Turkey
rkecili@anadolu.edu.tr

Chaudhery Ghazanfar Hussain is a research scholar in science and technology at the Department of Education, Punjab, Pakistan. His key areas of research include environmental modeling, analytical chemistry, data science, computer networks, and industrial development. He is the author of monographs on chemistry, sample pretreatments, nanotechnology, and software technology. He is a professional and affiliated with several organizations.

Chaudhery Ghazanfar Hussain
Department of Education Lahore
Science and Technology, Punjab
Pakistan
ur1947@gmail.com

https://doi.org/10.1515/9783111503325-014

Index

https://doi.org/10.1515/9783111503325-015